Unhealthy Pharmaceutical Regulation

Health, Technology and Society

Series Editors: **Andrew Webster**, University of York, UK and **Sally Wyatt**, Royal Netherlands Academy of Arts and Sciences, The Netherlands

Titles include:

Ellen Balka, Eileen Green and Flis Henwood (*editors*)
GENDER, HEALTH AND INFORMATION TECHNOLOGY IN CONTEXT

Courtney Davis and John Abraham (*editors*)
UNHEALTHY PHARMACEUTICAL REGULATION
Innovation, Politics and Promissory Science

Gerard de Vries and Klasien Horstman (*editors*)
GENETICS FROM LABORATORY TO SOCIETY
Societal Learning as an Alternative to Regulation

Alex Faulkner
MEDICAL TECHNOLOGY INTO HEALTHCARE AND SOCIETY
A Sociology of Devices, Innovation and Governance

Herbert Gottweis, Brian Salter and Catherine Waldby
THE GLOBAL POLITICS OF HUMAN EMBRYONIC STEM CELL SCIENCE
Regenerative Medicine in Transition

Roma Harris, Nadine Wathen and Sally Wyatt (*editors*)
CONFIGURING HEALTH CONSUMERS
Health Work and the Imperative of Personal Responsibility

Jessica Mesman
MEDICAL INNOVATION AND UNCERTAINTY IN NEONATOLOGY

Mike Michael and Marsha Rosengarten
INNOVATION AND BIOMEDICINE
Ethics, Evidence and Expectation in HIV

Nelly Oudshoorn
TELECARE TECHNOLOGIES AND THE TRANSFORMATION OF HEALTHCARE

Nadine Wathen, Sally Wyatt and Roma Harris (*editors*)
MEDIATING HEALTH INFORMATION
The Go-Betweens in a Changing Socio-Technical Landscape

Andrew Webster (*editor*)
NEW TECHNOLOGIES IN HEALTH CARE
Challenge, Change and Innovation

Andrew Webster (*editor*)
THE GLOBAL DYNAMICS OF REGENERATIVE MEDICINE
A Social Science Critique

Health, Technology and Society
Series Standing Order ISBN 978–1–4039–9131–7 hardback
(*outside North America only*)

You can receive future titles in this series as they are published by placing a standing order. Please contact your bookseller or, in case of difficulty, write to us at the address below with your name and address, the title of the series and the ISBN quoted above.

Customer Services Department, Macmillan Distribution Ltd, Houndmills, Basingstoke, Hampshire RG21 6XS, England

Unhealthy Pharmaceutical Regulation

Innovation, Politics and Promissory Science

Courtney Davis
Kings College London, UK

and

John Abraham
Kings College London, UK

palgrave
macmillan

First published 2013 by
PALGRAVE MACMILLAN

Palgrave Macmillan in the UK is an imprint of Macmillan Publishers Limited,
registered in England, company number 785998, of Houndmills, Basingstoke,
Hampshire RG21 6XS.

Palgrave Macmillan in the US is a division of St Martin's Press LLC,
175 Fifth Avenue, New York, NY 10010.

Palgrave Macmillan is the global academic imprint of the above companies
and has companies and representatives throughout the world.

Palgrave® and Macmillan® are registered trademarks in the United States,
the United Kingdom, Europe and other countries

ISBN 978-1-349-28417-7   ISBN 978-1-137-34947-7 (eBook)
DOI 10.1057/9781137349477

This book is printed on paper suitable for recycling and made from fully
managed and sustained forest sources. Logging, pulping and manufacturing
processes are expected to conform to the environmental regulations of the
country of origin.

A catalogue record for this book is available from the British Library.

Library of Congress Cataloging-in-Publication Data

Abraham, John, 1961– author.
    Unhealthy pharmaceutical regulation : innovation, politics and promissory science /
    Courtney Davis, Kings College London, UK and John Abraham, University of Sussex, UK.
        pages cm
    Summary: "European and American drug regulators govern a multi-billion-dollar
    pharmaceutical industry selling its products on the world's two largest medicines markets. This
    is the first book to investigate how effectively American and supranational EU governments
    have regulated innovative pharmaceuticals regarding public health during the neo-liberal era of
    the last 30 years. Drawing on years of fieldwork, the authors demonstrate that pharmaceutical
    regulation and innovation have been misdirected by commercial interests and misconceived
    ideologies, which induced a deregulatory political culture contrary to health interests. They
    dismantle the myth that pharmaceutical innovations necessarily equate with therapeutic
    advances and explain how it has been perpetuated in the interests of industry by corporate
    bias within the regulatory state, unwarranted expectations of promissory science, and the
    emergent patient-industry complex. Endemic across both continents, the misadventures of
    pharmaceutical deregulation are shown to span many therapeutic areas, including cancer,
    diabetes and irritable bowel syndrome. The authors propose political changes needed to
    redirect pharmaceutical regulation in the interests of health." –Provided by publisher.

        1. Pharmaceutical policy – United States. 2. Pharmaceutical policy – European Union
    countries. 3. Drugs – Law and legislation – United States. 4. Drugs – Law and legislation –
    European Union countries. 5. Pharmaceutical industry – Political aspects. I. Davis,
    Courtney, 1965– author. II. Title.

RA401.A1A27 2013
362.17'82—dc23

2013021799

To Cai, Manon and Jess

Contents

List of Illustrations

Tables

Figures

Preface and Acknowledgements

Our eight-year long investigation into the regulation of innovative pharmaceuticals in the two largest pharmaceutical markets in the world has taken us many places both in the literal spatial sense and intellectually, from the offices of the world's largest drug regulator just outside Washington DC to the company of consumer advocates lobbying the European Parliament. We have also learned about the workings of a whole range of institutions and drugs, from the FDA and the CHMP to the role of blood-glucose control and heart rhythm in drug evaluation. The perspectives of many disciplines have also needed to be brought to bear on getting to grips with the complexities of pharmaceutical innovation and regulation, such as clinical pharmacology, epidemiology, history of medicines, political economy, political science, science and technology studies, sociology, and toxicology.

This would not have been possible without the generosity of scores of people who have given up their time to be interviewed or helped in other ways with advice or document searches. Our thanks go to all of them for their co-operation and assistance, even if they may not agree with all of our findings, and if many cannot be identified. We are particularly grateful to our former colleague, Tim Reed, for helping with some early parts of the project, and to various colleagues on both sides of the pond for their encouragement over the years, notably, Joe Collier, Graham Dukes, Andrew Herxheimer, Joel Lexchin, Barbara Mintzes, Kenneth Oye, Larry Sasich, Rein Vos and Caroline Wilson, to name but a few. Special thanks go to Andrew Webster and Sally Wyatt, the series editors, for their immense patience and unswerving support for the completion of this lengthy endeavour. We also very much appreciate the sustained commitment to this book project shown by Andrew James and his colleagues at Palgrave. Last, but not least, we are grateful to the UK Economic and Social Research Council (ESRC) for supporting four years of the research underpinning this book, especially most of the international fieldwork, in a way entirely consistent with the important goal of truly independent social science.

List of Abbreviations

AACR	American Association for Cancer Research
ABPI	Association of the British Pharmaceutical Industry
ADOPT	A Diabetes Outcome Progression Trial
ADR	adverse drug reaction
ALLHAT	Antihypertensive and Lipid Lowering to prevent Heart Attack Trial
ASCO	American Society for Clinical Oncology
BfArM	Bundesinstitut fur Arzneimittel und Medizinprodukte
BGA	Bundesgesundheitsamt
CAST	Cardiac Arrhythmia Suppression Trial
CDER	Center for Drug Evaluation and Research (FDA)
CHMP	(EU) Committee for Human Medicinal Products (formerly CPMP)
COMT	catechol-O-methyl transferase
CPI	Critical Path Initiative
CPMP	(EU) Committee for Proprietary Medicinal Products
CSM	(UK) Committee on Safety of Medicines
DDMAC	Division of Drug Marketing and Communications (FDA)
DG	Directorate-General
DHHS	(US) Department of Health and Human Services
DREAM	Diabetes Reduction Assessment with ramipiril and rosiglitazone Medication
DTCA	direct-to-consumer advertising
EBE	European Biopharmaceuticals Enterprises
EFPIA	European Federation of Pharmaceutical Industry Associations
EGFR	epidermal growth factor receptor
EMA	European Medicines Agency
EMEA	European Medicines Evaluation Agency
EORTC	European Organization for Research and Treatment of Cancer
EPAR	European Public Assessment Report
EPPOSI	European Platform for Patients' Organisations, Science and Industry
ERMS	European Risk Management Strategy
EURODIS	European Patient Organization for Rare and Orphan Diseases

FDA	Food and Drug Administration
FDAAA	FDA Administration Amendments Act
FDAMA	Food and Drug Administration Modernization Act
FD&C	Food, Drug and Cosmetic
FOIA	(US) Freedom of Information Act
FR	(US) Federal Register
GAO	(US) Government Accountability Office (formerly General Accounting Office)
GDAC	Gastrointestinal Drugs Advisory Committee (FDA)
GSK	GlaxoSmithKline
HTA	health technology assessment
IBS	irritable bowel syndrome
IBSSHG	Irritable Bowel Syndrome Self-Help Group
IDEAL	Iressa Dose Evaluation in Advanced Lung cancer
IFFGD	International Foundation for Functional Gastro-intestinal Disorders
IFPMA	International Federation of Pharmaceutical Manufacturers' Associations
IMI	Innovative Medicines Initiative
IND	investigational new drug (US)
INTACT	Iressa Non-small-cell lung cancer Trial Assessing Combination Treatment
IoM	(US) NAS's Institute of Medicine
LAG	Lotronex Action Group
MCA	(UK) Medicines Control Agency
MiEF	Medicines in Europe Forum
MPA	Medical Products Agency
NAS	new active substance
NAS	(US) National Academy of Sciences
NBE	new biological entity
NBHW	(Swedish) National Board of Health and Welfare
NCI	(US) National Cancer Institute
NDA	new drug application (US)
NHLBI	(US) National Heart, Lung and Blood Institute
NHS	(UK) National Health Service
NIAID	(US) National Institute of Allergy and Infectious Diseases
NICE	National Institute for Health and Clinical Excellence
NIH	National Institutes of Health
NIHCM	National Institute for Health Care Management
NME	new molecular entity
NORD	(US) National Organization for Rare Diseases

NSC   non-small cell
ODAC   Oncologic Drugs Advisory Committee (FDA)
OTA   US Office of Technology Assessment
PhRMA   (US) Pharmaceutical Research and Manufacturing Association (formerly PMA)
PDUFA   Prescription Drug User Fee Act
PMA   (US) Pharmaceutical Manufacturers' Association
PROactive PROspective pioglitAzone Clinical Trial in macroVascular Events
RCT   randomized controlled trial
REMS   (US) Risk Evaluation and Mitigation Strategy
SPC   summary of product characteristics (EU 'product label')
TAG   Treatment Action Group
UKPDSG   UK Prospective Diabetes Study Group
US NAS   US National Academy of Sciences
USUDG   US University Diabetes Group

Introduction: The Health of a Political System

Until a few years ago, the political elites in the governments of Europe and North America were pursuing a neo-liberal approach to regulation of the banking industry and the financial sector. What that meant was that there was 'light-touch' regulation and a conviction that by making millions, and even billions, of dollars for themselves, bankers were also delivering good times for the rest of society via tax revenues and the like. The enormous wealth pocketed by some bankers and financiers, we were told, would trickle down to the rest of society. Apparently, wealth creation was the business of the financial sector, so governments should, for the most part, let them get on with it because what the banking industry was doing was in the interests of all of us. This neo-liberal era began in the 1980s on both sides of the Atlantic with New Right politicians, such as Reagan and Thatcher, but was continued under the Clinton and Blair governments, among others.

Neo-liberalism got its name from an emphasis on liberalization of markets, ostensibly to liberate the entrepreneurial spirit of the capitalist and consumer choice from the interference of government regulation. By the late 2000s, a widespread crisis of 'toxic' financial transactions produced an immanent collapse of the international banking system plunging most of the western world into its worst economic recession since the 1930s. The catastrophe for western economies has become so severe that the current generation of young people is thought to be the first since the beginning of the post-war period to have poorer prospects than their parents. It is now widely recognized by governments and political elites of all persuasions that the disaster resulted from inadequate regulation of the banking system and financial sector since the 1980s. In a climate of job losses, falling wages, and austerity in public services, economic growth stalled in the late 2000s in most western countries.

The reaction of most governments and many political commentators has been to call for less regulation of other sectors outside banking in order to stimulate economic growth. But what if the banking industry is not the only sector in which deregulation has produced toxic, and even catastrophic results? One should at least get to grips with answering this question regarding other sectors before embarking on further deregulation. This book is an invitation to start on that journey by exploring the regulation of just one such other sector, namely, the pharmaceutical industry during the neo-liberal era in Europe and the United States (US). While there are many opinions and much commentary about the conduct of pharmaceutical companies, there is far less social science investigation of the reasons for, and consequences of, that conduct.

The pharmaceutical industry is trans-national and vast, with some of its individual products fetching over a billion US dollars on the world market. The industry has prospered during the neo-liberal era. Between 1960 and the early 1980s, prescription drug sales were almost static as a percentage of Gross Domestic Product in western societies. However, from the early 1980s to 2002, prescription drug sales tripled to nearly US\$400 billion worldwide, and almost US\$200 billion in the US (Abraham 2010, p. 607). Between 2002 and 2006, US prescription drugs sales grew annually by 10 per cent on average, while global sales reached US\$600 billion by 2007 (Anon. 2008a). There is no doubt, then, that the drug industry has been able to grow, expanding its sales and profits in the process.

Of course, society's expectation of the pharmaceutical industry is not merely that it makes profits for shareholders and investors. Drug firms' products must also provide some health benefit. The pharmaceutical industry accepts that and contends that its growing sales reflect its success in creating products and innovations needed by patients. For the last 40–50 years (or more in the US and Scandinavian countries) governments have not been so naive as to accept that the pharmaceutical industry's commercial motives will always deliver new drug products in the best interests of patients, so the government drug regulatory agencies check drug companies' claims about their products before permitting them on the market. Yet there is a paradox at the heart of pharmaceutical regulation in the neo-liberal era. On the one hand, state regulation has been introduced and maintained on the assumption that the interests of the pharmaceutical industry and public health do not always converge. On the other hand, the last 30 years has seen a raft of deregulatory reforms, ostensibly to promote pharmaceutical innovation deemed to be simultaneously in the commercial interests of industry and the health interests of patients.

Drawing on fieldwork-based research in the US and Europe, combined with systematic analysis of a mass of documentary evidence, this book investigates how pharmaceutical regulation has evolved and operated during the neo-liberal era in order to determine whether the deregulatory reforms of that period can reasonably be regarded as being in the interests of public health, or alternatively if, like the banking system, pharmaceutical regulation has been festering in an unhealthy state. To do that we must explore a range of social scientific questions, of which we mention just a few here. Who are the key actors involved and what has been their relative influence on the trajectories of pharmaceutical regulation in the EU and the US over the last 30 years? Can the political convictions of government really determine how the therapeutic efficacy of an individual drug is evaluated? What is the relationship between deregulation, innovation and the availability of valuable therapies for patients? What role is played in shaping regulatory decision-making by public expectations and the assertions of what some scholars call 'promissory science'? Has pharmaceutical regulation and innovation during the neo-liberal era been an unwarranted misadventure or even mis-direction so far as health is concerned, or are they on the right track? It is only within the last decade that the full theoretical and empirical complexities of neo-liberal drug regulation in Europe and North America have become apparent. Based on extensive new international fieldwork and documentary/archival analysis, this book, for the first time, systematically links them together into what may be regarded as the evolution of new a social science discipline concerned with pharmaceuticals and public health policy.

Although this book is a social scientific investigation, it is written for academics and non-academics alike. Academic jargonizing is, therefore, kept to a minimum. While the real world is undoubtedly complex, our view is that the first job of the social scientist is to unravel those complexities systematically and logically, so that they can be explained in a relatively straightforward way. Strong social science should be able to make its case to both non-academic and academic audiences. Similar comments apply to pharmaceutical and clinical science – complex for sure, but, with sufficient effort and inclination, capable of clear explanation to the non-specialist reader. Having said that, we do introduce and develop some social science theories of drug development and regulation, which help the reader focus on the issues at stake in understanding the pharmaceutical sector during the neo-liberal era. The identification of these theories may be regarded as a way of expressing sets of claims about how the world of pharmaceutical regulation and innovation has operated, so that those claims can be set against evidence in a succinct way in order to build an ever-more accurate and illuminating picture of complex realities. The first chapter begins that task.

1

Putting Pharmaceutical Regulation to the Test: From Historical Description to a Social Science for Public Health

This book examines how innovative pharmaceuticals have been regulated in the US and the European Union (EU) since 1980 – a period which we refer to as the 'neo-liberal era'. Regarding the EU, our principal focus is on the period since 1995 because that is when a supranational EU regulatory agency and system became fully established with specific responsibilities for regulating innovative pharmaceuticals. Like many other writers, we refer to the post-1980 era in the US and western Europe as 'neo-liberal' because it was, and remains, a period in which the political project of minimizing state intervention, subjecting the state to competitive tests of 'the market', and elevating individual consumer choice above the state as a form of collective decision-making, all came to the fore. This has involved the 'liberalization' of markets, that is, relaxation of government regulations and controls believed to hamper business activity and the socio-economic signals of consumer demand (Fisher 2009).

In the US, neo-liberalism found its most committed and enthusiastic expression in the Republican Party's antagonism to 'big government' – a recurrent theme in the rhetoric and often policy objectives of the administrations of Ronald Reagan, George Bush (senior) and George W. Bush. However, the Democratic Administration under Clinton also accepted the political philosophy of 'market liberalization', including the view that the state should help business interests achieve economic success even if that meant retreat from their regulation by government. A similar trend has been evident in western Europe and latterly the EU.

Perhaps the most notorious European enthusiast for neo-liberalism was the UK's Conservative Prime Minister, Margaret Thatcher, who sought to 'free' private industry from state control and regulation – a sentiment deviated from only marginally by the 'New' Labour Governments of Blair and Brown. Until the recent banking crisis of the late 2000s, all of these politicians and governments either actively promoted, or were willing to be persuaded of, the idea that pro-business deregulation was not merely in the commercial interests of industry, but ultimately for the greater good – the 'public interest'.

In the pharmaceutical sector, a raft of pro-business deregulatory reforms ensued during the neo-liberal period (discussed in detail in Chapter 2). They included making the American and European government drug regulatory agencies largely and increasingly dependent on funds from the pharmaceutical industry; increasing the extent and flexibility of consultation between regulators and drug companies; reducing the amount and types of evidence that pharmaceutical firms had to collect to demonstrate the efficacy of particular categories of drugs in order to obtain marketing approval from regulators; and shortening the time taken by government regulatory agencies to grant marketing approval to drug companies for their products. Meanwhile, government regulation of pharmaceuticals maintained its legal responsibility and official democratic mandate to promote and protect public health. In that context, the crucial claim made by government and industry officials regarding the deregulatory reforms was that they would accelerate and increase pharmaceutical innovation, which was in the interests of patients and public health because they needed faster access to innovative drugs. Thus, one theory of pharmaceutical regulation since 1980, which we call 'neo-liberal theory', is that the pro-industry deregulatory reforms of that period were instigated by government in the interests of patients and public health. Implicit in the theory is the proposition that pharmaceutical innovations necessarily promise therapeutic advances for patients.

In this book, we put that theory, among others, to the test by examining both the macro-politics of regulatory change and the micro-sociology of individual drug development and regulation. We present the first social science research to provide an analysis of both American and supranational EU pharmaceutical regulation of innovative prescription drugs. Given that the direction of drug regulation has significant implications for patients, public health and healthcare systems, it is important to understand its socio-political and technical dynamics in order to learn lessons from the past and reflect on possible future policy options. We wrote this book because it advances those goals substantially beyond the limitations of existing literature.

Previous research on European pharmaceutical regulation is quite modest in extent and is almost always concerned with individual European countries or small groups of such countries. Dukes (1985) conducted an important survey of drug regulation across several different European countries, but it had no specific focus on innovative pharmaceuticals and pre-dated the neo-liberal era and the emergence of a supranational regulatory agency, the European Medicines Evaluation Agency (EMEA) – known as the European Medicines Agency (EMA) since 2010. In their valuable edited collection, Mossialos *et al.* (2004) provide a more recent overview of many different dimensions of pharmaceutical regulation in Europe, but attention to innovative prescription drugs is limited with most coverage ranging widely to pricing, over-the-counter drugs, 'alternative medicines', pharmacies and pharmacogenomics. While Abraham (1995a; 2009) and Daemmrich (2004) examine prescription drug regulation in-depth in the UK and Germany, respectively, by comparison with the US, and Hancher (1989) offers a similar type of comparison of the UK and France, those comparative studies, even combined, provide coverage of only three European countries. More significantly, their analyses are almost entirely confined to events before 1990 and make no attempt to consider supranational EU regulation. Wiktorowicz (2003) provides a more recent comparison of prescription drug regulation in the UK and France by comparison with Canada and the US, taking into account developments in the 1990s, but she also is little concerned with supranational EU regulation and innovative pharmaceuticals *per se*. Only Abraham and Lewis (2000) focus substantially on supranational EU pharmaceutical regulation since 1995, but their analysis is limited to its emergence and early years up to the late 1990s, rather than its effects, and addresses in only a preliminary way the nature of innovative pharmaceuticals within that regulatory system.

Much more has been published about American drug regulation. Temin (1980), Abraham (1995a) and Daemmrich (2004) produced broad analyses of US government control of pharmaceuticals spanning from the late nineteenth century and the origins of the American drug regulatory agency, the Food and Drug Administration (FDA). However, their studies do not stretch beyond the late 1970s or 1980s. Marks (1997) also made an important contribution to the twentieth-century history of US medicines regulation, particularly in relation to standards for drug trials and testing, but his investigation also terminates at 1990. Following in a similar tradition to Marks (1997), Greene (2007) examines post-war developments in the design, testing and promotion of diabetes and cholesterol drugs, though again the vast bulk of analysis is concerned with events before the 1990s, and most emphasis is given to drug development and marketing, rather than regulation *per se*.

None of those discussions of US drug development and regulation paid much attention to pharmaceutical product innovation within the American regulatory system. It has, however, been the concern of some other scholars. Most notably, Epstein (1996) offers an extensive account of how human immunodeficiency virus (HIV)/acquired immunodeficiency syndrome (AIDS) patients sought to influence the science underpinning pharmaceutical testing and regulation in order to facilitate faster and wider access to innovative AIDS drugs, but his analysis is entirely circumscribed by the HIV/AIDS field. In a quite different approach to pharmaceutical innovation, Angell (2004) examines the extent to which innovative pharmaceuticals across many therapeutic fields owe their origins to industrial research and offer value to patients and healthcare systems. Although the role of drug regulation and the FDA forms part of her discussion, the overwhelming majority of her critique is aimed at the activities of the pharmaceutical industry.

The two main recent analyses of US drug regulation are provided by Hilts (2003) and Carpenter (2010a).[1] Both take a historical approach, which includes retracing twentieth-century regulatory developments before the neo-liberal era. About two-thirds of Hilts (2003) and over three-quarters of Carpenter (2010a) are concerned with events before 1980, also previously investigated by Abraham (1995a), Daemmrich (2004), Marks (1997) and Temin (1980), which need not detain us here. However, the other parts of Hilts (2003) and Carpenter (2010a) make important contributions to any analysis of the FDA in the neo-liberal era, with which we shall certainly engage throughout this book, though only one chapter of Carpenter (2010a) is devoted to neo-liberal influences on the FDA proper. Hilts (2003) emphasizes the impact of 'deregulatory politics' and the New Right' on the FDA after Reagan's election to the presidency, though much of his discussion revolves around controversies over the agency's regulation of food, tobacco and medical devices, rather than drugs. Carpenter's (2010a) focus is fixed on pharmaceutical regulation throughout, but his investigation is less about US drug regulation as such, even less about the relationship between regulation and innovative pharmaceuticals, and much more an account of the organizational dynamics of the FDA within its social and political context. A limitation, therefore, of both Hilts (2003) and Carpenter (2010a) for our purposes of analysing US drug regulation is that they are, in effect, studies of the FDA. Indeed, Angell (2010) criticizes Carpenter (2010a) for neglecting to consider sufficiently industry influence on the FDA. Carpenter's (2010b) rebuttal accuses her of erroneously misrepresenting him, but also confirms that his investigations lead mainly elsewhere to the question of how the FDA has managed to maintain its power and influence, or found them diminished, in the face of wider neo-liberal politics in the US. Of course, as we embark on our analysis of US *and EU* drug regulation, a more fundamental limitation of all the major works by these American scholars, apart from Daemmrich (2004), is that their examination of pharmaceutical regulation and/or innovation is entirely confined to the US.[2]

Pharmaceutical studies becomes social science

Before 1990, pharmaceutical studies were highly fragmented and could scarcely be regarded as a 'field'. What was available generally took the form of a descriptive history of regulation, policy and/or the pharmaceutical industry, including its criminological activities (Braithwaite 1986; Dukes 1985; Liebenau 1981; Penn 1982; Temin 1980). That began to change with Abraham's (1995a) introduction of theories into the field from political sociology and political science. Most notably ideas put forward by writers, such as Bernstein (1955), Cawson (1986), Middlemas (1979), Miliband (1983), Offe (1973), Stigler (1971) and Wilson (1980), about how social and economic interests influenced governments' decision-making, and the regulatory state, in particular.

For instance, Bernstein (1955), writing from an American perspective, contended that government regulatory agencies, which typically formed in the aftermath of some public disaster associated with industrial activity, initially regulated the industry zealously in the public interest, but gradually over time became captured by the regulated industry, so that it eventually came to regulate primarily in the interests of the industry, rather than the public interest, until another public disaster when the cycle would restart. The school of thought derived from Bernstein's (1955) writings was to become known as 'capture theory'. According to capture theory, regulatory agencies created after some public disaster are given a legal mandate by the Legislature and the Executive arms of the state to regulate an industry in the public interest. When the regulatory agency shifts away from that mission due to capture, it is known as 'administrative drift' or 'bureaucratic drift' because the bureaucratic arm of government (the regulatory agency) has drifted away from the mandate of its Legislative and Executive arms.

Within capture theory, the focus is very much on the relationship between the regulatory agency and the regulated industry. Capture may occur because agency officials or experts, who sit on regulatory agencies' advisory committees, develop attitudes and obligations towards pharmaceutical firms resulting from hospitalities or consultancies. Capture is likely to be increased where informal consultation and meetings between government regulatory officials and industry is permitted and encouraged because the opportunities for regulated firms to lobby government officials are expanded.

One of the most instructive books written about the dynamics of regulatory capture was compiled by Owen and Braeutigam (1978) as a 'how to' manual for industry that recommends techniques with which to manipulate government regulatory official and expert advisers. On lobbying regulatory agencies, they provide the following advice to regulated firms:

> Effective lobbying requires close personal contact between the lobbyists and government officials. Social events are crucial to this strategy. The object is to establish long-term personal relationships transcending any particular issue. Company and industry officials must be 'people' to the agency decision-makers, not just organizational functionaries. A regulatory official contemplating a decision must be led to think of its impact in human terms. Officials will be much less willing to hurt long-time acquaintances than corporations. Of course, there are also important tactical elements of lobbying, of which not the least is information gathering at low levels of the agency staff. Each contact must be carefully tailored to the background and personality of the official being lobbied. For this reason it is useful to keep files on the backgrounds of agency officials. (Owen and Braeutigam 1978, pp. 6–7)

Recognizing that regulatory decisions are often influenced by government agencies' expert advisory committees, firms are also advised to co-opt those experts (often academics), as follows:

> This is most effectively done by identifying the leading experts in each relevant field and hiring them as consultants or advisors, or giving them research grants and the like. This activity requires a modicum of finesse; it must not be too blatant, for the experts themselves must not recognize that they have lost their objectivity and freedom of action. At a minimum, a programme of this kind reduces the threat

that the leading experts will be available to testify or write against the interests of the regulated firms. (Owen and Braeutigam 1978, p. 7)

Regulatory capture may also occur in a much more passive, structural way, without any lobbying by industry. For instance, the 'revolving door' phenomenon may foster capture. This refers to a subculture within leading organizations in the regulatory process in which officials begin their careers as regulators, but then move on to join the regulated industry; or they begin their careers in industry, then work for some years in the regulatory agency until they are promoted back into the higher echelons of industry. The 'revolving door' can contribute to capture in at least two ways. If regulators have a background of training in industry, then they may be more likely to bring values to the agency which are sympathetic to the regulated industry than if they received training outside industry. More significantly, if regulators view their career development in terms of future promotion into the regulated industry, then they may be unduly concerned to maintain 'friendly relations' with industry at the expense of public interest regulation (Abraham 1995a, p. 73).

In the decades following Bernstein's articulation of capture theory, political scientists, especially in Europe, began to theorize governance and political power in terms of relations between organized interests and the state. By analysing the influence of the trade union movement on UK Labour governments, Middlemas (1979) drew attention to the importance of organized interests in gaining privileged access to the state, above and beyond other interest groups, to the extent that the organized interests governed in partnership with the state, including the delegation of governing powers to those interests in the form of self-regulation. Middlemas (1979) referred to this arrangement as 'corporate bias' and its proposition subsequently became known as 'corporate bias theory'.

Corporate bias theory differs from capture theory particularly because it suggests that regulation and regulatory decision-making needs to be located in a broader political context than solely the relations between regulator and regulatee. Specifically, the wider constituents of the state must be taken into account, not only the 'bureaucracy' (regulatory agencies). The politics of the Executive (the Administration in the US, and the Council of Ministers and national European governments in the EU) and the Legislature (the Congress in the US and the European Parliament in the EU) are also regarded as highly significant in corporate bias theory. For corporate bias theory, the influence of an organized interest, such as the pharmaceutical industry, may extend to lobbying the top strata of government within the Executive and the Legislature. Representatives of the organized interest may even establish themselves as key advisers to the Executive or sit on high-level joint committees with government Ministers/Secretaries of State setting the policy agenda for regulation of that interest-group/industry. Hence, corporate bias theory allows that a possible mechanism by which industry can drive regulation in its own interests is via the Executive and Legislature without necessarily effecting direct capture of regulatory agencies because the bureaucracy (the regulatory agencies) may be made responsive to industry interests by its constitutional masters in the Executive and Legislature.

Unlike, capture theory, corporate bias theory does not hypothesize a cyclical process of regulatory change. Nor does it postulate that government agencies necessarily begin life with high ambitions to regulate industry vociferously in the public interest. A further difference is that capture theory assumes that pro-industry (de)regulation is associated with the capture phase (of the regulatory cycle) during which the government agency is relatively passive and powerless. By contrast, corporate bias theory allows for the possibility of a relatively strong, pro-active state, which may encourage pro-business (de)regulation in collaboration with industry. Conversely, it follows that corporate bias theory does not assume that the state is zealous in its goals for business regulation only when wishing to regulate strongly in the public interest – an assumption made by capture theory.

After the AIDS crisis, a quite different theoretical perspective from those introduced into discussions of the pharmaceutical sector by Abraham (1995a) began to emerge, especially among American scholars in the aftermath of AIDS patient activism. In particular, Epstein (1996) showed how AIDS treatment activists in the US affected some aspects of new AIDS drug development and regulation. Epstein treated his work as a self-contained ethnography of a social (patient) movement. However, other analysts, such as Daemmrich, Edgar, Krucken and Rothman, read much more into the implications of AIDS patient activism for understanding regulatory change. They took the view that such patient activism had altered drug regulatory philosophy in the US, and that that was part of a wider phenomenon in which changing attitudes of patients, and specifically 'disease-based' patient groups (e.g. cancer or Alzheimer's patient groups) had come to drive regulatory developments and change (Daemmrich 2004; Daemmrich and Krucken 2000; Edgar and Rothman 1990). For instance, Daemmrich writes:

...during the 1980s and 1990s ...[t]he American 'patient' evolved from needing state protection from industry and physicians to a free-market consumer who deserved access to still-experimental drugs. As a consequence, FDA placed fewer demands on manufacturers for lengthy testing and redesign of clinical trials than in the past. (2004, p. 81)

Similarly, Carpenter declares:

Before the 1980s it was rare for the public's attention to be drawn to a drug that the FDA had not approved or was reviewing slowly. The AIDS epidemic changed this ...Yet AIDS was only the beginning of a much larger story of disease-based political mobilization in the United States. To a degree never before witnessed, disease-specific lobbies now press Congress for medical research funding, insurers and state governments for favourable coverage rulings, and the FDA for quick approvals. (2004, p. 57)[3]

On this view, in post-AIDS America at least, drug regulation had become responsive to patient activism and its associated interests. Grander claims implied that a new disease-based politics had taken centre-stage, displacing the old structures of industry interests, on the one hand, and the 'public interest', on the other. This became known as 'disease-politics theory', which added a new dimension to the social scientific nature of the field, and one whose claims are sufficiently clearly articulated to be scrutinized against empirical evidence. We are generalizing this theory to western countries but, in fairness to its proponents, we should point out that they assert it only in relation to the US. Nonetheless, the US is, of course, a major focus of this book.

There can be, what we call, 'hard' and 'soft' versions of disease-politics theory. The 'hard' version is that the new disease-based politics has been in the *interests* of patients and public health; the 'soft' version is that it has resulted from patient activism/demands/pressure, but whether it has been in patients' interests is left open. A variant on the 'soft' version is the idea that US drug regulators have responded to patient activism and media pressure in order to protect their reputation in the public sphere. This reputational theory, which is most associated with Carpenter (2004; 2010a), is essentially instrumentalist because it implies that public image is paramount for regulatory agencies, rather than that they are making decisions in the best interests of public health – though, of course, the two may coincide from time to time. For this reason, Carpenter's

reputational theory does not necessarily claim that regulation is responsive to the health interests of patients, merely to the demands of patient
activism in order to preserve the regulatory agency's image of serving
patients' interests. He does not, however, take the additional step of
attempting to ascertain whether that reputational strategy is *really* in the
interests of patients and public health. That may be a result of his methodological constraints. Although the broad institutional and historical
sweep of Carpenter (2010a) is very impressive, like Hilts (2003), he does
not undertake any in-depth analysis of the techno-scientific basis for
regulatory decisions regarding specific drugs, so it may have been difficult for him to comment with confidence on whether regulatory judgements were in the interests of patients and public health.

The techno-scientific aspects of regulation are particularly important
when investigating decision-making about innovative pharmaceuticals.[4]
In that respect, an important contribution to pharmaceutical studies
during the 2000s has been the sociology of expectations applied primarily to technological innovation, especially in the areas of biotechnology
and medical technology (Brown and Michael 2003; Brown and Webster
2004; Pollock and Williams 2010). Hedgecoe's (2004) work stands out as
an application of this 'expectations theory' to pharmaceuticals, specifically pharmacogenetics. The principal contribution of this theory is the
idea that innovations, including pharmaceutical innovations, do not
progress in development and/or reach the market solely, or perhaps
even primarily, because of compelling techno-scientific logic, but rather
because various social actors, such as drug manufacturers or particular
laboratory scientists, make promissory claims about the social/health
value of the new technology/drug, which create powerful expectations
about (and hence demand for) that technology within wider society,
including patients. This is what we refer to as 'promissory science'.

In the pharmaceutical sector, 'expectations theory' maps most directly
on to drug promotion and marketing. Consequently, it has much in
common with another strand of research that also developed during
the 2000s, namely studies of pharmaceutical marketing (Applbaum
2007; Fishman 2004; Lakoff 2005; Sismondo 2008). The way in which
pharmaceutical companies promote some scientific studies during drug
development and recruit medical professionals to act as 'opinion leaders'
to support the marketing of new products has been written about for
decades (Abraham 1995b; Collier 1989; Relman 1980). Nonetheless,
before the 2000s, the overwhelming approach to pharmaceutical
marketing was to analyse promotion and advertising of drug products
once they had reached the market. The recent studies of pharmaceutical

marketing have given renewed emphasis, and drawn particular attention, to the role of marketing strategies in promoting clinical trial results and even medical conditions, long before an associated drug product actually reaches the market. Those studies can be seen to dovetail with 'expectations theory' because the purpose of such marketing strategies, which amplify promissory science, is to influence how medical professionals, patients and regulators view a forthcoming pharmaceutical.

Expectations theory and marketing studies help to build up a social scientific picture of what may be happening within pharmaceutical innovation and regulation, especially at the interface between industrial science and medical professionals. However, they differ from the other theories we have discussed in this section in a number of respects that point to some limitations in how they have developed to date. Expectations theory tends to concentrate its study on the social *processes* of *early-stage* technological innovations and much less on outcomes, such as regulatory decisions or health outcomes in use. In particular, while identifying how social actors create expectations via various promissory claims about new medical technologies, expectation theorists rarely, if ever, follow through with an analysis of whether or not those promissory claims are valid – and by implication what the truth-value of that promissory science tells us about whether or not the new technologies are in the best interests of patients' health. Similarly, while it is interesting to learn about how transnational pharmaceutical firms go about constructing their marketing strategies, analysts of such marketing rarely engage in any substantive investigation to determine whether such marketing claims are true, and, therefore, whether the claims represent legitimate dissemination of scientific information or commercial bias, together with what then follows for public health.

In addition, pharmaceutical studies focused on industry marketing strategies tend to ignore the regulatory dimension, as if to suggest that understanding the promotional activities of companies is sufficient to explain how new drugs reach markets. However, such a suggestion is mistaken because, no matter how extensive and sophisticated the marketing strategy of a drug company is, the ultimate decision about whether or not a new drug is permitted on to the market rests with the relevant regulatory authority. Hence, the role of regulatory agencies is absolutely crucial to understanding how pharmaceuticals reach the market. While expectations and marketing theory have been a valuable auxiliary to the development of pharmaceutical studies as social science, these limitations have undermined their potential to inform policy.

The emergence of this rich array of theoretical perspectives on pharmaceuticals in society has never been brought together before. We suggest that it represents the beginning of a social science (sub)discipline concerned with the sociology of pharmaceuticals and public policy. That is to say, the scientific study of the socio-political relations of pharmaceutical production, development and consumption. Evidently, our focus in this book is on regulation, innovation and health.

The theoretical and methodological approach of this book

Following Abraham (1995a; 2008), we take an empirical realist, interests-based approach. That is, we presuppose that within regulated capitalism, such as exists in the pharmaceutical sectors of Europe and the US, drug firms have objective commercial interests in maximization of profits for their shareholders and investors, while patients and the wider public have objective health interests in the maximization of the benefit–harm and benefit–risk ratios of pharmaceutical products. The pharmaceutical industry often argues that it is a highly profitable industry because it manufactures products that patients and healthcare systems need, with the implication that the commercial interests of drug companies and the health interests of patients and the public coincide. Sometimes those interests do converge, but as Abraham (2008) has pointed out, the very existence, and historical development, of government intervention to regulate the pharmaceutical industry logically implies that those interests can often diverge or conflict, and that consequently pharmaceutical manufacturers cannot be trusted to be the sole arbiters of whether their products are in the health interests of patients. If drug companies could be so trusted, then the existence of government regulation to check the safety and efficacy of pharmaceutical products would be unjustifiable. Thus, government drug regulatory agencies have been established ostensibly to regulate the pharmaceutical industry in the interests of patients and public health. Moreover, both the EU and US drug regulatory agencies accept that it is their legal responsibility to protect and promote public health.

Like the well-known and centuries-old philosophy known as 'positivism', our realist methodology is committed to the pursuit of truth and the identification of mechanisms and causes to explain objective phenomena. For many readers, this statement may seem like 'common sense', yet it is surprising how many scholars from the late twentieth century onwards have become uncomfortable with, and eschewed, the idea that a crucial role for social scientists is to the discover the truth about our world. A notable exception is Carpenter, who recently applied a 'positivist' approach to his investigation of the FDA (2010a, p. 28).

Although our realist approach has much in common with positivism, it also differs from it in some important respects. Throughout the ages, one of the hallmarks of positivism has been the conviction that the best way to get at the truth is to (attempt to) adopt a standpoint of 'value-freedom' or 'value-neutrality', except, of course, for a value-commitment to pursue truth itself ((Hammersley 1995; Keat 1981; Weber 1949). This 'neutrality standpoint' has become so pervasive within academic research that it has infected huge areas of social science well beyond positivism.[5] Yet it is misguided, serving more as an academic ideology that mistakes 'neutrality' for objectivity, than a principle of (social) scientific endeavour (Lukes 1973).

Instead of taking the problematic notion of 'neutrality' as our starting point, we apply a modified version of the transcendental philosophy[6] utilized by the realist philosopher of science, Roy Bhaskar (1975), by asking the question: what must the pharmaceutical sector be like in order for its raison d'etre to make sense? This is not an arbitrary, subjective, or utopian question. All parties in the sector, including the pharmaceutical industry, agree, publicly at least, that the intelligibility of producing prescription pharmaceuticals (and other medical drugs) in society is to improve health. While the pharmaceutical industry may fulfil other important socio-economic objectives, such as employment and tax revenues, such objectives are secondary, not least because they could be achieved by expansion of other industries without a pharmaceutical industry. Thus, in order for the existence of the pharmaceutical industry to make sense, its products should improve health. It follows, therefore, that in analysing European and American drug regulation, it would be very strange, and indeed make little sense, to adopt a neutral standpoint about whether such regulation should be in the interests of public health, as that is the ostensible raison d'etre of such regulation. Rather, our pursuit of truth and explanation is informed by an objective[7] value-commitment to determine how well such regulation meets health interests.

Our approach also differs from the historical constructivism of Epstein (1996), Marks (1997) and Daemmrich (2004), and the sociological constructivism to be found in much of expectations theory, which tend to limit analysis to how agents create and act upon their beliefs, networks and goals, falling short of relating such agency to a common framework of objective health interests. While such constructivism can provide some valuable insights, as far as it goes, we contend that social science and policy analysis needs to incorporate not merely what industry, government, professionals and patient/consumer organizations are doing in 'their worlds', and what their agendas might be, but also how their activities relate to the primary purpose of objective health improvement via drug treatment. Following Lukes' (2005) conceptualization of power, our objectivist realism has the further implication that patients' desires and demands are *not necessarily* consistent with either the interests of public health or even their own health interests because patients *may* lack the requisite power (of, say, comprehensive knowledge) to realize their health interests. Furthermore, within our theoretical framework, a powerful drug regulatory agency is one which protects and promotes the interests of health effectively, given that that is its raison d'etre, rather than merely maintaining its reputation with pertinent audiences – the apparent foundation of Carpenter's (2010a) instrumentalist (not realist) conceptualization of power. While a regulatory agency may establish instrumental power by furthering its reputation with particular audiences, we ask the deeper question: does that instrumental, reputationally based power translate into real power to maximize the interests of public health? Similarly, while a pharmaceutical firm may generate expectations about the therapeutic value of a drug through promissory science, we scrutinize the validity of the claims inherent in those expectations by setting them alongside the techno-scientific data supposed to support them.

We have spent the last eight years investigating these issues with respect to the regulation of innovative pharmaceuticals in the EU and the US in the neo-liberal era. That has included several years of field-work in the US and across Europe between 2003 and 2008. Much of our research involved the collection and analysis of documents while in the field and when based at 'home-desks'. Documentary and archival analysis was complemented by some 50 semi-structured interviews, out of 109 sought, with scientists and managers from the pharmaceutical industry, current and former EU and US drug regulators, expert science advisers, relevant legislators and clinical investigators, and representatives of American and European patients' groups, consumer organizations, and public health advocacy bodies. Respondents' requests for anonymity have been respected throughout this book.

We chose to investigate drug regulation in Europe and the US because they are the two largest pharmaceutical markets globally and homes to the two largest and best-resourced regulatory agencies in the world. Japan, the third largest pharmaceutical market in the world is beyond the scope of this book and deserves a separate study. The decision to research both the EU and the US was made as much because we are interested in both regions as a desire to compare the two. Nonetheless, we are fully aware that international and inter-regional comparison can sharpen social scientific analyses, so we certainly utilize it in this book but it is not the sole or pivotal analytical device. Moreover, often when social scientists compare countries or regions, they focus only on contrasts. In this book, we are as interested in evidence of convergent regulatory trajectories and outcomes between the two regions as we are in contrasts.

Having said that, significant contrasts can sharpen the dimensions of analysis and even aid explanation by providing 'natural political experiments'. For instance, at the macro-political level, if a particular regulatory development exists in one region, but not in the other, upon comparing outcomes in the two regions, one can draw on that underlying contrast to evaluate that particular regulatory development. At the micro-sociological level one can compare contrasting regulatory decisions about specific innovative pharmaceuticals to help to uncover the reasons for a particular regulatory approach. For us international comparison is not an end in itself, but a method to be applied in the service of a larger social scientific investigation.

In Chapter 2 we concentrate on the macro- and meso-levels of political analysis over time. Specifically, we examine longitudinally, and at an organizational and ideological level, how neo-liberal reforms in Europe and the US since 1980 have affected pharmaceutical legislation, drug regulatory agencies, techno-regulatory standards for drug approval, pharmaceutical industry strategies, and patients. To do that, in addition to academic writings, we draw on our analysis of official documents from: the US Congress; American Government Administrations and oversight bodies; the US Department of Health and Human Services (DHHS); the US drug regulatory agency – the FDA; the American National Academy of Sciences (NAS); the US Federal Register; the European Parliament; the European Commission; the EU's supranational drug regulatory agency – the EMEA[8]; EU's Committee for Human Medicinal Products (CHMP)[9]; the European Commission's Directorate-General for Enterprise (DG Enterprise); the European Commission's Pharmaceuticals Unit; the international pharmaceutical trade press; the American Pharmaceutical Research and Manufacturing Association (PhRMA)[10]; the European Federation of Pharmaceutical Industry Associations (EFPIA); medical and pharmaceutical professional organizations; patient groups; and consumer organizations, among many others.

Regarding many background historical events and facts discussed in Chapter 2, we draw heavily on the specialist international pharmaceutical press, known as *Scrip*. It is funded by its subscribers, many of whom are pharmaceutical companies or executives, but some are academic libraries. It is independent of the drug industry as such and provides a window on significant developments within the sector in innovation, regulation and public policy. On the whole, we have used *Scrip* judiciously to provide non-contentious background information, rather than using it to advance a particular position associated with *Scrip*. Our extensive use of it reflects our interest in the industry's activities, rather than any bias in evaluation of those activities. Some articles in *Scrip* are authored by individuals, but many are anonymous and referred to in our bibliography as 'Anon'. In addition, we interviewed individuals with specialist knowledge and/or experience of the key legislative and regulatory developments during the 30-year period involved.

Our longitudinal analysis in Chapter 2 identifies several key changes to the regulation of innovative pharmaceuticals during the neo-liberal era in both the EU and the US. Although such macro- and meso-level analysis can be highly suggestive of the effects of neo-liberalism on pharmaceutical regulation vis-à-vis the interests, organization and ideological representations of the stakeholders involved, a full understanding of those effects must also explore the techno-science, regulatory processes and outcomes regarding innovative pharmaceuticals themselves.[11] That requires the shift to a micro-sociological level of analysis, which we undertake in Chapters 3–6. Those four chapters examine a number of case-study innovative pharmaceuticals, which have been regulated in ways specifically relevant to the neo-liberal period, such as approved quickly under priority review, approved on the basis of non-established surrogate markers of efficacy, or maintained on the market using risk management strategies. The case-study drugs were selected on a number of criteria. First, they were all officially defined as innovations, and indeed most were officially regarded as offering modest or significant therapeutic advance. Second they were all approved on to the market in either the EU or the US or both after 1995 when the supranational EU drug regulatory agency, the EMEA, was established. Third a selection of regulatory scenarios was chosen deliberately to sharpen comparative insights into the regulatory decision-making process along the dimensions of type of illness for which pharmaceutical intervention was introduced, and the inter-regional dimension. Thus, the case studies include drugs to treat life-threatening conditions (e.g. Iressa for cancer), serious conditions (e.g. the glitazones for diabetes, Tasmar for Parkinson's disease, and Orlaam for opiate addition), bacterial infections, which vary in seriousness (e.g Trovan), and what we think most readers would acknowledge are usually less serious conditions (e.g. Lotronex for irritable bowel syndrome). Inter-regionally, the case-study scenarios include drugs approved on to the market in one territory, but never approved in the other; drugs approved in both territories and maintained on both markets; and drugs approved in both territories, but subsequently withdrawn from the market as unhealthy in one territory, while permitted to remain on the market in the other territory. The case studies were not chosen because they were publicly controversial – less than 30 per cent of them could be regarded as such at the time they were selected for study, though subsequently two of the drugs (the glitazones) became publicly controversial, four years after we had chosen them.

To investigate the regulation of these drugs, we undertook extensive documentary and interview research about each one in the EU and the US. The academic techno-scientific literature about each of the drugs, and associated information about the conditions they were developed to treat, was reviewed. In addition, we drew upon many non-academic sources. For example, freedom-of-information requests were made to the US government to obtain all the publicly available regulatory data about the drugs. We also scrutinized numerous regulatory documents produced by the EMEA, the CHMP, the FDA, and FDA advisory committee hearings that were pertinent to the drugs. Specifically, we analysed publications about the drugs by their manufacturers; the FDA's reviews of the drugs' application for approval on to the US market; the EMEA's European Public Assessment Reports for the case-study drugs, which provide summary bases for approval of drugs across the EU; and American and European technical analyses of the drugs' post-marketing studies. We interviewed regulators, scientists and other key actors (including some patient activists), who were knowledgeable about the regulation of the case-study drugs. The expertise and relevance of the interviewees was determined by our systematic review of the techno-scientific literature and documentary data analysis about each drug.

No table can do justice to the complexity of issues involved in our analyses of the case-study drugs because the roles of institutional and political context over time are so significant in piecing together the relationships between theoretical understanding and empirical evidence. Nonetheless, Table 1.1 shows some of the key comparative dimensions salient to the case studies found in Chapters 3–6, though, as we noted above, it should be borne in mind that the EU/US comparison is only one aspect of this book, which is concerned holistically with regulation of innovative pharmaceuticals in both regions.

Table 1.1   Some key comparative EU/US dimensions

Comparative Dimension	Iressa	Glitazones	Lotronex	Tasmar, Trovan, Orlaam
Policy Context	Extent of accelerated approval	Commitment to placebo-controlled trials	Flexibility of regulations	Commitment to post-market risk-assessment
Decision-making Process	Public participation and patient–industry complex	Expectations of innovation	Public participation and patient–industry complex	Reducing risk to individual patients versus number of patients at risk
Regulatory Outcomes	Approval versus non-approval	Strictness of approval	Approval versus non-approval	Market withdrawal versus risk management
Temporal Trends in Regulatory Policy	Convergence – EU follows US	Mutual Convergence of EU and US	Non-convergence	Convergence – EU follows US

Pharmaceutical product innovation and regulation

In this section, we explain some of the basic background to the workings and definitions of pharmaceutical regulation and innovation. While the term 'innovation' is often used to mean many things, or frequently unhelpfully used with no clear definition at all, within the pharmaceutical sector, drug product innovation has a specific meaning. Similarly, too often the term 'regulation' is discussed with little or no precision, sometimes for ideological reasons, rather than to aid systematic analysis of regulations' objectives and achievements. It is crucial that all parties understand that they are using the same fixed definitions of terms, so that intelligent analysis can proceed. We begin by explaining the well-established understanding of pharmaceutical product innovation within industry and policy literature, and then outline how it is related to the basic principles of drug product regulation in the US and the supranational EU.

When a pharmaceutical firm or laboratory discovers a new compound with some therapeutic potential its novelty status is defined by its technical uniqueness from other molecules. The discovery of such technical novelty may also be regarded as a form of intellectual property, especially because work and resources have been involved in the discovery process. In western countries, drug companies typically make applications to patent their discoveries of new compounds. Patent offices provide authoritative determination and record of the technical uniqueness of the compound together with the party who made the discovery. The patent protects the intellectual property rights of the company/ laboratory that made the discovery by preventing any other enterprise from copying the discovery for commercial purposes. Such protection is afforded by patent and intellectual property laws usually for 15–20 years from the date the patent is issued.

The new, technically unique, drug compound is known as a 'new molecular entity' (NME) – the preferred term in North America. In Europe, NMEs are often known by the term, 'new active substance' (NAS). Within the class of NMEs/NASs, there is a further distinction sometimes made between biologically active substances (e.g. vaccine compounds), known as 'new biological entities' (NBEs) and all the other chemical compounds, known as 'new chemical entities' (NCEs). Although the number of NBEs has become more significant in the last 20 years, it remains the case that most NMEs are NCEs. Understanding these terms is important because they are used in official statistics released by governments and the industry about pharmaceutical product innovation. As far as possible we use the term, 'NME' or 'NAS', distinguishing further only when the statistics demand it.

If an NME reaches the market, then it is defined as a drug product innovation (or an innovative drug product). As this book is concerned only with innovative drug products, it follows that we will be discussing only NMEs/NASs. For stylistic reasons, we will sometimes refer to 'new drugs' or 'new drug products', but this is just another way of expressing NMEs. In order to reach the market, the NME must gain regulatory approval from government agencies responsible for the markets in question. In this book, that means the EU and the US. When a new drug is granted permission on to the market by the regulatory agency in the US, it is referred to as a 'new drug approval', whereas in the EU, it is referred to as 'marketing authorization'. In this book, to simplify matters, we refer to 'marketing approval' for both regions.

Evidently, pharmaceutical product innovation is, in effect, defined by technical and commercial (intellectual property and market) criteria, except for the requirements imposed by regulatory agencies to meet health-related criteria. Those health-related criteria, which exist in both EU and US pharmaceutical regulation, are drug quality, safety and efficacy. By 'quality' is meant the purity of the drug compound – that it is not adulterated with other 'ingredients', and that it meets very high standards of purity. In the US, the 1906 Pure Food and Drugs Act was passed enabling the federal government to regulate drug quality to standards set by government chemists (Abraham 1995a). In most European countries, similar legislation and regulatory controls were introduced in the late nineteenth or early twentieth century (Abraham and Lewis 2000). For some decades now, laboratories in the mainstream pharmaceutical sector in Europe and North America have been able to assure high standards of drug quality with relative ease (Abraham and Sheppard 1999). Although fraudsters involved in drug adulteration and the production of counterfeit 'medicines' still exist and present a danger to public health, their impact on medicines in western countries since the 1950s has been comparatively small, and different in nature from the science and politics of drug efficacy and safety.

We presume that the meaning of drug 'safety' is self-evident in broad terms, while 'efficacy' relates to a drug's effectiveness in treating the illness or condition for which it has been approved for use, or is hoped to be approved for use. Hence, in the context of pharmaceutical regulation, it is possible for a drug to be of very high quality, but to be unsafe or ineffective or both. This book is almost entirely concerned with drug efficacy and safety, primarily efficacy. The issue of drug quality will not be discussed further.

In the US, the 1938 federal Food, Drug and Cosmetic Act (FD&C Act), as amended in 1962, requires that manufacturers wishing to market an NME must demonstrate that the product is safe and effective for use (as well as meeting adequate standards of quality). The legislation established an implicit public right to protection from unsafe and ineffective drugs. Similarly, in the EU, by the time the EMEA began work in 1995, Articles 11(1) and 68(1) of Regulation (EEC) 2309/93 and Articles 26(1) and 126(1) of Council Directive 2001/83/EC had established that marketing approval required evidence of drug efficacy, safety and quality (European Council 1993, p. 1; 2004, pp. 67–128). Indeed, such regulatory requirements existed in all major western European countries from 1980 (Abraham and Lewis 2000).

Thus, within the basic drug regulatory system that existed before and throughout the neo-liberal era in Europe and the US, a pharmaceutical product innovation was, and is, defined as an NME that has gained marketing approval by demonstrating that it is safe and effective. The basic system does not require new drugs to offer therapeutic advance for patients or public health by, for example, being more efficacious than drugs already on the market to treat the same condition. As the forgoing discussion reveals, the official accounting of innovation in the sector means that a new drug may be counted as an 'innovation', but offer no therapeutic advance. An 'innovative drug', in this official sense, might offer therapeutic advance or even be a major therapeutic break-through, but equally, it might be less effective than drugs already on the market. Logically, therefore, innovation does not necessarily imply ther-apeutic advance, and systematic analysis of the pharmaceutical sector must keep 'product innovation', on the one hand, and 'therapeutic advance', on the other, conceptually distinct. It is, then, a matter of empirical evidence whether an individual drug innovation also provides therapeutic advance.

This is reflected in the nature of the key clinical testing required by regulators to demonstrate drug efficacy for the purposes of marketing approval. To appreciate that, it is necessary to understand some of the principles and terminology of clinical drug testing design. When a new drug is being developed various types of clinical experiments may be conducted with patients and healthy volunteers, known as clinical trials. Such trials may be 'controlled' or 'uncontrolled'. If 'uncontrolled', then they involve merely observation of the patient (or healthy volun-teer) while taking the experimental drug over some period of time. By 'controlled' is meant that the trial is designed to compare the experi-mental drug with some 'control'. If the 'control' is placebo, then the experiment is known as a 'placebo-controlled trial', but if the 'control' is another therapy, such as a drug, then the experiment is said to be an 'active-controlled trial'. In controlled trials, some patients (or healthy volunteers) are allocated to a group taking the experimental drug, while others are allocated to a group taking the 'control'. This can be done in many different ways. If the clinical investigators and the patients (or healthy volunteers) both know who is taking what, then the trial is referred to as 'open'; if the clinical investigators know, but the patients (or healthy volunteers) do not know, then it is a 'single-blind' trial; and if neither investigators nor patients/healthy volunteers know who is taking the experimental drug and who the 'control', then the trial is said to be 'double-blinded'. The precise way in which each patient is allo-cated to the experimental or control group can also vary, but probably the most common technique is for the allocation to be done randomly, in which case the trial is said to be 'randomized'. The double-blinded, randomized controlled trial (RCT) is generally regarded as the most scientifically rigorous type of design.

Since the 1960s in the US, and the 1970s in Europe, the regular process of drug development for the purposes of gaining marketing approval from regulatory agencies has comprised laboratory and animal tests followed by three sequential phases of human trials, known as phases I, II and III (see Figure 1.1). Phase I trials mark the first occasion that the new drug is given to people. Such trials are conducted with a small number of healthy volunteers to ascertain basic human safety and toxicity. If the drug seems fairly safe when taken by healthy people, development proceeds to phase II, which involves medium-sized clinical trials, sometimes placebo-controlled, with a view to collecting preliminary evidence about the drug's safety and efficacy in patients with the condition the product is intended to treat. Typically, phase III studies are larger RCTs, whose purpose is to provide statistically significant evidence of a drug's effects on the appropriate patient population. Phase III studies are the most costly component of drug development for manufacturers (Office of Technology Assessment 1993, p. 56). Clinical trials conducted *after* a drug is on the market (post-marketing), are known as Phase IV. The full range of design possibilities for trials (e.g. double-blinded or placebo-controlled, etc.) may be applied to trials of any phase.

To be granted marketing approval by regulators, clinical drug testing must, of course, progress from healthy volunteer trials to trials with patients suffering from the condition the drug is being developed to treat. However, regulatory agencies typically require only that new drugs are more effective than placebo in order to meet the drug efficacy approval standards, though there are some exceptions, such as antibiotics. Indeed, an NME can be approved on to the market even if it is therapeutically inferior to available medicines. To discover whether a new drug is a therapeutic advance on existing therapies, then it is necessary to require active-controlled trials. In the US, section 505(d) of the FD&C Act requires that manufacturers must provide 'substantial evidence' of a drug's efficacy 'consisting of adequate and well-controlled investigations'. The FDA has interpreted that phrase to mean at least two adequate and well-controlled trials. For some decades, randomized, placebo-controlled trials and active-controlled trials have both been listed as acceptable types of study-design in the FDA's regulations, but there is not, and never has been, a general requirement for active-controlled trials where proven treatments already exist (Code of Federal Regulations 2010, pp. 151–3; Temple 1997).

In the EU, Annex 1 of Council Directive 2001/83/EC states that:

> In general, clinical trials shall be done as controlled clinical trials if possible, randomized and as appropriate versus placebo and versus an established medicinal product of proven therapeutic value; any other design shall be justified. The treatment of the control group will vary from case to case and also will depend on ethical considerations and therapeutic area; thus it may, in some instances, be more pertinent to compare the efficacy of a new medicinal product with that of an established medicinal product of proven therapeutic value, rather than with the effect of a placebo. (European Council 2004)

Like the FDA, the EMEA has interpreted this to mean that active-controlled trials are not required for marketing approval, though the EU regulatory agency has the discretion, on a case-by-case basis, to recommend that placebo-controlled studies are not sufficient. Hence, during the neo-liberal era, an NME granted marketing approval, thereby becoming an innovative pharmaceutical product, has not necessarily offered any therapeutic advance over drugs already on the market because, typically, comparison of the NME's therapeutic value with those other drugs has not been required. In the EU, this was underlined by Articles 11(1) and 68(1) of Regulation (EEC) 2309/93 and Articles 26(1) and 126(1) of Council Directive 2001/83/EC, which established that a marketing approval could only be refused on the basis of safety, quality or efficacy, with the clear implication that failure to demonstrate therapeutic advance over existing drugs could not be a basis for denial of approval (European Council 1993, p. 1; 2004, pp. 67–128).

Nonetheless, in some therapeutic contexts, ethical issues regarding the circumstances of patients has led regulatory agencies to apply their discretion. For instance, the FDA has required active-controlled, rather than placebo-controlled, trials when existing treatments have been shown to improve survival or reduce severe morbidity because, under those circumstances, the consequence of not treating patients (by giving them placebo) would be serious – affecting their long-term health (Temple and Ellenberg 2000, p. 455). As a concrete example, FDA scientists told us that the regulatory agency has always required an active control in clinical trials to test a new antibiotic where effective therapy exists.[12,13] Furthermore in such circumstances, it has been considered unethical by both the FDA and the EMEA to approve a drug that was 'globally inferior' to existing treatments, with respect to *safety and efficacy* (EMEA 2001a).[13]

There are circumstances, therefore, in which regulators require clinical trials involving active controls, which are designed to compare the experimental drug with another existing treatment for the same condition (the 'comparator drug'). Active-controlled trials generally represent a higher standard than placebo-controlled trials and are more challenging for manufacturers because it is more difficult to prove that an experimental drug is better than an existing therapy than to prove that it is better than nothing. When active-controlled trials are performed, they can take one of two forms: either a superiority trial or a 'non-inferiority' trial. A superiority trial is designed to test whether the experimental drug is statistically significantly better than the comparator along some dimension, usually efficacy. In fact, placebo-controlled trials are a type of superiority trial because they are designed to test whether the experimental drug is better than placebo. An active-controlled superiority trial investigates whether the experimental drug is more effective than an existing treatment. The active-controlled superiority RCT design is the most demanding standard against which to test an experimental drug.

Regulators, however, frequently permit pharmaceutical firms to conduct 'non-inferiority' active-controlled trials. In 'non-inferiority' trials the company investigates whether the new drug achieves efficacy that is equivalent to some pre-specified, statistical fraction of the comparator drug's efficacy. In other words, to pass the test of a 'non-inferiority' trial, the new drug need not demonstrate that it is of equivalent efficacy to the comparator drug, but only that it is no worse than the comparator by more than a particular amount. That amount is known as the 'non-inferiority margin' or, in the techno-scientific jargon, as 'delta' (EMEA 2004a, p. 2). The 'non-inferiority' design may be valuable in some situations, such as when a new drug is less effective than existing treatments, but because it has safety advantages over them, it could nonetheless be a valuable therapeutic option. On the other hand, as a regulatory standard, 'delta' is often allowed to be as high as 20 per cent – a statistical margin that could be clinically important (Li Bassi *et al.* 2003, p. 249). Evidently, the 'non-inferiority' trial design is a lower test standard than the active-controlled, superiority design.

These complex machinations of new drug development and regulation explain why it is possible that an NME, which reaches the market, may be defined as safe and effective, but offer no therapeutic advance over existing therapies or even be less effective than them. As we will explain in Chapter 1, that reality is made explicit by the FDA, which classifies NMEs according to those that do, and those that do not, offer therapeutic advance. Clearly, it is in patients' interests that innovative drugs offer therapeutic advance, while drug innovations providing no therapeutic advantage are unlikely to be in the interests of public health. By contrast, all pharmaceutical innovations, irrespective of therapeutic advance, are in the commercial interests of their manufacturers, with the possible exception of safety disasters.

One important dimension linking patents/intellectual property rights, innovation, regulation and commercial interests is time. As noted above, patents and intellectual property rights, granted soon after NMEs are discovered, prevent competitors, known as generic companies, from copying NMEs discovered by the research-based pharmaceutical industry for about 15–20 years. After the patent expires, the generic firms are permitted to make copies of the NMEs, selling them at relatively low prices that usually undercut the original brand-name version. The generic companies are able to charge lower prices partly because they have not had to invest in any of the research and development needed to discover and bring the drug to market. Consequently, the research-based pharmaceutical firms, which generate drug innovations, have strong commercial interests in maximizing the amount of time their innovative products are on the market while patent-protected; and conversely, in minimizing the development and regulatory review time needed to attain marketing approval because drug development and regulation eat into patent-protected time. Rapid drug development and regulatory review-times enable fast returns on companies' investments. To get some idea of the scale of commercial interests involved, with respect to the time dimension, it has been estimated that a pharmaceutical firm could lose on average over a million US dollars for each day's delay in gaining marketing approval from the FDA (Montaner *et al.*, 2001). In 2003, the FDA Commissioner, Mark McClellan, claimed that cutting FDA review times by 10 per cent would reduce drug companies' average development costs by US$12 million per drug (Ault 2003).

In the last decade, the problem of access to new patented high-priced anti-retroviral drugs to treat HIV/AIDS in poor countries has generated a debate, especially within bio-ethics, about whether the patent system should be retained or not because it permits monopoly pricing (Mannan and Story, 2006; Selgelid and Sepers, 2006). However, during the neo-liberal era, the protection of intellectual property rights has expanded internationally, albeit with significant attempts to create opportunities for poor countries to access cheaper versions of drugs to treat life-threatening conditions (t'Hoen, 2009). For better or worse, the patent system, therefore, has been a cornerstone of pharmaceutical research, development and innovation. As long as commercial enterprise and goals remain a significant part of new drug development, some form of protection of industrial investment in new drug discovery and development from commercial competitors – who have made no such investment – is likely to be required to maintain incentives for the pharmaceutical research-based firms. The existence of such an intellectual property regime does not, however, preclude regulation of pricing to ensure adequate access to essential drugs for the poor. Whatever the future holds for the pharmaceutical patent system and associated intellectual property rights regimes, within our analysis of the pharmaceutical sector in the neo-liberal era, patents are part of the mechanics of the drug development process along with the regulatory requirement to meet quality standards. Patents and their effects on pricing, about which much has been written elsewhere, are not a major focus of this book.

When a new drug is granted marketing approval by a regulatory agency, it is approved to treat a specific disease or condition in particular ways, which are known as 'indications'. For example, if a drug is said to be 'indicated for non-small-cell lung cancer', then that means it has been granted marketing approval to be prescribed for non-small-cell lung cancer. Every prescription drug is approved with a list of information about the new drug for doctors (as well as a patient information leaflet with which readers will be familiar). That list of information for doctors is known as the product 'label' in the US, but within the EU regulatory system it is known by the obscure term, 'summary of product characteristics (SPC)'. In this book, we will use the much more user-friendly, American term, 'label'. The label contains all kinds of information about the drug, including its indications, adverse effects, dosage, associated precautions and warnings and so on.

Product labelling is produced by the manufacturer, but its contents must be approved by the appropriate regulatory authorities. The labelling defines the legal boundaries within which the drug may be promoted to doctors by the manufacturer. However, doctors may prescribe drugs to patients in ways not indicated on the label – so-called 'off-label' use. It is legal for doctors to do that, but they are much more exposed to hostile drug injury litigation if the patient is harmed by 'off-label' use. Drug injury litigation is typically aimed against the drug manufacturer if the drug was prescribed according to the label, but with off-label use the pharmaceutical company may argue that the drug itself was not at fault, but the doctor was for improper prescribing. Some estimates put off-label use of prescription drugs as high as 20 per cent, or even 60 per cent in some therapeutic areas (Light 2010, p. 21). Off-label use can be a particularly significant issue for drug regulation when government authorities decide to leave a product on the market, but with new labelling restrictions imposed because of the emergence of major safety problems with the drug. In that context, doctors who do not share the government's level of concern about the drug's safety (perhaps because in their experience it has been benign) may continue to prescribe it as before in an unrestricted fashion (and hence off-label).

Once a drug is on the market, it becomes subject to post-marketing surveillance regulations. The manufacturer is required by law to report to the appropriate regulatory authorities all adverse reactions to its drugs in a timely manner. The timing and extent of such reporting typically varies according to the seriousness and unexpected nature of the adverse reactions. In addition, doctors are asked to report all suspected adverse drug reactions (ADRs) they encounter in their clinical practices. Such reporting is voluntary for doctors. Estimates of the extent to which doctors report ADRs range from 1 to 10 per cent. Such reporting is often referred to as 'spontaneous' because, unlike the collection of pre-market clinical trial data, there is no experimental design and no placebo/comparator group. In the last five years or so, regulatory systems in Europe and the US have also developed post-marketing surveillance systems enabling patients to make reports of ADRs directly to regulatory agencies. In addition, regulators may ask manufacturers to conduct post-market clinical trials or epidemiological studies either as a condition of marketing approval or in response to some post-marketing safety problem that arises.

Conclusion

From the discussion above, we suggest that five central theoretical perspectives on pharmaceutical regulation during the neo-liberal period since 1980 can be identified. They are not all mutually exclusive, but they are distinctive in their explanatory claims.

(1) *Neo-liberal Theory* – that the pro-industry deregulatory reforms were instigated by government to promote innovation that was not only in the interests of the pharmaceutical industry, but also in the interests of patients and public health. Embedded in this theory is a 'coincidence of interests' thesis – that the pharmaceutical industry operates largely in the interests of patients and public health, so requires only 'light-touch' regulation. Typically, this is the position taken by officials who speak for governments and the top managerial echelons of regulatory agencies and the drug industry.

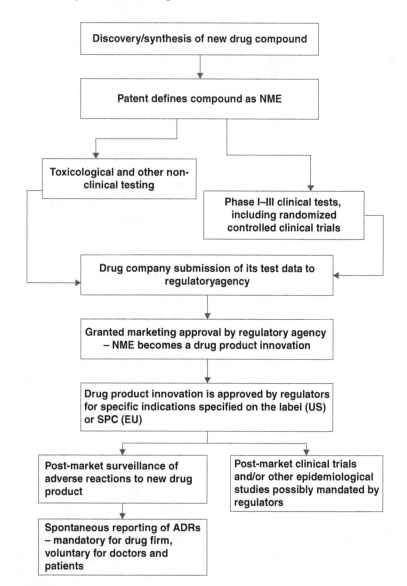

Figure 1.1 The development and regulation of pharmaceutical product innovation

It overlaps with expectations theory inasmuch as the therapeutic promise of techno-scientific innovation is implicated in the claim that the deregulatory reforms of the neo-liberal era have been in patients' interests. However, expectations theory is very different because it does not assume that promissory science and innovation is in the interests of patients and public health, and its focus is located around micro- and meso-processes of innovation, rather than macro-political drivers.

(2) *Capture Theory* – that government regulatory agencies, whose legal mandate defined and created by the Legislature was to regulate industry in the public interest, gradually became captured by the regulated industry, so that, over time, during the neo-liberal era, it has actually regulated primarily in the interests of the industry, instead of the public interest. The main thrust of capture theory is, of course, diametrically opposed to neo-liberal theory because the focus of the former is on how shifts towards 'light-touch' regulation by the bureaucracy ('administrative drift') reflect capture by industry interests at the expense of the public interest, whereas the neo-liberal theory proclaims those shifts to be embraced by government and consistent with the public interest.

(3) *Corporate Bias Theory* – that some organized interests (for instance, the pharmaceutical industry) gained privileged access to the state, over and above other interest groups at the highest levels of the government (the Executive and Legislature) as well as regulatory agencies (the bureaucracy) to the extent that those organized interests have governed in partnership with the state, setting the agenda for regulation so that it has been biased in favour of those interests at the expense of diverging or conflicting interests. Although we have emphasized the differences between capture theory and corporate bias theory, it is possible for them to mutually reinforce each other. For instance, in the context of corporate bias, an industry might persuade the Executive and/or Legislature to require the regulatory agency to interact with that industry in a way that increases the likelihood that the agency will be captured. Once the regulatory agency is captured that might provide the industry with more opportunities to extend corporate bias in its favour.

(4a) *Disease-politics Theory (Hard)* – that drug deregulatory reforms have been the result of patient activism and changes in patient culture towards consumerism in the aftermath of the crisis over access to AIDS treatment, thus reflecting patients' interests. This theory shares with neo-liberal theory the idea that the deregulatory reforms of the neo-liberal era have been in patients' interests. However, unlike neo-liberal theory, which claims that patients' interests have been met as a coincidental effect of releasing more industrial innovation through deregulation, 'hard' disease-politics theory regards the emergence of such deregulation as a direct response to patient activism and demand.

(4b) *Disease-politics Theory (Soft)* – that drug deregulatory reforms have been the consequence of patient activism and changes in patient culture towards consumerism in the aftermath of the crisis over access to AIDS treatment, which altered regulatory agencies' approach to maintaining their public credibility/reputation. In terms of identifying the primary drivers of deregulatory reform, this theory is at odds with capture theory and corporate bias theory. However, it could be compatible with them in the sense that patient activism could facilitate regulatory capture, corporate bias, or both, as secondary effects.

 (5) *Expectations/Marketing Theory* – that the creation of expectations and promissory science around new drug/medical innovations among networks of medical professionals, research scientists and patients has grown during the neo-liberal period and put pressure on regulatory agencies to provide access to those innovations even if that has undermined existing regulatory standards. If the promissory science in question is consistent with patients' interests, then this theory can complement 'hard' disease-politics theory, as expectations fuel patient activism. On the other hand, if the promissory science is contrary to patients' interests, then expectations theory is likely to be more compatible with capture theory or corporate bias theory because the expectations may perform the function of facilitating regulatory capture or industry influence in the political sphere.

A notable aspect of this book is that we relate these theories to the pharmaceutical sector in order to determine in whose interests drug regulation in the EU and the US has developed and operated since the 1980s. Such a determination is not merely of academic or theoretical interest because citizens expect, and the law mandates, that drug regulatory agencies operate primarily in the interests of public health. Hence, the evolution of pharmaceutical studies into social science marks not only an important intellectual step, but also a significant implication for how the field might inform socio-political change and attendant policy improvement. We are now ready to embark on our investigation of how neo-liberalism has shaped the regulation of innovative pharmaceuticals in the last few decades. That journey sweeps far and wide, from the macro-politics of who has controlled the US Congress to micro-social phenomena, such as a drug company's choice of comparator treatments during trials for innovative diabetes drugs in Europe. It is an investigation of how politics, economic interests, ideology, regulatory culture and promissory science are inter-related and embedded in a sector, whose purpose, most agree, should be to improve health.

2
The Political Economy of 'Innovative' Drug Regulation in the Neo-Liberal Era

Introduction

As explained in Chapter 1, the basic legislation governing pharmaceutical regulation in Europe and the US was established by the 1970s. That legislation did not require, as a condition of marketing approval, that new drug products should provide therapeutic advance for patients over drugs already on the market, and nor has it been revised to do so. Nonetheless, other far-reaching institutional, administrative, policy, and legislative reforms concerned with expediting the development, regulatory review, and marketing approval of new drugs have been introduced since 1980. Most of those reforms have been aimed at speeding up patient access to new drugs, product development times, and regulatory review times. European and American governments have justified the reforms on the grounds that they benefit patients and public health by increasing and accelerating new drug innovations that patients need, though, as we show in this chapter, such arguments and rationalizations have rested on a particularly limited conceptualization of patient 'need', namely the need for quicker access to new drugs.

The argument that accelerated drug development and review was a response to patients' demands and interests has been made significantly more widely, and apparently been more persuasive, in the US than in Europe because of the exceptionally active role played by the American AIDS patient movement to expedite access to drug treatment for HIV/AIDS. For instance, upon winning the 1997 'Innovations in American Government Awards Program' in recognition of far-reaching changes to expedite its drug approval process, the FDA (1997a) claimed:

Speeding the delivery of new drugs to Americans, while preserving the FDA's high standards for quality, efficacy, and safety, has always been the primary goal inspiring the drug review program innovations.

This illustrates the way in which the question of how the FDA should respond to the therapeutic needs of patients in relation to its regulation of pharmaceutical products has been framed primarily in terms of speed of access. In fact, the 1997 Food and Drug Administration Modernization Act (FDAMA) created a new mission for the FDA stressing that it should 'promote the public health by promptly and efficiently reviewing clinical research and taking appropriate action on the marketing of regulated products in a timely manner' (FDAMA 1997, section 406b).

In attempting to account for this emphasis on speed, some FDA officials contend that the agency was forced to rethink its assumptions about the purpose of drug regulation as a consequence of encounters with the AIDS patient advocacy movement. One FDA official claimed that, before the mid-1980s, the message received by the agency from the US Congress and public was 'that it is better to take more time and be certain, than to move quickly and possibly make a tragic mistake', but following the 1988 demonstrations outside FDA offices by AIDS activists, 'the traditional relationship between FDA, the active protector, and the passive, protected patient, had to change' (Holston 1997).

Another senior FDA official elaborated and rationalized the situation as follows:

The fundamental debate was: yes, you are a consumer protection agency, but you can protect people to death, and that's what's happening. You've got to live with this tension ... There are some situations where you promote health by protecting people and other situations where you promote health by being a facilitator and making sure they get what they need, and making sure you're not being a roadblock ... And it was *that* piece of the puzzle that the agency had always been blind to. They were so concentrated on [their public health protection role] that they didn't realize they were protecting people to death. And that's really what the HIV population woke this agency up to, philosophically.[1]

As we discussed in Chapter 1, a number of academic scholars, who we call 'disease-politics theorists' have also argued that the demands of patient advocacy groups – and in particular the AIDS advocates – for quicker access to new drugs have driven the procedural and philosophical changes in US drug regulation which occurred during the 1980s and 1990s (Smith and Kirking 1999; Carpenter 2004; Carpenter and Fendrick 2004; Daemmrich 2004).

In this chapter, we trace how the macro-political landscape of the EU and the US influenced drug regulatory reforms. In so doing, we investigate and explain which actors and institutions were most significant in shaping the orientation of regulation of innovative pharmaceuticals in both regions. In particular, we identify the full range of regulatory reforms introduced since 1980 ostensibly in order to stimulate pharmaceutical innovation by accelerating drug development and marketing approval. The evidence presented compels one to appreciate that drug regulatory reform in the period from 1980 is much more complex than the official view that it should be characterized as quickening delivery of innovative pharmaceuticals needed by patients. Our analysis in this chapter reveals the ideological motivations, material interests, and institutional outcomes of such reforms.

Neo-liberal reforms and corporate bias in European nation-states

The UK election of Prime Minister Margaret Thatcher in 1979 and her New Right Conservative Party in three subsequent elections marked the beginning of the neo-liberal shift. The Thatcher Government was elected with a positive commitment to reduce state intervention in the economy; pharmaceutical regulation was to be no exception. At that time, drug regulation was conducted by a section within the UK government's Department of Health called the Medicines Division, also known as the UK Licensing Authority. During the 1980s, the Association of the British Pharmaceutical Industry (ABPI) badgered the British Government to organize drug regulation more efficiently according to the industry's desire for faster marketing approvals. The Thatcher Administration was already planning reform of the civil service as part of its neo-liberal agenda so it was sympathetic to the pharmaceutical industry's claims that state regulation was insufficiently responsive to the needs of business and innovation because it did not approve new drugs on to the market fast enough (Abraham and Lewis 2000, pp. 60–2). However, an alternative perspective of the situation, rarely heard at the time, was that the Thatcher Government had starved the Medicines Division of resources.

In the ten years to 1986, new drug licence applications of all types had increased by 87 per cent, while staff levels in the Medicines Division grew by only nine per cent. In the late 1980s, the FDA had six times as many staff handling drug applications as the UK Licensing Authority (Anon. 1989a, p. 3). The Medicines Division may not have been structurally inefficient, but rather under-resourced. Further doubt is cast on this 'inefficiency thesis' by the fact that from 1961 to 1985 more new 'innovative' drugs (NASs), were first marketed in the UK than in Austria, the Benelux countries, Italy, Scandinavia, Spain, Switzerland, the US, or, what was then, 'the Eastern Bloc' of Europe (Andersson 1992, p. 68). Indeed, in 1988, the UK was found to have the fastest approval times for new drugs in the EU, then 'the European Community (EC)' (Anon. 1988a).

Nonetheless, the neo-liberal reforms went ahead. A new regulatory authority, known as the Medicines Control Agency (MCA), was established with a new director, who came from the pharmaceutical industry. The Medicines Division had recouped about 60 per cent of its annual running costs in fees from pharmaceutical firms for regulatory work involved in the licensing process, while 40 per cent came from the Treasury via taxation. By contrast, under the MCA, the entire running costs of UK drug regulation were to become dependent on fees paid by pharmaceutical firms (Anon. 1989b).

Negotiations over the licensing fees laid bare the 'exchange' underpinning the new arrangements. For example, in 1989, the industry objected to paying a licensing fee as large as £50,000 without assurances that their drugs would pass more quickly through the UK regulatory system (Anon. 1989c). On arrival, the new director of the MCA announced that the agency aimed to reduce the net processing times for new drugs by 24 per cent within a year (Anon. 1989a, p. 2). Between 1989 and 1993, new drug processing times fell by more than half, from 154 working days to just 67, while the number of licences granted for new drugs increased from 57 in 1989 to 77 in 1993 – results for which the pharmaceutical firms paid handsomely, making the MCA one of the richest regulatory agencies in Europe (Abraham and Lewis 2000, pp. 65–6). Moreover, the agency increased its consultation with companies and integrated industry interests into its mission statement, which promoted the perspective that the interests of industry and public health coincided:

> Overall, the agency aims to provide an efficient, cost-effective service that protects the users of medicines while not impeding the effectiveness of the pharmaceutical industry. (MCA 1991, p. 1)

Notably, there is no evidence of patient activism or demands for any of these reforms at that time in the UK. Rather, in 1991, the British Government invited the industry to join a board of experts to advise the

Department of Health on the scope of the MCA's targets and performance (Anon. 1991a). Patient activism or even awareness about such
matters were so absent that government ministers made little attempt
even to present their reforms as if they were a response to patient demands
(Abraham and Lewis 2000, pp. 64–5).

These neo-liberal reforms in the UK were to be influential on other
European countries and the framework for Europeanization of pharmaceutical regulation that ultimately informed the nature of the supranational EU regulatory system established in 1995. For instance, soon
after these UK reforms, one can observe similar developments in Sweden
and Germany. In both countries, a shift to the political right was taking
place, though we are not suggesting that the neo-liberalism of Germany,
Sweden, and the UK were identical. Nevertheless, in the context of pharmaceutical regulation, the commonalities were unmistakeable and far
outweighed subtle differences.

In the context of the growing power of the political right elected to
government in Sweden in 1991, the Swedish Audit Office received relentless complaints from the pharmaceutical industry that drug approval
times were too long (Abraham and Lewis 2000, p. 67). They were certainly
longer than in many other countries – between 1972 and 1983 NASs came
to the market in Sweden slower, on average, than in France, Germany,
Italy, the UK, or the US (Andersson 1992, pp. 62 and 68). In response to
industry complaints, pharmaceutical regulation was removed from the
government's Department of Drugs at the National Board of Health and
Welfare (NBHW), in 1990, and taken over by a newly established Medical
Products Agency (MPA) to improve 'efficiency'. This changed the political culture of Swedish drug regulation. The MPA derived higher licensing
fees from industry on the understanding that the regulatory agency
would deliver faster drug approval times, which were indeed cut by more
than half between 1989 and 1993, together with increased consultation
with pharmaceutical firms. As in the UK, neo-liberal tendencies afforded
the pharmaceutical industry more influence over regulatory policy. For
example, in response to industry requests, in 1993, the MPA agreed to
no longer publish the fact that an application had been rejected, thus
denying the public and wider scientific community access to such knowledge (Abraham and Lewis 2000, pp. 68–70). There is no evidence that
organized patient activism drove or demanded these changes in Sweden,
or even had significant awareness of them.

In Germany too, the election of the neo-liberal Christian-Democrats-led
right-wing coalition in the 1980s and 1990s made the Government more
responsive to pharmaceutical industry complaints about regulation.

In this period, the pharmaceutical industry consistently pressed the German drug regulatory authority, the Bundesgesundheitsamt (BGA) to accelerate new drug approvals in the administrative courts. In fact, from 1961 to 1985, more new drug innovations (NASs) were first introduced in Germany than nearly any other industrialized country, and between 1972 and 1983 more NASs reached the German market faster than in France, Italy, Sweden, or the US (Andersson 1992, pp. 62 and 68). Senior staff at the BGA blamed the poor quality of many industry applications for the slowness of the regulatory review process. For instance, of the 62 new drugs approved in 1988, only two fell into the BGA's category of outstanding significance (Abraham and Lewis 2000, p. 73).

Nevertheless, against this background of dissatisfaction with the BGA in industry and, in 1993, a public scandal concerning HIV-infected blood transfusions to some 350 haemophiliacs, the German Health Minister, who blamed the BGA for the scandal, vowed to dissolve the agency. In 1994, it was replaced by the Bundesinstitut fur Arzneimittel und Medizinprodukte (BfArM). The BfArM was organized into 'business units', which were to be 'customer-oriented', meaning industry-friendly, and dedicated to quickening drug approval times, which were halved by 1996. The new agency also established extensive mechanisms of consultation with pharmaceutical companies to meet industry needs (Abraham and Lewis 2000, pp. 73–5). As with Sweden and the UK, organized patient groups existed in Germany throughout the 1980s and early 1990s, but there is little evidence that they were active in seeking the regulatory reforms visited upon the BGA and BfArM (Daemmrich and Krucken 2000, pp. 517–18 and 526).

Meanwhile, in France, during the 1980s and early 1990s, a corporatist and co-operative partnership between the government regulatory agencies and the pharmaceutical industry remained dominant. The closed and centralized institutional power of this governing system meant that regulatory decision-making was protected from public scrutiny and other groups, such as consumer or patient organizations (Hancher 1990). Unlike Germany, Sweden, and the UK, there is little evidence of sharp neo-liberal government reforms to drug regulation in France in response to industry demands probably because the corporate bias already in place met with considerable industry satisfaction. The absence of a well-staffed centralized bureaucracy, which led the French regulatory agency to delegate responsibilities to external experts sympathetic to pharmaceutical firms, limited the requirements the agency could impose on the industry (Wiktorowicz 2003, pp. 638, 643 and 646). There is even less evidence of patient activists driving regulatory reforms in France in this period (Barbot 2006; Callon and Rabeharisoa 2008).

Thus, in four European countries, reforms (or arrangements in the case of France) aimed at ensuring that drug regulation could deliver rapid marketing approval were established during the 1980s and early 1990s. In each case, those reforms (or arrangements) were established in response to industry demands and to accommodate industry interests. Few scholars dispute this account of regulatory change in Western Europe. However, as we have noted, several scholars in the field contend that the situation was quite different in the US, where it is claimed that pharmaceutical development and regulation were transformed by patient activism and demands, especially AIDS patient activism. For some such scholars, that contrast between the US and, at least Germany, if not the rest of Western Europe, is an explicit part of their international comparative thesis (Daemmrich and Krucken 2000; Daemmrich 2004).

The drug lag mythology and early neo-liberal shift in the US

During the 1960s through to the 1980s, the FDA gained the reputation for having among the most stringent drug regulatory standards in the world, along with Norway. In particular, it prevented the sale of many unsafe or ineffective drugs in the US, including some drug disasters, which found their way on to markets in other countries (Abraham 1995a; Abraham and Davis 2005; 2006; 2007). Unimpressed, throughout the 1970s and into the 1990s, the pharmaceutical industry and some researchers, such as the conservative economists from the industry-funded Tufts University Center for the Study of Drug Development (Tufts Centre), persistently accused the FDA of being unnecessarily cautious and bureaucratic about approving NASs. Such 'over-regulation', as they characterized it, resulted, they insisted, in important new drugs reaching markets and patients in the UK and other European countries while remaining under review in the US. The consequent 'drug lag', they argued, delayed American patients' access to important medicines, negatively impacting on their health (Wardell 1974; 1978; Kaitin *et al.* 1989; Kaitin and Brown 1995).

In 1981, the critics of the FDA's supposed 'over-cautious' regulation from industry and conservative 'think-tanks', like the Tufts Centre and the American Enterprise Institute, were boosted by the election of Ronald Reagan as US President. A New Right Republican, he believed in minimal regulatory restrictions on business interests, and began what was to be a run of 12 years in the White House for the Republicans. Nonetheless, disease-politics theorists, contend that the FDA did not respond to pressure from these quarters, including the Reagan Administration, to accelerate new drug review and approvals until such reforms were also called for by AIDS activists in the late 1980s. For example, Edgar and Rothman (1990) assert that, before the mid-1980s, FDA's 'heavy-handed paternalism' was 'heavily biased in favour of caution' about approving new drugs and denied patients the right to their own risk–benefit calculus about pharmaceuticals. AIDS patient activism, they argue, reversed this. Endorsing that view, Daemmrich and Krucken (2000, pp. 514 and 519) argue:

> Whereas complaints about slow drug approvals in the US had little impact on FDA policies during the 1970s, the aggressive tactics of AIDS activists brought about visible policy outcomes in the 1980s. ... Preventive risk-taking by the FDA was criticized by economists and physicians in the drug lag debate. The setting for risk-taking in regulatory decision-making, however, only changed significantly during the AIDS crisis. Hidden consequences of preventive policies, first articulated by drug lag critics, only became politically salient in the 1980s. Patients now demanded the right to assume medical risks and political responsibility for adverse effects of rapidly approved drugs.

Reinforcing this perspective, Carpenter (2004, p. 52) claims that 'patients, more than pharmaceutical firms, shape the political costs to FDA of delaying drug approval'.

One can see from the writings of Edgar and Rothman (1990) and Daemmrich and Krucken (2000) that they have significantly accepted the premise of the drug lag critics that the FDA was overly cautious in the 1970s and early 1980s, with regard to patients' interests. By implication, therefore, they claim that regulatory reforms to accelerate drug approvals were not only driven by patients' demand, but they also reflected patients' interests because AIDS activists' exposure of the ostensible adverse consequences of FDA's over-cautious approach made the US government recognize the 'legitimacy of demands for more rapid drug approvals', vindicating the drug lag criticisms of the FDA made during the 1970s and 1980s (Daemmrich and Krucken 2000, pp. 512–19). This is what we referred to as 'hard' disease-politics theory in Chapter 1.

While Carpenter (2004; 2010a) subscribes to the theory that AIDS activism drove regulatory reforms to speed drug approval, he is more neutral, indeed silent, on the question of whether such reforms reflected patients' interests. His thesis is rather that accelerated regulatory review and approval was embraced by the FDA in the face of patients' demands because it was in the reputational interests of the agency not to be seen delaying patients' access to innovative medicines. Within his thesis, it would seem to remain an open question whether or not the FDA's reputational interests map on to patients' health interests, which is why we refereed to it as 'soft' disease-politics theory in Chapter 1.

Yet the validity of the claims made by the drug lag critics is highly questionable. Daniels and Wertheimer (1980) evaluated 198 drug innovations that were commercially available in countries outside the US at the end of 1976 and concluded that only 14 per cent offered a potential therapeutic advance. Later, Schweitzer *et al.* (1996) analysed the approval dates of 34 pharmaceuticals, which were marketed in the G-7 countries plus Switzerland between 1970 and 1988, and were designated by panels of doctors and pharmacists to be very therapeutically significant (at the time of their approval). They found that the FDA approved more of these therapeutically significant drugs before the UK regulatory authorities, and ranked third out of the eight countries in approving the drugs on to the market. Moreover, there is evidence that pharmaceutical firms themselves were, in large part, responsible for delays in FDA drug review times. A study by the US Congress General Accounting Office (GAO) in 1980, investigated 27 new drug applications that had been with the agency for more than three years without an approval/ non-approval decision. The GAO concluded that this was because the applications were incomplete – a finding confirmed by the applicant companies themselves (Hilts 2003, p. 277).

While anti-regulation critics seeking less state intervention in the market complained that over-regulation at the FDA stifled pharmaceutical innovation and acted as an obstacle to patients' access to medicines, it was in fact industry's unwillingness to develop 'unprofitable' drugs that explained the lack of innovation for treatment of 'rare' diseases, rather than excessive regulation. Consequently, in 1983, Democrats and Republicans working together in the US Congress felt the need to pass the Orphan Drug Act. This involved the state providing tax and (protectionist) market exclusivity incentives for manufacturers to develop drugs otherwise of limited commercial value because they were for ('rare') diseases affecting less than 200,000 Americans and/or they incurred development costs unrecoverable from sales. The legislation was initially opposed by the US Pharmaceutical Manufacturers' Association (PMA), now known as the Pharmaceutical Research and Manufacturers'

Association (PhRMA), which was loath to admit publicly that private enterprise was failing to meet the needs of those patients (Anon. 1982a) However, pharmaceutical firms soon realized that the Orphan Drug Act was a good deal for them and their institutional commercial interests quickly overcame 'misplaced' ideological embarrassments (Meyers 1997). Notably, the Orphan Drug Act also permitted access to still-experimental drugs, before marketing approval, for patients who could not be satisfactorily treated by available alternative medicines – a provision often wrongly accredited to AIDS patient activism.

Regarding the general claim that the FDA was overcautious in the 1970s and early 1980s, Abraham (1995a) conducted an in-depth analysis of FDA's approval of five new non-steroidal anti-inflammatory drugs in that period. He discovered that none offered any significant therapeutic advance over similar drugs already on the market, such as aspirin and/ or ibuprofen. Indeed, it was questionable whether some of them met the FDA's efficacy standards, while all had very significant drawbacks in terms of toxicity. For example, regarding one of the drugs, Suprol (suprofen) the FDA's reviewing medical officer stated in November 1985, just one month before approval, that 'suprofen will be the first drug to be approved [by the FDA] in about ten years that has demonstrated no benefit, efficacy or safety, over aspirin' (cited in Abraham 1995a, p. 237). Subsequently, three of the five drugs (including Suprol) had to be withdrawn from the market on very serious safety grounds, two of which (Oraflex and Zomax) were so deadly that they can reasonably go down in history as drug disasters. By reference to the scientific and regulatory standards of the time, Abraham (1995a) concluded that the FDA *had not been cautious enough in the 1970s and early 1980s*, even though the UK and some other European countries were even less cautious.

Although, the drug lag thesis was largely misleading, it is not true that the FDA did not respond to it.[2] From 1974, the FDA began to classify NDAs according to a drug's therapeutic potential as a way of prioritizing agency resources during the review process. Drugs received an 'A' classification if considered an important therapeutic gain over existing therapies, a 'B' denoted a modest therapeutic gain, and 'C' implied little or no therapeutic gain. These classifications were an attempt to reduce the amount of time the agency took to review 'A' and 'B' rated drugs (FDA Oral History Program 1997). This classification system changed to the current 'priority' or 'standard' review system in 1992, within which the 'A' and 'B' ratings were combined under 'priority' (Anon. 1993d).

The FDA's regulatory response to calls for acceleration of drug approvals, however, took on a whole new flavour and scale after the

Reagan Administration entered the White House in 1981. That same year, the Republicans gained control of the Senate. The nation's shift to the right occurred in the context of a large federal budget deficit and rising inflation – reaching 12 per cent in 1980 (Hilts 2003, p. 210). Simultaneously, according to the US National Academy of Sciences (NAS), the international competitive position of key US industries, such as automobiles, steel, textiles and consumer electronics, had declined and there was 'a clear relative deterioration in the foundations of the pharmaceutical industry's competitive position – the research efforts necessary for discovery and introduction of new patented drugs' (NAS 1983, p. 1). As the New Right within the Republican party believed that the cause of America's economic and industrial decline was federal government interference in the private sector, concerns about 'drug lag' caused by 'excessive regulation' of the pharmaceutical industry were likely to receive a sympathetic hearing from the Reagan Administration, which had declared itself committed to a radical deregulatory agenda (Hilts 2003, pp. 210–11). The Republicans controlled the White House continuously from 1981 to 1993 and the Senate from 1981 to 1987.

In January 1981, just weeks after Reagan's election, the chairman of his health policy advisory group sent the FDA a warning that a 'change of attitude' at the agency towards the drug industry was 'essential', while Joseph Stetler, another member of the group, and a former president of the Pharmaceutical Manufacturers' Association (PMA), told a New York conference that the congressional climate was receptive to proposals that would provide 'regulatory relief' for industry (Anon. 1981a). This was not just empty rhetoric. In July 1981, a Commission on the Federal Drug Approval Process was convened at the behest of the Republican congressman, James Scheuer, and Democrat Al Gore. The Commission's aim was to 'reform and restructure' the FDA because, according to Scheuer:

> regulatory overkill at the FDA has made life-saving new drugs unavailable to American patients for years on end while superfluous tests are run and other needless and costly delays are imposed. Thousands of Americans suffer and die pointlessly while the regulatory machine lumbers on. (quoted from Anon. 1981b)

The Commission's recommendations in 1982 sought reforms to promote more rapid FDA review of new drugs. A year later, the NAS (1983, pp. 81–7) advocated adoption of the Commission's recommendations on the grounds that they would boost innovation.

The similarity in perspective of the New-Right Republicans and the drug lag critics is unmistakeable. In particular, proponents of the drug lag thesis and the New Right implied that faster regulatory reviews were not only in the interests of the pharmaceutical industry, but also in the interests of patients. In this respect, neo-liberalism was not only a political stance to protect the sectional interests of business, it was also an ideological discourse, which sought to fuse in the public mind the commercial interests of industry in product innovation with the needs of patients for therapeutic advance and health. That ideology was frequently articulated by proponents of FDA reform throughout the 1980s – and well before the demands of AIDS activists for quicker access to treatment.

For example, regarding proposals to accelerate the FDA's drug review process, the NAS (1983, p. 86) concluded:

To the extent that the system is improved, the pharmaceutical industry and therefore the public will gain immeasurably. Equally important, as incentives to invest in new pharmaceutical research are increased, greater gains can be expected in the discovery of new drugs that are effective in reducing the public burden of serious diseases that still remain to be conquered. Thus, the very economic incentives that will help return the U.S. pharmaceutical industry to its former stature will have important public health benefits as well.

While in 1986, the US Health Secretary, Otis Bowen reassured an open meeting of the PMA that 'our aims are your aims' and that 'this Administration will remain committed to working with you for positive change, for competition that promotes innovation in the marketplace and, above all, for better health for the American people' (Anon. 1986a).

Concrete reforms at the FDA followed in the early and mid-1980s, which aimed to reduce regulatory review times and data requirements for all new drugs, not only those that promised significant therapeutic advance. Specifically, changes to the regulations governing NDAs and the testing of new drugs, referred to in the US as investigational new drugs (INDs), were 'accelerated and intensified at the request of' Reagan's Task Force on Regulatory Relief (Federal Register 1985, section 1). Those changes became known as the 'NDA and IND rewrites' of 1985 and 1987, respectively (Federal Register 1985; 1987). They adopted several of the recommendations made by the 1982 Commission on the Federal Drug Approval Process, including: narrowing the scope of the FDA's review of phase I studies during the IND process; replacement of individual case-report forms with summary presentations of data in NDAs; acceptance of foreign data from countries then with less stringent regulatory standards; and new FDA guidelines allowing closer and more frequent contact between industry and agency staff.

The replacement of individual case-report forms with summary presentations of data in NDAs meant that manufacturers no longer had to provide the FDA with as much detailed evidence about the effects of their drugs during clinical trials, except when patients had died or dropped out of the study. FDA managers keenly predicted that this measure would reduce drug review times by six months (Anon. 1982b) While this might have helped to reduce review times, it also required the regulators to place more trust in pharmaceutical firms' reporting, characterization, and quantification of trial events. As we will see later in this book, that shift in trust has, time and time again, strained even the most generous limits of credibility.

The FDA management's focus on streamlining the drug approval process in this period is further illustrated by the transfer of staff and resources into the drug review divisions, to reduce review times, and out of other departments, even during overall budget cuts at the agency (Anon. 1986b; 1989d). Contrary to the assertions of some FDA officials and academics that it was AIDS patient advocates who forced the agency to reassess its regulatory philosophy, before the AIDS crisis, FDA management was attempting to do just that in the 1985 NDA rewrite by reorienting the agency's goals towards 'facilitating the approval of important new safe and effective therapies' (Federal Register 1985, section 1).

Evidently, such policies filtered down the agency to affect FDA reviewing. One reviewer from the FDA's oncology drugs division testified before Senate that management's 'pressure' to expedite the drug approval process could 'compromise the scientific integrity of reviews' because agency scientists were tempted to ignore deficiencies in NDAs, rather than request additional information from pharmaceutical firms (Anon. 1981c). In addition, a Congressional staffer during the mid-1980s told us:

> There was a premium placed on the numbers of new drugs that were approved on a monthly basis. And there were all these December approvals. So there was pressure back then to be able to say: 'we're approving a lot of drugs'.[3]

Indeed, a Congressional Committee found that 30 per cent of innovative new drugs (NMEs) approved from 1980 to 1985 were approved in

December and over 50 per cent were December approvals in 1985 (US Congress 1987). Subsequently, Carpenter (2010a, p. 532) has shown graphically how the steepest growth in this 'December effect' occurred from the early to late 1980s and then persisted until the mid-1990s. If it is true that a rush of approvals in December is indicative of pressure within the agency to increase the annual number of recorded drug approvals, then it certainly operated throughout the early and mid-1980s before AIDS patient activism.

It is clear, then, that acceleration of new drug review at the FDA was driven by demands from industry and their neo-liberal allies in US government before AIDS patient activism even emerged. Moreover, the Reaganite neo-liberal reforms did not distinguish between innovative pharmaceuticals representing therapeutic advance and those offering *no* therapeutic advantage to patients, let alone drugs needed to treat life-threatening diseases. Indeed, after all the reforms of the Reagan period to quicken drug approvals, the FDA reported that, between 1988 and 1991 inclusive, only 17 per cent of NMEs approved by the agency were judged to offer an important therapeutic advance to patients, while 47 per cent offered little or no therapeutic gain (Anon. 1989e; 1990a; 1991b; 1992a).

AIDS, AZT and accelerating patient access to experimental drugs in the US

The HIV virus that causes AIDS was identified in late 1984, but there were no approved drug treatments for AIDS until 1987 (FDA 2009a). However, this was not due to 'regulatory overkill'. In fact, what is notable about the FDA's response to the AIDS crisis is the speed with which the agency acted to facilitate patient access to early investigational drugs to treat the disease. The first drug to receive marketing approval was zido-vudine, more commonly known as AZT. The NDA for AZT was reviewed by the FDA in just three and a half months (FDA 2001a) and approved on the basis of very preliminary data. The evidence for AZT's efficacy was based on only one double-blind, randomized, placebo-controlled phase II trial, involving 281 patients treated for six months, and with incomplete data on optimal dose, optimal duration of therapy, and longer-term adverse events. No phase III studies were required. Instead, a phase IV post-marketing study was agreed (Anon. 1988e; 1989l; Kessler 1989). In addition, all patients, who had been on the trial, including those taking placebo, were put on AZT, and the FDA, together with the company, began to distribute the drug to over 4,800 patients (at least a third of AIDS victims in the US at that time) on a 'compassionate use' basis – a mechanism that allowed patients access to experimental drugs before marketing approval (FDA 2009b; Hilts 2003, p. 244).

Meanwhile, the FDA was also developing what became known as 'treatment IND' regulations – partly within the IND rewrite and partly in response to the AIDS epidemic (Code of Federal Regulations 1987). In April 1987 the agency first published its proposals for new 'treatment IND' regulations which would allow still-experimental drugs intended to treat life-threatening and serious diseases to be distributed to patients before general marketing at a price charged by manufacturers. The new rules, finally approved later in 1987, applied where (1) there were no comparable or satisfactory therapies or drugs available to treat the condition; (2) the drug was under investigation in adequately controlled studies or awaiting evaluation; and (3) the patients were ineligible for enrolment in clinical trials. The evidence needed for FDA approval of a 'treatment IND' could be sufficient in phase II trials for life-threatening conditions, but would not normally be available before phase III for a 'serious' (but not life-threatening) disease (Kessler 1989, p. 284).

Thus, in the early years of the AIDS epidemic the FDA demonstrated its capacity to act quickly and flexibly in the face of a clear public health emergency. Nevertheless, many patients developed resistance to AZT so there remained a pressing need for new therapies and in October 1988, AIDS patient activists protested outside FDA headquarters, accusing the FDA of 'blocking the delivery of promising therapies to people with AIDS' (Hilts, 2003, p. 249). However, a bitter irony of the early confrontations between the AIDS patient activists and the FDA during 1988 was that there were no new AIDS drug applications for the FDA to hold up at that time. The results of the phase I trial of AZT in combination with a new promising AIDS drug, didanosine (ddC) were not presented until June 1990 (Epstein 1997, p. 694; Feigal 1999, p. 33). By late 1988, manufacturers had developed a small number of drugs to treat the opportunistic infections associated with AIDS, and the agency had approved some of these in addition to authorizing, through use of the new 'treatment IND' regulations, pre-approval distribution of others (Holston 1997; FDA 2005a).

As our earlier discussion of the IND and NDA rewrites demonstrates, the FDA was developing and implementing considerable policy changes that were well underway before the agency was targeted by AIDS patient activism in late 1988. However, it seems clear that the concept behind 'treatment IND' regulations concerning access to drugs before marketing approval was driven by the epidemiological reality of the AIDS health crisis and, to a lesser extent, associated AIDS patient demand, including that which led to the FDA's 1988 policy allowing US citizens to import unapproved AIDS drugs from abroad in small quantities for personal use (Anon. 1988c). As for the pharmaceutical industry, it was deeply ambivalent about these special AIDS measures, fearing they were diverting too much of the FDA's attention away from accelerating new drug approval more generally (Anon. 1987a). However, once it became clear that the concept behind 'treatment IND' was going ahead at the FDA, the industry began to lobby for these special measures regarding access to AIDS drugs to be formalized and *extended* to other disease categories, as occurred with the final 'treatment IND' regulations.

Turning AZT into deregulation: expediting development and approval of drugs for serious and life-threatening conditions in the US

In July 1988 following a meeting of the President's Task Force on Regulatory Relief, Vice-President George Bush, who headed the task force, urged FDA Commissioner Young to develop regulations to 'speed the availability of new drugs for AIDS and other life-threatening conditions for which adequate therapies are not available' (Anon. 1988d). The outcome of the FDA's consequent deliberations with the PMA, patient groups, and others was the 'Interim rule on Procedures for Drugs Intended to Treat Life-Threatening and Severely Debilitating Illnesses', also known as 'Subpart E' regulations. These extended the application of the procedures that the agency had used to expedite approval of AZT (Anon. 1988e).

The 1988 'Subpart' E regulations permitted marketing approval of a product for the treatment of life-threatening or severely debilitating diseases on the basis of phase II controlled clinical trials that provided adequate data on the drug's safety and effectiveness. Life-threatening diseases were defined as those with potentially/likely fatal outcomes, while 'severely debilitating diseases were defined as 'conditions that cause major irreversible morbidity' (Federal Register 1988). The FDA's rapid approval of AZT had been an exceptional situation – it was the first, and only drug application to treat a fatal disease reaching epidemic proportions, with 'highly persuasive' phase II data (Kessler 1989, p. 287). To be sure, AIDS patient activists had pressed for these regulations to accelerate approval of further AIDS drugs, but it is notable that Vice-President Bush's Task Force on Regulatory Relief enthusiastically extended those concerns to other life-threatening conditions, such as cancer, as well. Moreover, the PMA then successfully argued for the extension of expedited development and review not only to other life-threatening illnesses where no alternative treatment existed, but also to 'severely debilitating' diseases, including Alzheimers, osteoporosis, and rheumatoid arthritis (Anon. 1988f). Many types of innovative drugs with potentially large markets now fell within the framework of expedited regulatory review. The pharmaceutical industry's ambivalence about the FDA's response to the AIDS crisis evaporated as companies began to see how the context of AIDS and AIDS patient activism could enhance their longstanding and long-term strategy to accelerate drug regulatory review at the agency (Anon. 1988g).

These new regulations permitted approval based on one 'particularly persuasive multi-centre trial (as occurred with AZT), though the FDA stated that typically two 'entirely independent' studies would be required at Phase II (Anon. 1988f). According to the agency, under Subpart E regulations, the same standards for drug efficacy would be required, but the aim was to 'meet the regular standards earlier' (Anon. 1988h). Yet, there would be less safety data than normal and there would be less information on drugs' optimal therapeutic use. That is implicit in the regulations' acknowledgement of the FDA's 'need to answer remaining questions about risks *and benefits*' when coming to its risk–benefit judgement (Code of Federal Regulations 1988, para 312.84a, emphasis added).

Initially, the FDA proposed that approval under Subpart E would have conditional status, with 'remaining questions' to be answered by the drug company through *compulsory* post-marketing studies. However, the industry opposed both those proposals and they were dropped from the final regulations, though of course 'remaining questions' could be answered via *voluntary* agreements between the agency and company on post-marketing studies (Anon. 1988g). No wonder the PMA was so satisfied. While it is true that the idea behind the 'Subpart E regulations' arose and gained credibility from the AIDS crisis, the main impetus behind their implementation as formal procedures within the regulatory state came from the President's Task Force for Regulatory Relief, which encouraged a context enabling the pharmaceutical industry to shape much of the regulations' substantive content, especially widening applications and making oversight arrangements voluntary (Anon. 1990b).

Surrogate endpoints: AIDS, cancer and opportunistic deregulation

Efforts to reform the FDA by industry and the Republicans under President Bush (senior) continued. By the late 1980s, AIDS activists, industry and some government scientists wanted the FDA to approve new AIDS drugs based on what are known as *non-established surrogate measures* of treatment efficacy that could be assessed earlier in the drug development process (Epstein 1997). In phase III clinical trials the standard scientific method for evaluating a new drug is to measure its effect on an event or symptom that is therapeutically meaningful to patients, known as a clinical endpoint, such as death, loss of vision, organ rejection in transplant patients, pain, or other events reducing quality of life – that is, direct clinical efficacy (Fleming and DeMets 1996). However, where death or morbidity caused by disease progression are typically delayed for many years (e.g. diabetes or heart disease), clinical trials with mortality and morbidity as clinical endpoints may require large numbers of patients, take several years, and be very expensive.

Consequently, stretching back decades before the 1980s, to save time and costs, government scientific and/or regulatory agencies had sometimes approved the efficacy of new drugs based on their effects on *established surrogate* measures of efficacy, that is, laboratory/physical measures that substitute for clinical endpoints. For example, a drug's capacity to lower blood pressure is established (among the medical, scientific, and regulatory communities) as a valid predictor, and hence surrogate measure, of clinical efficacy to prevent strokes. Thus, a trial designed to test the efficacy of a drug to lower blood pressure would be a trial with blood pressure as the surrogate endpoint, without directly investigating the drug treatment's effect on the clinical endpoint, strokes. It is then assumed that the treatment's effect on the surrogate endpoint will predict the treatment's effect on the true clinical outcome. A *nonestablished* surrogate endpoint is a measure 'reasonably likely' to predict clinical benefit, but not demonstrated to be a valid substitute for clinical endpoints (Code of Federal Regulations 1992, section 314.510).

Clinical trials using surrogate endpoints are in manufacturers' interests because they require fewer patients, have shorter duration, and are cheaper, than trials tracking direct clinical effects. A drug approved on the basis of evidence regarding surrogate endpoints will reach the market faster for both the manufacturer and patients than if approved using clinical endpoints. However, approval of a drug based on an *established* surrogate endpoint is a lower standard of efficacy proof than evidence of drug effectiveness based directly on clinical endpoints because of the extrapolative uncertainty introduced by having to predict true clinical efficacy from the surrogate measure. Drug approval based on a *non-established* surrogate endpoint is an even lower standard because there is little or no evidence, even correlative evidence, that prediction from surrogate measure to clinical outcome is valid (Fleming and DeMets 1996; Sobel and Furberg 1997).

Nevertheless, in 1989, President Bush established the Committee to Review Approval of New Drugs for Cancer and AIDS, with Dr Louis Lasagna (a long-standing proponent of the drug lag thesis) as Committee Chair. Lasagna immediately announced that the Committee would examine the role of surrogate endpoints in expediting drug approval (Anon. 1989g). Although AIDS patient activists pressed for CD4(T)-cell counts as surrogate endpoints for approval of AIDS drugs, cancer patient activism about such matters was, by contrast, largely absent. It is notable that it was the Bush Administration and industry that drove the extension of concern about surrogate endpoints beyond AIDS to include cancer as well.

In September 1990 the Lasagna Committee recommended that the FDA should approve AIDS and cancer drugs based on the non-established surrogate endpoints of CD4(T)-cell counts and tumour shrinkage, respectively (Anon. 1990c) and AIDS activists – anxious to speed patient access to two promising AIDS drugs in development at that time – lobbied the FDA to accept the Lasagna Committee recommendations (Anon. 1990d). The Lasagna Committee made its recommendation in spite of the fact that several studies published between 1988 and 1990 demonstrated the failure of surrogate endpoints in a number of therapeutic areas to predict the effect of specific treatments on clinical outcomes. Cases of 'surrogate endpoint failure' leading to inappropriate, ineffective or harmful (sometimes deadly) treatment of patients included the use of: encainide and flecainide for cardiac arrythmias (CAST Investigators 1989); quinidine and lidocaine for arrhythmia suppression (Coplan *et al.* 1990; Hine *et al.* 1989; MacMahon *et al.* 1988); calcium channel blockers in acute myocardial infarction and unstable angina (Held *et al.* 1989); and sodium fluoride to treat osteoporosis in post-menopausal women (Riggs *et al.* 1990).

As it turned out, CD4(T)-cell counts proved to be a reasonable surrogate measure of AIDS survival. AIDS patient activism and NIAID may be credited with accelerating its recognition as such, which was in the interests of HIV/AIDS patients' health. However, acceptance of Lasagna's recommendations also meant permitting marketing approval of cancer drugs based on tumour shrinkage, and the situation with cancer drugs was very different from that with the AIDS drugs even though the Bush Administration and the Lasagna Committee had thrown them together. In the 1970s and early 1980s, the FDA had approved cancer drugs based on tumour response rate (ability to shrink tumours) as a surrogate for prolonged survival. However, there were numerous types of cancer in which tumour shrinkage did not consistently correlate with either increased survival or any other clinical benefit. Consequently, in 1985, the FDA's Oncologic Drugs Advisory Committee (ODAC) recommended that tumour response rate should not be a sole basis for approval of cancer drugs (Johnson *et al.* 2003, p. 1404).

Despite a clear rejection by FDA and its expert oncology advisers that tumour response rate could serve as a surrogate endpoint in a cancer setting, the Lasagna Committee recommended in 1990 that FDA should approve cancer drugs that produced tumour regression/shrinkage in more than 20–30 per cent of patients in phase II trials. Moreover the Committee recommended that more work should be done by FDA, in consultation with other groups, to identify potential surrogate endpoints in other disease settings (Anon. 1990c). This would become a prime site for the generation of promissory science and expectations about innovative pharmaceuticals.

Thus by 1991, considerable political momentum had accumulated to shorten development and review times for drugs to treat life-threatening diseases. Surrogate endpoints provided an obvious means to that end. In addition to the Lasagna Committee recommendations and the demands of the AIDS treatment activists, the White House Council on Competitiveness, headed by Vice-President Quayle, which was committed to the removal of 'regulatory burdens' on US business, also endorsed the idea that non-established surrogate measures of drug efficacy could be a basis for marketing approval. According to Quayle, the Council on Competitiveness operated on two premises: first, that a free market and a competitive economy 'are the best allies of the American people', and second, that 'the less regulation – the less government intrusion into peoples' lives – the better' (Anon. 1991c).

The regulatory outcome of these political pressures was the 1992 Accelerated Approval (Subpart H) regulations. The regulations had been negotiated in November 1991 between the FDA, the Secretary of State for Health, and Quayle's Council, but it was the Council on Competitiveness's proposals that dominated. The FDA wanted marketing approval based on surrogate endpoints after completion of Phase II trials to be restricted to drugs for life-threatening diseases where no alternative therapy existed (Anon. 1991d). However, following industry intervention, accelerated approval was extended to drugs to treat not only life-threatening or severely debilitating, but also 'serious', illnesses – the formal definition of which included reversible morbidity (Anon. 1991e; Code of Federal Regulations 1992; FDA 1998a, p. 4; Willman 2000a).

According to Jeffrey Nesbit, chief of staff to Commissioner Kessler in 1991, that loosening of the criteria for accelerated approval to include 'serious' illnesses occurred as a direct result of industry lobbying, with little or nothing to do with patient activism or demands:

> The pharmaceutical companies came back and lobbied the agency and the Hill for that word, 'serious'...Their argument was, 'Well, OK, there's AIDS and cancer. But there are drugs [being developed] for Alzheimers. And that's a serious illness'. They started naming other diseases. They began to push that envelope. (cited in Willman 2000a)

A measure of how important this was to the pharmaceutical industry can be gleaned from the comments of a Washington drug industry lawyer and US regulatory specialist, Peter Hutt, who argued that the accelerated approval provisions were the 'big bang' item of the Quayle reform programme and, if the FDA could be made to interpret them broadly, could cut three years from the drug development process – more than all the other reforms combined (Anon. 1991f)

The FDA published the final Accelerated Approval Regulations in the *Federal Register* in December 1992. They permit marketing approval of new drugs intended to treat serious or life-threatening illnesses that appear to provide meaningful therapeutic benefits to patients compared with existing treatments on the basis of adequate and well-controlled clinical trials establishing that the drug product has an effect on a surrogate endpoint that is 'reasonably likely' to predict clinical benefit (Code of Federal Regulations 1992). As one senior FDA scientist, Robert Temple, subsequently made clear at an Oncological Drugs Advisory Committee Meeting:

> The accelerated approval rule specifically accepted a lower than usual standard. Usually you are supposed to show that there is clinical benefit or have a surrogate that everybody believes is fully acceptable. [The accelerated approval rule] said we can use surrogates that are not of that quality that are more iffy than that, for a particular reason to serve an unmet medical need. (ODAC 2003, p. 80)

Since the true effect on morbidity and mortality of products granted this type of accelerated approval was unknown, manufacturers were required, under the regulations, to conduct 'adequate and well-controlled post-marketing (phase IV) studies to validate the surrogate endpoint or otherwise confirm the effect on the clinical endpoint'. Section 506(b)(3) of the existing 1938 Food, Drug and Cosmetic Act provided for expedited withdrawal of marketing approval by the FDA if: (1) the company failed to conduct the required post-marketing studies with 'due diligence'; (2) post-marketing studies failed to verify clinical benefit of the product; (3) other evidence demonstrated that the fast-tracked product was not safe or effective under conditions of use; or (4) the firm disseminated false or misleading promotional material about the product.

When Reagan was first elected to the White House, American neo-liberalism was largely confined to the New Right in the Republican Party. However, by the time Clinton became President in 1992, neo-liberalism was no longer the sole preserve of the Republicans – even if Democrats tolerated as much as embraced it. Indeed, the two Clinton Administrations of the 1990s did not reverse any of the major regulatory policy shifts instigated during the Reagan and Bush (senior) era. Instead the new 'third way' agenda of the Party's leadership was in many respects characterized by an approach to business regulation and government 'reform' that was consistent with the assumptions and ideology of neo-liberalism. Consider the initial reluctance of the FDA's oncology division to approve cancer drugs under subpart H due to disagreement among oncologists about which surrogate endpoints were appropriate (Clinton and Gore 1996, p. 3). Once again, there is little evidence that pressure for the FDA to accept tumour shrinkage as a surrogate endpoint in cancer trials came from patient advocates (Anglin 1997, pp. 1407–9). Rather, the Clinton Administration's 'Reinventing Government' initiative was decisive, within which President Clinton and Vice-President Gore directed the FDA to accept tumour shrinkage as a valid surrogate marker for accelerated approval (Clinton and Gore 1996). The principal explicit goal of the 'Reinventing Government' initiative was to cut regulatory 'red-tape' for the benefit of business.

The neo-liberal model of resourcing regulation hits the FDA: the Prescription Drug Users Fee Act (PDUFA)

As we have noted, prior to the AIDS crisis, and well before AIDS patient activists protested against the FDA in late 1988, the pharmaceutical industry and its neo-liberal allies in the Reagan Administration and the

Senate sought reforms to reduce regulatory review times for *all types of drug innovations*, irrespective of whether they were to treat serious or life-threatening illnesses, or judged to promise therapeutic advance. It would be wrong to say that the AIDS crisis was a side show so far as that reform programme was concerned, because the challenge of AIDS treatment affected the shaping of some of those reforms, as we have seen. Nonetheless, the neo-liberal reform of pharmaceutical regulation more generally continued throughout the debate about accelerated access to, and approval for, new drugs to treat serious or life-threatening conditions.

As FDA management attempted to decrease regulatory review times across the board, in response to the Reagan government's demands, it did so in a context of severe cuts to its budget, which was provided and set by Congress. Agency staff numbers were reduced from 8,200 in 1979 to just over 7,000 in 1987. Over the same period Congress had passed 20 new laws giving the FDA new responsibilities in both the food and drug area (Anon. 1989i). The FDA budget continued to be suppressed during the Bush (senior) Administration (Hilts 2003, p. 255).

While warning the Reagan and Bush Administrations that the FDA's lack of resources threatened its efforts to expedite new drug development and review, FDA Commissioner Young nevertheless attempted to keep reductions in the agency's drug review times on track by increasing NDA staff at the expense of other (already overstretched) parts of the agency (Anon. 1986b; 1989i) Consequently, other FDA regulatory activities, such as enforcement, were drastically scaled back. The number of seizures, injunctions and prosecutions undertaken by the agency dropped from 500 in 1980 to 173 in 1989 and, according to Hilts, by 1990 the FDA was 'an organizational disaster waiting to happen':

> The industries the FDA regulated were booming. The number of applications for approval of drugs, devices and other products shot up from 4,200 in 1970 to 12,800 in 1989. Over the same period, the reports of serious reactions to drugs increased from about 12,000 to 70,000 per year, not including the 16,000 reports of problems with medical devices, for which the FDA was now responsible. In 1970 the Freedom of Information Act (FOIA) did not exist [or had barely got off the ground at the FDA]; by 1989, as a result of its passage [in 1967], the FDA was fielding 70,000 consumer enquires, 40,000 FOIA requests, 3,000 queries from members of Congress, and 180 citizens petitions for action. In the decade between 1980 and 1990 alone, Congress passed twenty-four laws giving the FDA new responsibilities

and requiring diversion of at least 675 staffers from other tasks. And the AIDS crisis forced the agency to divert 400 people to deal with it, from blood transfusion issues to the testing of viral drugs. (Hilts 2003, p. 255)

Remarkably, the agency had managed to reduce its average new drug approval times from 33 months in 1987 to 19 in 1992 (US GAO 1995, p. 4). However, since the late 1980s, Vice-President Quayle's industry-friendly Competitiveness Council had demanded that drug review times should be reduced to 12 months (Anon. 1991g). The FDA responded by arguing that lack of staff made that target impossible (Anon. 1990e). A growing awareness that the FDA was inadequately resourced emerged in Congress and beyond. The idea of charging the pharmaceutical industry fees for the FDA's drug regulatory work had been suggested as a solution to the agency's funding problems from the mid-1980s by both the Bush and Reagan Administrations but had never been authorized by Congress (Anon. 1991h). In August 1992, the PMA indicated publicly that it would not oppose fee charges on condition that the fees, which became known as 'user fees', were wholly concentrated on new drug review and, that the FDA undertook to implement 'specific improvements' in the regulatory review process (Anon. 1992b). Later that year, Congress enacted the 1992 Prescription Drug User Fee Act (PDUFA).

The new PDUFA legislation authorized the FDA to collect fees from pharmaceutical firms for each NDA, together with an annual fee for each product on the market and each manufacturing plant in operation. In 1993, it was estimated that over the next five years companies would pay over US$300 million in user fees enabling the FDA to hire an additional 600 staff to review new drug applications (Anon. 1993a) However, in order to collect and spend user fees under PDUFA, each year the FDA had to spend from its Congressional annual appropriations at least as much on the drug review process as it had spent in 1992 (adjusted for inflation). The Act stipulated that user fees could only to be used to fund the drug review process (US GAO 2002, p. 7).

In return for user fees, the FDA agreed to meet increasingly demanding review performance goals regarding all innovative pharmaceuticals, irrespective of whether they offered therapeutic advance. For example, that by 1997 it would review 90 per cent of applications for new products given a standard review classification within 12 months and 90 percent of those accorded priority review status within 6 months (FDA 1997b). Such timeframes were goals rather than deadlines, however the particular character of the user fee legislation meant that FDA was under pressure to treat the timeframes as deadlines. PDUFA contained a 'sunset' provision whereby the user fee legislation had to be renegotiated every five years, allowing the pharmaceutical industry to refuse continued user fee funding if it felt that the FDA was not fulfilling its obligations under the legislation, or if industry and Congress could no longer agree on specific performance goals. Thus in the US, receipt of user fees became, in effect, conditional upon negotiated performance measures dictated by Congress and the regulated industry. The implicit threat by industry to terminate user-fee funding embedded in PDUFA, in the context of a government (Executive and Legislature) no longer willing to fully fund the FDA, acted as a lever forcing the agency to comply with the performance goals. It also allowed industry considerable *de facto* influence over FDA priorities and policies.

In fact, PDUFA has been re-authorized every five years (known as PDUFA II in 1997, PDUFA III in 2002, and PDUFA IV in 2007) and the FDA has become increasingly dependent on revenues from user fees which had grown to fund over 40 per cent of the total costs of the agency's new drug review activities by 2000 (Federal Register 2000, p. 47994). Under PDUFA II, the agency agreed to year-on-year improvements until 2002 when the goal was to review 90 per cent of standard NDAs within ten months. PDUFA II also added procedural goals requiring the FDA to: respond to industry requests for meetings; meet with industry; provide industry with meeting minutes; and resolve major disputes appealed by industry (FDA 2011a). Those new performance goals were explicitly intended to increase FDA's responsiveness to, and communication with, companies during drug development (Kaitin and Di Masi 2000).

As a result of user fees and the associated increase in FDA staff numbers, new drug review times did indeed decline. FDA review times (not including the time taken by companies to respond to queries by the authorities) for priority and standard NMEs more than halved between 1992 and 2008 – from 13.9 to 6 months, and from 27.2 to 13 months, respectively (see Table 2.1).

According to a study by the US GAO (1995, p. 11), by 1994 the 'drug lag', insofar as it ever existed, was starting to reverse, with UK approval times lagging behind those of the US. For the 12-year period ending 30 September 1994, the UK Medicines Control Agency (MCA) reported that median approval time for applications for new active substances (NASs) was 30 months, while the FDA's median approval times for NMEs in 1994 was 18 months. In 1988, only 4 per cent of new drugs introduced onto the world market were first approved by the FDA. By 1998, 66 per cent of the new products entering the market had received their

Table 2.1   CDER approval times for priority and standard NMEs and new BLAs calendar years 1993–2008

Year	Priority			Standard		
	Number approved	Median FDA review time (months	Median total approval time (months)	Number approved	Median FDA review time (months)	Median total approval time (months)
1993	13	13.9	14.9	12	27.2	27.2
1994	12	13.9	14.0	9	22.2	23.7
1995	10	7.9	7.9	19	15.9	17.8
1996	18	7.7	9.6	35	14.6	15.1
1997	9	6.4	6.7	30	14.4	15.0
1998	16	6.2	6.2	14	12.3	13.4
1999	19	6.3	6.9	16	14.0	16.3
2000	9	6.0	6.0	18	15.4	19.9
2001	7	6.0	6.0	17	15.7	19.0
2002	7	13.8	16.3	10	12.5	15.9
2003	9	6.7	6.7	12	13.8	23.1
2004*	21	6.0	6.0	15	16.0	24.7
2005*	15	6.0	6.0	5	15.8	23.0
2006*	10	6.0	6.0	12	12.5	13.7
2007*	8	6.0	6.0	10	12.9	12.9
2008*	9	6.0	6.0	15	13.0	13.0

*Beginning in 2004, figures include BLAs.

Source: FDA. Available at: http://www.fda.gov/downloads/Drugs/DevelopmentApprovalProcess/HowDrugsareDevelopedandApproved/DrugandBiologicApprovalReports/UCM123959.pdf (Accessed 15 October 2011).

first approval by the agency (Willman 2000a). As well as reviewing new drug applications more quickly, the FDA also began to approve a higher percentage of them than before. Before PDUFA, the agency approved around 60 percent of all applications submitted, but by 2000, that figure had risen to about 80 percent (Federal Register 2000, p. 47994).

PDUFA formalized and significantly accentuated the neo-liberal agenda of measuring the FDA's performance in terms of the speed with which new drugs were approved on to the market. Supporters of this agenda, including FDA managers required to implement it, asserted that it was progressive for health because the 'consumer' received 'more products, more quickly' (Federal Register 2000, p. 47994). However, it follows from the distinction between pharmaceutical innovation and therapeutic advance, explained in Chapter 1, that faster marketing approval only benefits patients and public health if the risk–benefit profiles of the new drugs placed on the market sooner represent an advance over drugs already available. Moreover, even if this is the case, there must also be enough scientific data on the drugs' advantageous risk–benefit profiles to allow realization of optimal use by patients and physicians, and those benefits must not be outweighed by public health 'dis-benefits' produced indirectly by PDUFA – for example, a decline in the proportion of funding available for post-marketing safety surveillance activities.

Although the neo-liberal reformers and those in the FDA management, who had resigned themselves to the ideology of their political masters, sought to represent the definition of FDA performance evaluation as speed of approval as good for patients and consumers, some patient/consumer organizations were not convinced. In evidence to FDA public meetings on PDUFA, the American National Organization for Rare Diseases (NORD) commented:

> We believe the overriding success of the agency must be measured not by the speed of its work, but by the completeness and scientific soundness of its work in order to protect the health and welfare of the American public. (Diane Dorman, NORD cited in FDA 2001b, p. 174)

While, the US Center for Medical Consumers remarked poignantly:

> Speedier drug approval clearly benefits industry; in some cases it may benefit patients or specific populations, but in most cases there is no such evidence. It is, therefore, inappropriate for the agency to so narrowly define the objective of PDUFA... Its self-evaluation and how it defines performance measures should focus on how well its activities improve the well-being of those who are sick and disabled. (Arthur Levin, Center for Medical Consumers cited in FDA 2000a)

Regarding policies pertaining to 'standard' NDAs, patient and consumer groups complained that there was no public health justification for the stringent review timeframes for such drugs as they offer little or no therapeutic advantage. Yet, when PDUFA was re-authorized as PDUFA II in 1997, the review timeframe for 'priority' drugs remained unchanged, but the goal for 'standard' drug review time was reduced from 12 to 10 months. Patient and consumer groups were concerned that, with agency resources stretched, stringent review-time goals for 'standard' (probably 'me-too') drugs forced the FDA to devote inadequate resources to review drugs that are important to public health. Moreover, they pointed out that, even in relation to drugs for life-threatening or serious diseases where no therapeutic alternatives exist, FDA reviewers should have the flexibility to exercise some scientific judgement on whether the nature and complexity of the data present demanded a longer review time than that allowed by the agency's performance goals (Travis Plunkett, Consumer Federation of America cited in FDA 2001c, p. 26).

These comments by patient and consumer organizations highlight the potential divergence between the interests of public health and the neoliberal reform programme, which sought to define the FDA's performance solely in terms of speed of marketing approval. The New Right, who advocated the virtues of 'free markets' with limited state regulation of business, entrepreneurship, and associated innovation, promoted this ideology as convergent with consumers' interests. However, such promotion should not be mistaken for fact or even acceptance by patient and consumer organizations. Scholars and other commentators on pharmaceutical policy, who argue that the acceleration of access to AIDS drugs for patients was indicative of a more general shift in American culture towards a desire for faster and more risky pharmaceutical products with less protection by state regulation, appear to have made just that mistake. They have reported an ideological framing of patients' interests by the neo-liberal reformers in government as if it were reality. The responses of patient and consumer advocates to PDUFA provide further evidence that the AIDS drug scenario was an exception in a sea of regulatory reform designed to assist business and industry, with the interests of patients and public health a residual concern. In this respect, a debate that arose in 1994 in the context of a GAO investigation into FDA performance measures under PDUFA is telling. The pharmaceutical trade press, *Scrip*, reported the debate as follows:

In the belief that the ultimate aim of the 1992 Prescription Drug User Fee Act is to improve public health by enabling patients to receive new drugs sooner, Representative Edolphus Towns, chairman of the US House government operations/intergovernmental relations subcommittee, has questioned whether the data to be collected under the act will be sufficient to determine how far it is achieving this goal. The Department of Health and Human Services, however, has stated that the main purpose of the Act is to find FDA the resources to review applications within certain time frames and thus the programme should be evaluated on the basis of whether the FDA is successful in accelerating its review, not on a specific improvement in public health ... DHHS [pointed out] first, that evaluation of improvement in public health 'requires a complex multi-factorial analysis' that goes beyond an assessment of whether new drugs are reaching the public earlier, and second that *'the main aims of [PDUFA] are to give the FDA the necessary resources to review applications within specified time frames and allow drug sponsors [firms] to predict more accurately when product reviews will be completed'*. (Anon. 1994a, emphasis added)

There is evidence to suggest that PDUFA and associated political shifts in the nature of the relationship between the FDA and industry towards one of 'collaboration' in the drug development process have affected the culture of drug regulation within the agency. According to FDA management, the agency agreed to expand its performance goals without requesting additional user fees because it assumed that there would be a consistent increase in the number of NDAs received by the agency between 1997 and 2002 providing sufficient fees to cover the added workload (FDA 2001c, pp. 142–3). Instead the number of NDAs fell (FDA 2004). Hence, the costs of PDUFA commitments exceeded the fees collected resulting in significant increases in FDA reviewers' work-load (FDA 2001b, p. 7; KPMG Consulting 2002; US GAO 2002). Between 1999 and 2001, FDA reviewers scheduled nearly 4,000 meetings with pharmaceutical firms under PDUFA II obligations. According to the FDA, a typical meeting could involve 17 reviewers, and the time require-ments for a meeting involving all FDA review disciplines could range from about 125 to 545 hours per meeting (US GAO 2002, p. 20).

An investigation by the Inspector General of the US Department of Health and Human Services (DHHS) (2003) also found evidence that increased workload and inflexible PDUFA timeframes might be under-mining the integrity of the review process. Forty percent of the FDA reviewers responding to the study, who had been at FDA at least five years, indicated to the investigation that the review process had wors-ened during their tenure in terms of allowing for in-depth, science-based reviews, while 58 percent thought that the six-month review timeframe for priority NDAs was an inadequate period of time in which to conduct sufficiently substantive, science-based reviews.

Aside from altering the drug review process, PDUFA affected other FDA activities. As PDUFA required the agency to spend a large, infla-tion-adjusted, amount each year from its annual appropriations on drug approval activities in order to collect user fees, when appropriations from Congress were limited, FDA officials had to reduce resources spent on other activities to ensure that enough appropriated funds were spent on the review process to enable collection of user fees. By 2000, non-drug-review programmes had to be cut by US$234 million to cover mandatory pay increases, and to ensure that sufficient appropriated funds were funnelled into the drug review process (FDA 2000a). The FDA's workforce and resources for programs other than PDUFA contracted each year from 1992 to 2000 (Federal Register 2000, p. 47994). Thus, the dramatic decreases in approval times were bought at the expense of post-marketing monitoring of adverse drug reactions, tracking of manufacturers' phase IV study commitments, oversight of post-launch advertising campaigns by drug companies, and inspections of industry facilities and conduct (FDA 2000a; FDA, 2001c; Public Citizen 2002; US GAO 2002).[4,5,6]

In 1992, funds obligated for new drug review constituted 17 per cent of the FDA's total budget. By 2000 this figure had increased to 29 per cent. This change was reflected in the distribution of agency staff. In 1993, 14 per cent of full-time employees, or 'full-time equivalents' (FTEs), were dedicated to new drugs review. By 2000, it was 26 per cent. Conversely, the percentage of FTEs employed on other FDA activities dropped from 86 to 74 per cent over the same period (US GAO 2002, pp. 14–18). The activities that were 'down-sized' were important for protecting public health and safety.

Reduction in the funds for key FDA activities after PDUFA is all the more remarkable when we consider that resources for those activities had *already* been severely reduced due to budget cuts and prioritization of the drug review divisions during the 1980s. An FDA needs-assessment study conducted by the agency's Office of Planning and Evaluation in 1991 estimated that the FDA would need to double its workforce between 1991 and 1997 in order to meet its regulatory responsibilities. The study indicated that more staff was needed to boost activities in a number of areas – not only in the review divisions. For instance, during the 1980s, the number of compliance staff at the FDA fell by 30 per cent. Consequently, in 1989, the agency conducted 35 per cent fewer inspections than in 1980 (Anon. 1991i) By 2010, the US DHHS Office of Inspector General reported that only 0.7 per cent of foreign clinical trial sites were inspected by the FDA, while the figure for American sites was just 1.9 per cent (Anon. 2010a). Despite this demonstrated need for *more staff* across all FDA activities, there has been a *decrease* under PDUFA – not only in the proportion, but also in the absolute number, of FTEs working on non-drug-review activities, such as enforcement and surveillance. Between 1992 and

2000, the number of FTEs working on non-drug-review activities fell from 7736 to 6571.

By the FDA's own admission, after PDUFA, post-market drug safety regulation became particularly underfunded, and 'severely challenged' with 'critical new drug safety work...not getting needed funding' (cited in NORD 2002). PDUFA 'stressed' the agency's system for monitoring adverse drug reactions directly, because of the need to prioritize the review process, and indirectly because more drugs entered the US market in absolute terms than before. Also, more drugs than previously reached US patients before patients in other countries. Accordingly, American patients became more likely to be the first exposed to the risks of new drugs placing greater demands on the agency's post-marketing surveillance systems without commensurate resources (FDA 2001c, pp. 65–7). Following an unprecedented number of withdrawals of prescription drugs from the US market on safety grounds in the mid-to-late 1990s and increasing public concern that the FDA was approving unsafe products, the FDA convened open meetings in 2000 and 2001 to discuss PDUFA (Friedman *et al.* 1999; Willman 2000b). Announcing the 2000 meeting in the *Federal Register*, the agency management admitted:

> We are increasingly concerned that spending enough appropriations on the drug review process to meet the statutory conditions [of PDUFA] makes FDA less able to manage the resources available in a way that best protects the public health and merits public confidence. Just one example of an area we have not been able to fund adequately is responding to reports of adverse events related to the use of prescription drugs. (Federal Register 2000, p. 47994)

Subsequently, Congress, the FDA, and the pharmaceutical industry agreed that some user fees could be collected for 'post-approval risk management activities' under the 2002 PDUFA III re-authorization (FDA 2009c). However, objectives to protect public health, as defined by the FDA, in these tripartite negotiations were severely compromised. Under PDUFA III, user fees could be spent on 'collecting, developing, and reviewing safety information' and adverse event reports, but only for drugs approved after October 2002, and typically only for up to two years after marketing approval – three years for drugs considered particularly dangerous (FDA 2005b). Those limitations made it highly questionable that the PDUFA III agreement could adequately meet the drug safety and public health concerns raised by FDA management since research had demonstrated that half of all drug safety withdrawals in the

US occurred *after* the first two years of marketing, and that half of the major drug safety labelling changes occurred after the first *seven* years of marketing (Lasser *et al.* 2002).

Moreover, virtually all of the FDA Office of Drug Safety's proposals that user fees should be spent to supplement the passive adverse events reporting system with a number of pilot projects for active surveillance 'failed to make it past industry-FDA negotiations' for PDUFA III (Patient and Consumer Coalition 2002, p. 1). The increase in funding for drug safety regulatory activities under PDUFA III amounted to US$2.2 million for drug safety, but fell well short of the US$100 million that FDA officials claimed was needed for the agency to undertake proper post-marketing surveillance (NORD 2000).

The process established for negotiating PDUFA and its subsequent re-authorizations enabled the pharmaceutical industry to bargain for specific performance goals prioritizing speed of new drug review and responsive interactions with industry in exchange for FDA funding. As we have seen, that skewed the agency's regulation towards activities that were in the commercial interests of industry – sometimes coinciding with patients' interests, but at the expense of activities that were primarily or solely in the interests of public health. The extent of such industry influence was accentuated by the exclusion of patient and public health organizations from the PDUFA negotiations (FDA 2000a; 2001c). As one former patient advocate recounted:

> The point is that they had cut the deal at the table, excluding people like us. And so the essential deal, about how much money and how to use it, it was cut behind closed doors between the FDA and industry... These were deals largely done out of the earshot of patient and consumer groups.[7]

For PDUFAs I, II and III, patient organizations and public health advocacy groups were merely asked to comment on measures already agreed between industry, the FDA, and Congress – an unsatisfactory state of affairs that did not go unnoticed according to consumer organizations:

> But the complaint generally is... that the public is invited in to comment late in the game when things are already fairly well formed and that puts the advocates at an unfair advantage. Not only do we have less money and less influence and less power to influence decisions but, if things have pretty well formed, it's really hard to change them in any substantive way.[6]

As a consequence of complaints by patient and consumer advocates about their exclusion from participation, a commitment to consult with patient groups from the beginning of the negotiations for PDUFA IV was written into the PDUFA III agreement.[8]

What is particularly revealing about our analysis of PDUFA is that when one moves away from the rather exceptional context of AIDS treatment to US pharmaceutical regulation more generally, one finds that, since 1992, for 15 years or more, patient and consumer health advocates were effectively excluded from the consultative role that industry was accorded in shaping the FDA's goals and priorities.[5] Patients, consumers, and associated organizations did not drive or demand PDUFA. Political and economic interests of government and industry drove the PDUFA processes, which profoundly altered American drug regulation and the FDA. Disease politics, though present at times, seems only marginally relevant to this highly significant chapter in contemporary American pharmaceutical policy.

The creation of supranational EU pharmaceutical regulation

Meanwhile, the Europeanization of pharmaceutical regulation across western European nation-states was gathering pace like never before. The transnational research-based pharmaceutical industry saw advantages in the Europeanization project if it harmonized regulatory standards across Europe in ways conducive to greater market access for drug products. Reflecting those transnational priorities, in 1978, the industry established the European Federation of Pharmaceutical Industry Associations (EFPIA), which was a conglomerate of the national industry associations in Europe, and became the official representative of the European industry in negotiations with the European Commission and Parliament. In theory, at least, increased European harmonization of regulatory standards could mean faster access to a transnational, if not pan-, European market. Conversely, separate and distinct national regulatory regimes, with different technical standards and divergent safety regulations tended to add to the costs of transnational firms. Thus, industry interests converged with those of the European Commission and Europhiles in national governments and the European Parliament, who were committed to European integration.

In 1988, a Commission report on the 'Single European Market' (successor of the ' Common Market') accepted the argument put forward by EFPIA that the European pharmaceutical industry was significantly constrained by the 'lengthy and differing drug registration [approval] procedures' of the different EU Member States. Moreover, according to the European Commissioner in 1995, the previous 20 years had seen the EU's share of all the world's new medicines developed decline from half to a third, while the US held four times as many patents in the biotechnology sector as the EU. Against this background, Vogel (1998, p. 5) notes:

> An important objective of the creation of a single European drug approval procedure was to promote more European-wide drug research and development, thus helping the industry to confidently continue to hold its place on the world stage in the foreseeable future.

In fact, transnational European harmonization of drug regulation dates back to 1965 in the form of policy statements (or 'directives') from Brussels, which were intended to provide a common framework agreed by the Member States of the EC and (subsequently) the EU. Unlike the FDA's single-minded historical mission to protect public health under the 1938 Food, Drug and Cosmetic Act, together with its 1962 Amendments, the roots of the European pharmaceutical regulation reveal the dual principles of the Commission's perspective – the protection of public health *and* the free movement of products within the EC (European Commission 1965; 1995, p. 35).

Ten years later Directive 75/319/EEC established a European expert body, namely the Committee for Proprietary Medicinal Products (CPMP) 'to facilitate the adoption of a common position by the Member States with regard to decisions on the issuing of marketing authorizations' (European Council 1975). That same directive created the first transnational marketing approval system in the EC based on the expectation that Member States would voluntarily endorse each others initial evaluations – an endorsement known as 'mutual recognition'.

In theory, such transnational mutual recognition could save drug companies and regulators time compared with multiple separate national applications and regulatory reviews. However, this voluntary procedure failed because Member States frequently refused to mutually recognize each others assessments in terms of safety and/or efficacy. Consequently, pharmaceutical companies became and/or remained sceptical about the advantages of using the EC procedures over national regulatory systems. This led to poor uptake by the industry, with most new drug applications between 1975 and 1995 being submitted via national, rather than EC, routes (Cartwright and Matthews 1991).

In 1987, recognition of the increasing technical complexity of some new drugs led to a separate European regulatory procedure mandatory for biotechnology-based drugs and optional for other some new drugs, including NASs. A product was assessed by one Member State (called the 'rapporteur') on behalf of the EU. After all the other Member States and the pharmaceutical manufacturer had commented on the assessment, if the CPMP approved, then marketing authorization was permissible in all Member States. However, *CPMP opinions were not binding on Member States*, so this procedure was not sufficient to achieve a single EU pharamaceutical market (Sauer 1997, p. 3).

Conscious of the limitations of these early voluntary procedures with respect to Europeanization, in the late 1980s, the pharmaceutical industry set out its 'blueprint for Europe'. The industry stressed that any new Europeanized system should enable companies to obtain a single, uniform marketing authorization quickly (within 210 days). While the industry wanted regulators to submit to stringent timeframes and guidelines, it also argued for the maintenance of three approval tracks (national, decentralized, and supranational-centralized) so that firms had maximum flexibility in deciding within which regulatory context to place their drug applications. Specifically, the industry proposed that there should be a centralized EU drug regulatory agency to handle biotechnologically-based drugs, while companies could retain the option of submitting non-biotechnological, innovative pharmaceuticals to either a decentralized EU authority or national regulatory agencies within a mutual recognition procedure. Crucially, the industry recommended that if national regulatory agencies failed to mutually recognize each others assessments within a specified time, then the EU drug regulatory agency could impose a decision on the Member States (Anon. 1988i, p. 7).

As we have seen, by the early 1990s, three key EU Member States with major pharmaceutical sectors (Germany, the UK, and France), together with a fourth important Member State in waiting (Sweden) had governments committed to less state regulation of private industry and sympathetic to further European integration.[9] The shaping of the EU supranational drug regulatory system, which was becoming established during the early and mid 1990s, was strongly influenced by those neo-liberal developments taking place in many of its most important Member States (Abraham and Lewis 2000). Furthermore, within the Commission, pharmaceutical regulation fell under the brief of the Directorate-General (DG) Enterprise, rather than DG Health,[10] until 2010 when it was decided by the President of the European Commission,

Jose Manuel Barroso, to transfer responsibility for EU pharmaceutical policy to DG Health (Anon. 2009a). DG Enterprise was the department responsible for promoting trade and industry, so it was particularly receptive to the concerns of the drug industry. For all of these reasons, combined with the fact that alternative voices from patients, health professionals, consumer organizations, and public health advocacy groups were relatively weak in Europe, a new EU-wide drug regulatory system was established in 1995 largely consistent with the demands of the industry.

On 1 January 1995, the European Medicines Evaluation Agency (EMEA) was created, with headquarters in London, to administer the new systems of EU drug regulation. The CPMP, which changed its name to the Committee for Human Medicinal Products (CHMP) in 2004, remained the key expert EU regulatory body, comprising 30 members – two regulators from each of the 15 Member States at that time. Together the EMEA and the CPMP formed the core of supranational EU pharmaceutical regulation, though formally the European Commission and European Council, both had to ratify the CPMP's opinions into decisions. Simultaneously, the supranational EU 'centralized procedure' was established. It was mandatory for technologically advanced medicinal products, including most biotechnology-derived products, and optional for other innovative pharmaceuticals. Assessment under the centralized procedure was via a single application to the EMEA, and crucially CPMP opinions *became binding on Member States* (European Council 1993; 1996, pp. 31–49). Hence, marketing approval via the centralized procedure provided a company with an immediate, single European licence to sell its pharmaceutical product across the whole of the EU (bypassing entirely national regulatory agencies), just as an FDA approval enabled a firm to market its product across the US.

As the industry had requested, this supranational regulatory agency and system was complemented by a new mutual recognition system and a national route available for non-biotechnological pharmaceutical products that could be innovative or non-innovative. Within the new mutual recognition system, failures between Member States to mutually recognize each others assessments became disputes to be settled at the EU level by CPMP 'arbitration', which was *binding on Member States* (Anon. 2005a). Thus, the decentralized procedure enabled pharmaceutical firms to gain simultaneous marketing approval across part of the EU, say three, four, five or more Member States. As Abraham and Lewis (2000) found, the industry wanted this flexibility in case companies felt that they would have a better chance of marketing some products by avoiding particular Member State regulatory agencies or the EMEA, or that they lacked the marketing capacities/incentives to launch some products across all parts of the EU.

A drift towards increasing supranational Europeanization has continued since 1995, though by the early 2000s, only 60 per cent of NASs went through the centralized procedure probably because of the high rejection rate recorded in the late 1990s (Anon. 2001a; 2002a). Under Regulation 726/2004, however, from November 2005 it became compulsory for orphan drugs and innovative drugs to treat AIDS, cancer, diabetes, and neurodegenerative diseases, as well as biotechnology-derived pharmaceuticals, to be assessed by EMEA and the supranational centralized procedure, with the expectation that EMEA and the CHMP (the CPMP's successor) within the supranational centralized procedure would continue to expand its reach (Anon. 2005b). In May 2008, innovative drugs to treat viral diseases, autoimmune diseases, and other immune dysfunctions were added to the list fuelling the expectation that EMEA and the centralized procedure would, in time, take over the marketing approval of all innovative pharmaceuticals in the EU (Anon. 2008b). Given our focus on innovative pharmaceuticals in this book, henceforth, our discussion of EU drug regulation will be almost entirely concerned with the supranational level and the centralized procedure, involving the EMEA, CPMP/CHMP, European Commission, European Parliament, and European Council. In 2004, the EMEA changed its name to the European Medicines Agency and adopted the abbreviation, 'EMA' in 2010.

Congress and the conservative 'FDA reform movement' 1995–1997

The election of President Bill Clinton to the White House for two consecutive terms from January 1993 to December 2000 interrupted the New Right's control of the US Administration, but not neo-liberalism. By November 1994 the Republican Party, which provided a home for most of the New Right politicians in the US, for the first time in decades, controlled both houses of Congress. The Republicans had substantial majorities, although not the two-thirds majority needed to override a Presidential veto. Nonetheless, neo-liberal reform of the FDA returned to the political agenda in Washington. Within a few months of the 1994 Congressional election results, Newt Gingrich, speaker of the House of Representatives and leading New-Right politician, announced that he was enlisting the help of conservative Washington-based think-tanks to work on proposals to 'overhaul' the FDA (Anon. 1994b) Later that year, a former legislative assistant to Senator David Durenberger, told an industry meeting that it would see 'enormous responsiveness' to its frustrations with the FDA amongst the new Congressional majority and suggested that industry had a unique opportunity to lobby for change at the FDA (Anon. 1994c). Industry responded with alacrity to these invitations. By February 1995, the Pharmaceutical Research and Manufacturers' Association (PhRMA) was working on FDA reform proposals (Anon. 1995a).

PhRMA's plans for FDA reform took the form of detailed, draft legislation circulated to key government figures on Capitol Hill. The association's proposals included amendments providing for: (1) a new FDA mission statement emphasizing timely availability of safe and effective products; (2) the establishment of a permanent DHHS panel to evaluate the impact of FDA regulation on the competitiveness of the US drug industry; (3) new NDA requirements allowing submission of more safety and efficacy data as summary tables; (4) following the French system, the use of contract reviewers, external to the FDA (known as third-party review), whenever agreed by the pharmaceutical company; (5) a new drug approval standard that could consist of data from only one well-controlled clinical trial instead of the two, which had been required since the 1962 Drug Efficacy Amendments to the 1938 Food, Drug and Cosmetic Act; (6) a presumption of approval for drugs already approved by the EMEA or the UK MCA, so that the FDA would have to decide on NDAs within six months, based on the original submission without being able to request any further data and rejection would only be allowed if the drug had been demonstrated to be unsafe or ineffective (Anon. 1995b).

Some New-Right think-tanks also published proposals. One, Newt Gingrich's Progress and Freedom Foundation, drafted recommendations to privatize the drug regulatory review process by using contract reviewers outside the FDA (Anon. 1995c; 1996a). The pharmaceutical industry, however, was less enthusiastic about plans for wholesale privatization of the drug review process because they were concerned that such a move would undermine consumer confidence in prescription drugs. Commenting on the proposals of Gingrich's Foundation, PhRMA stated its preference that the FDA should retain responsibility for NDA reviews, assisted by occasional third-party review when felt necessary. The PhRMA sought reduction of regulation in ways likely to maximize industry interests, rather than ideologically-driven privatization (Anon. 1995c). Congress's first draft bills in summer 1995 largely reflected the proposals put forward by PhRMA. One of the bills even repeated verbatim the language of PhRMA's draft legislation (Anon. 1995d; 1995e; 1995f; 1995g). A consumer advocate during negotiations about this legislation claimed to have seen a draft of the bill, parts of which were left blank with the words: 'Industry inserts language here'.[5] Whether or not that is correct, it is clear that New-Right reformers in Congress were highly influenced by industry preferences and moderated their legislative proposals at the behest of PhRMA.

The New-Right foundations, especially the Washington Legal Foundation, paid for a series of anti-FDA advertising throughout 1995 and 1996 accusing the agency of killing Americans with over-cautious drug regulation. PhRMA arranged to fly in 140 patients to Washington to talk with the legislators about how the FDA had deprived them of life-saving drugs. However when PhRMA provided the *Washington Post* with the names of some of those people for the *Post* to contact, the newspaper reports did not include one story where a patient's problems could have been addressed by the proposed FDA reforms. Indeed, according to Hilts (2003, pp. 309–11), their problems were not even within the control of the FDA.

The New-Right's campaign for FDA reform had political clout in the mid-1990s, but little evidence to support it. By 1994, FDA drug review times had declined dramatically and were as fast as those of the UK's MCA, and faster than those in France, Spain, Germany, Australia, Japan, Italy and Canada (FDA 1997c; Hilts 2003, pp. 316–18). Furthermore, the agency was the first regulatory authority in the world to approve five out of the six available AIDS therapies and a number of drugs considered to be 'breakthroughs', such as Taxol, Fludarabine, Pulmozyme, Riluzole, Cognex and Betaseron. By the mid-1990s, the 'drug lag thesis' still peddled by the Washington Legal Foundation and the editorial page of the *Wall Street Journal,* was little more than ideologically driven fantasy (Anon. 1995h; Hilts 2003, pp. 297–303).

Aware of this, some patient groups became alarmed at the Congressional reform proposals and increasingly annoyed that none of the organized interests pressing for reform had bothered to ask the patient groups what they thought. [8] Consequently a number of patient advocacy groups collaborated to form the Patients' Coalition in 1995.[11] By 1997, it comprised over 100 organizations representing patients with serious, rare and life-threatening disorders (Patients' Coalition 1997). This was the largest mobilization of patient groups in the US since AIDS activism, and arguably more significant because it crossed a large number of disease categories. The Patients' Coalition argued against many of the proposals for FDA reform put forward by both the House and the Senate, such as statutory review deadlines which, if exceeded, could trigger either a 'European hammer' whereby drugs approved by the EMEA or MCA would automatically be approved in the US. The Patients' Coalition also opposed the suggestion that NDA approval should be based on a single clinical trial, though the Coalition was willing to support a 'single-clinical-trial' rule in exceptional circumstances where the single study was very persuasive in demonstrating a direct therapeutic benefit on a clinical endpoint (Anon. 1996b; 1996c; Patients' Coalition 1996).

According to the Coalition, patient groups speaking in favour of the reform proposals were either 'astroturf' organizations or relied on industry for funding.[5,7,8] Moreover, as part of their campaign, the Patients' Coalition commissioned an opinion poll, which indicated that only 22 per cent of those polled believed that the FDA was a 'big government bureaucracy' (Patients' Coalition 1997, p. 2). This was despite a barrage of prominent newspaper and television advertisements and 'info-mercials', paid for by industry and the right-wing foundations, claiming that FDA regulation was responsible for the deaths of thousands of American patients (Hilts 2003).

Thus, there is little evidence from the mid-1990s that patients groups or the general public were pushing for further acceleration of the drug review process or further lowering of regulatory standards in the way that the AIDS activists had done in the late 1980s. On the contrary, many of the early AIDS patient advocates from the 1980s played key roles in *opposing* the Congressional reform proposals in 1996 precisely because of what they had learnt from their close involvement with the regulatory science of AIDS treatment. In particular, all of the American AIDS patient groups began to stress the importance of trials that generated knowledge concerning the optimal use of the available drugs (Epstein 1997).

Analysis of documents from the most active AIDS patient groups reveals that they opposed further lowering of regulatory standards despite claims by the proponents of the reforms that the new legislation would 'facilitate the timely availability of new products' (US Senate 1996). In urging rejection, AIDS patient advocates noted that the US had fewer drugs withdrawn for safety reasons than other countries (Link 1995). Evidently, *safety*, not merely speed of approval, was of primary importance even for these patients with life-threatening conditions. Significantly, the AIDS patient advocates argued against many of the proposals that were ostensibly designed to quicken patient access, for instance: third party reviews, shortening the time FDA had to review critical safety data before allowing trials in humans to go ahead, and the introduction of rigid new drug review timeframes (AIDS Action Council 1996).

Many of the New Right reform proposals of 1996 were also opposed by the FDA and the Clinton Administration, which held the power of presidential veto over Congressional proposals (Anon. 1996b). In 1996, the drug industry urged Republicans in Congress to drop some of the more 'extreme' deregulatory proposals of 1995 in order to secure greater bipartisan support in Congress, so that a Presidential veto by Clinton's Democratic Administration could be avoided and the legislation could be passed quickly (Anon. 1996d; 1996e). Towards the end of that year, industry and the FDA began negotiations on the reauthorization of PDUFA I, due to expire in 1997. PhRMA's plan was to secure the agency's agreement on a number of FDA reforms as conditions of such reauthorization. PhRMA and the FDA agreed that the controversial proposal on mandatory third-party (privatized) reviews should be discarded. However, the industry now sought timeframes for meetings with FDA management as well as for drug review; the PDUFA I drug review timeframes to be further reduced; and continued to press for only one pivotal clinical trial to be required to demonstrate efficacy for all new drugs (Anon. 1996f). By February 1997, the FDA agreed to most of these conditions (Anon. 1997a; 1997b).

The industry managed to secure concessions from the FDA on off-label promotion, even though the Patients Coalition asserted vigorously that off-label promotion was a hazard to the American public and presented a direct threat to the nation's health (Anon. 1997c; Patients Coalition 1997, p. 5). By contrast, with respect to accelerated approvals, a provision supported by consumer and patient advocacy groups, which would have imposed fines on pharmaceutical firms that failed to conduct confirmatory post-marketing studies, was watered down to requirements on companies to submit progress reports on such studies each year until completion (Anon. 1997c; 1997d). The House and Senate bills were reconciled on the 8 November 1997 and passed by Congress as the Food and Drug Administration Modernization Act (Anon. 1997e). Despite all their campaigning efforts, and in sharp contrast to the pharmaceuticals industry's influence on the Food and Drug Administration Modernization Act (FDAMA), proposals from patient groups that significantly conflicted with industry's interests were not adopted into FDAMA.[8]

FDAMA and fast-track drug development and review

As we have seen, like PDUFA, and unlike the 'Subpart E' and 'Subpart H' rules, FDAMA was mainly about general FDA reforms affecting all types of drug product regulation. However, within that, it codified the 1992 Accelerated Approval (Subpart H) rule into the statute. Relatedly, section 112 of FDAMA added a new section to the 1938 Food, Drug and Cosmetic Act. The new section 506 allowed companies to seek 'fast-track' designation for their product development programmes concurrently with, or at any time after, submission of an IND application. Largely reiterating the criteria for eligibility under the 1992 Accelerated Approval rule, a drug could qualify for fast-track designation, if it was intended for the treatment of a 'serious' or life-threatening condition *and* its development programme evaluated the product's potential to address 'unmet medical needs' for such a condition. The programme became referred to as 'a fast track drug development program' (FDA 1998a, p. 3). A 'serious' condition was defined as one impacting on survival or associated with morbidity substantially affecting day-to-day functioning 'but the morbidity need not be irreversible, providing it is persistent or recurrent' (Federal Register 1992; FDA 1998a, p. 4). Thus rheumatoid arthritis or depression could meet the definition of 'serious' for the purpose of fast-track designation. An 'unmet medical need' was defined as a medical need not addressed adequately by existing therapies, including products that demonstrate potential advantages over existing alternative therapies in terms of improved effects on serious outcomes or avoidance of serious toxicity (FDA 1998a).

The FDAMA specified some ways in which product development and review could be expedited. For instance, it permitted 'rolling' NDA submission and FDA review of discrete portions of a new drug application before the company submitted a complete application. FDAMA stipulated that manufacturers of fast-track products had a right to request meetings at various stages of the drug's development. Following this legislation, FDA's guidance to industry stated that 'appropriately timed meetings between the regulated industry and FDA are a critical aspect of efficient drug development...to ensure that the evidence necessary to support marketing approval will be developed and presented in a format conducive to an efficient review' (FDA 1998a, p. 10). For drugs with fast-track designation, meetings between the manufacturer and the agency to discuss product labelling issues were to be scheduled early in the review process.

Following the neo-liberal trend: funding and review times in the EU

By the time EMEA was created in 1995, many countries had taken the neo-liberal route of funding their drug regulatory agencies via industry fees. In Europe, this varied from 50 per cent funding in Germany to 100 per cent in the UK, while under PDUFA I, industry fees made up 36 per cent of the FDA's spending on new drug review in 1995, rising to 62 per cent in 2010 (see Figure 2.1). From its inception, supranational EU drug regulation accepted the principle that the EMEA should be partly funded by fees from pharmaceutical firms reflecting the dominant neo-liberal perspective within the Member States of the EU.

Like the European Member States, fees collected from drug companies for regulatory work by the EMEA contributed to the overall revenues of

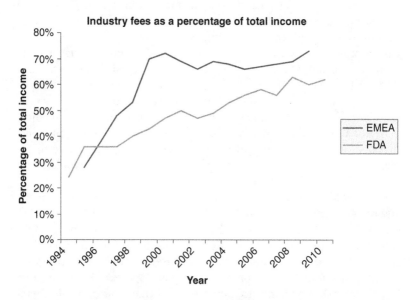

Figure 2.1   Percentage of industry fee contribution to total EMEA budget and to FDA spending on new drug review

Figures for FDA for 1994–1999 were compiled from 65 Federal Register 47994 and for 2000–2010 were compiled from Annual Financial Reports to Congress, available at: http://www.fda.gov/AboutFDA/ReportsManualsForms/Reports/UserFeeReports/FinancialReports/PDUFA/default.htm (Accessed 15 October 2011).

Sources: Figures for EMEA for 1995–2000 were compiled from EMEA Annual Reports. Figures for 2000–2009 taken from Ernst & Young 'European Commission Evaluation of the European Medicines Agency – Final Report – January 2010'. Available at: http://ec.europa.eu/health/files/pharmacos/news/emea_final_report_vfrev2.pdf (Accessed 15 October 2011).

the agency. Unlike the FDA, the EMEA was not restricted to using fees collected solely for new drug review or other specific purposes. In its first year, the EMEA collected industry fees to the tune of just 28 per cent of its total annual review. Thereafter, however, it pursued industry fees vigorously, quickly overtaking the FDA in this respect, and reaching over 70 per cent funding from companies in the late 2000s. As shown in Figure 2.1, the proportion of funding contributed by industry towards total EMEA revenues and towards FDA spending on new drug review has increased over the years. In 2002, a group of CPMP members unofficially published an open letter stating that the independence of the EMEA should be strengthened by reducing the proportion of its budget financed by industry fees, but that argument fell on deaf ears, as the agency's financial dependence on industry fees subsequently increased (Anon. 2002b).

According to Sauer (1997, p. 2), the first executive director of the EMEA, the centralized procedure was intended to operate as a 'fast track' system for obtaining a marketing authorization in Europe. It was expected to have a major impact on shortening new drug review times (Thomas *et al.* 1998, p. 788). Accordingly, in line with demands from industry and managers of national drug regulatory agencies in Europe who subscribed to the neo-liberal regulatory state, the supranational centralized procedure was established with strict and fairly rapid time-frames. The total amount of time allowed for the CPMP to arrive at its opinion was set at 210 days (European Commission 1996, p. 43). However, unlike the FDA's system, which differentiated between innovative pharmaceuticals receiving priority review (due to their promise of modest or significant therapeutic advance) and drug innovations receiving standard review (offering little or no therapeutic advance), the EMEA introduced no formal mechanisms for distinguishing between products that were innovative in a merely technical sense and those that promised therapeutic advance.

Nevertheless, the EMEA operated an *informal* accelerated evaluation arrangement for products intended to treat heavily disabling or life-threatening diseases, where there was an absence of appropriate alternative therapies, and where there was anticipation of 'exceptional high therapeutic benefit'. In such cases a final CPMP opinion could be adopted within 120 days, but that was at the discretion of the CPMP and EMEA (CPMP 1996). That arrangement was much more restrictive than the FDA system granting priority review status, which was open to life-style pharmaceuticals and drugs to treat less serious diseases with no requirement that expected clinical benefit must be very high. Whereas the FDA's priority (or standard) review 'track' allocation was routine practice for all innovative pharmaceuticals in the US, according to both European regulators and industry, the informal 120-day track in the EU centralized procedure was intended to be, and was, used only rarely (CMS Cameron McKenna and Andersen Consulting 2000, p. 174). In 2004, this accelerated assessment procedure was formalized in legislation by the European Parliament with a new time-frame of 150 days (Anon. 2005c; 2006a).

Table 2.2 shows that every year (1995–2010) the EU timelines were adhered to, with 13 out of these 16 years showing an assessment time of less than 190 days (range 157–203 days), compared to a target time of 210 days. Following the FDA's consultative approach to industry, the EMEA explicitly advised companies planning to submit NDAs through the centralized procedure to meet with EMEA staff beforehand for regulatory advice to help ensure quicker validation of applications (Anon. 1998a). Those suggestions may have been a response to the growth in EU non-approvals in the centralized procedure from 11 per cent in 1995 to 41 per cent in 1998. Subsequently, the rejection rate fell to 23 per cent in 1999, 21 per cent in 2000 and 27 per cent in 2001 (Anon. 2002c).

Table 2.2   EMEA approval times for centralized procedure products 1995–2010*

Calendar year	EMEA assessment phase (days)	EMEA post-opinion phase (days)	Decision process (days)	Total (days)	Total (months)
1995	189	45	119	353	11.6
1996	169	40	79	288	9.5
1997	178	32	86	296	9.7
1998	185	42	83	310	10.2
1999	183	38	70	291	9.6
2000	178	45	71	294	9.7
2001	170	32	76	278	9.1
2002	192	31	61	284	9.3
2003	190	48	60	298	9.8
2004	187	7	94	288	9.5
2005	203	56	41	300	9.9
2006	171	36	31	238	7.8
2007	171	25	33	229	7.5
2008	184	24	45	253	8.3
2009	157	29	42	228	7.5
2010	167	20	59	246	8.1

Sources: compiled from data on review times contained in EMEA Annual Reports.

Europe's innovation quest and the legislative review: from 'exceptional circumstances' to 'conditional marketing'

Article 71 of Regulation 2309/93 required the European Commission to report on the progress of the EU's supranational drug regulatory system within six years of the EMEA's creation. Consequently, in 2001, the Commission initiated a sweeping review of EU pharmaceutical legislation, which culminated in 2004 in a new Regulation (Regulation 726/2004) and a new Directive (Directive 2004/27/EC) amending Directive 2001/83/EC. This major review of EU pharmaceutical regulation confirmed concerns that the European pharmaceutical industry was losing competitive ground to the US in terms of research activity and product innovation (Gambardella *et al.* 2000, pp. 83–4). As we shall see, this broader context, and the fact that DG Enterprise was tasked with promoting the competitiveness of the European pharmaceutical industry, largely framed the responses and activities of the Commission throughout the review.

While any new legislative proposals arising out the review would, ultimately, have to be approved by the European Parliament and the European Council, the role of the Commission was particularly significant since it is the Commission alone of all the EC institutions that has the power to propose and draft new legislation. In theory, Commission proposals are meant to be consistent with the broad directions laid down by the Council. However, the Commission's initial draft proposals for new pharmaceutical legislation ignored specific Council recommendations where these were seen as being inconsistent with industry interests. For example, at the 2281st Council meeting of Health Ministers in June 2000, the Council invited the Commission, as part of the 2001 legislative review, to address the Council's resolution that 'identification of medicines with significant added therapeutic value is of great importance to promote innovation, which is vital not only from a health-protection perspective but also from an industrial policy viewpoint' (Europa, 2000). The Commission appeared to disagree, commenting that regulation was not the appropriate way of achieving this aim (La Revue Prescrire, 2002, p. 11) and neither the new Directive nor the new Regulation included comparative clinical data requirements that would allow the EMEA to differentiate between medicines on the basis of whether or not they offered added therapeutic value.

In contrast, the Commission was quick to include proposals that promoted industry interests, such as accelerated regulation of innovative pharmaceuticals within the centralized procedure (Bardelay and

Kopp, 2002; European Commission 2001a, p. 14). The Commission proposed two solutions: first, the formalization in legislation of an accelerated assessment process; and second, a provision which would permit the 'conditional' authorization of innovative drugs within the centralized procedure. This conditional marketing approval would allow for the early marketing of new drugs based on less trial data than normal (European Commission 2001b, p.4). As we have seen, in 1988 and 1992, the FDA introduced regulations (the accelerated approval regulations) pertaining to drugs intended to treat serious, debilitating or life-threatening conditions, which permitted the agency to approve drugs on to the market on the basis of less evidence of safety and efficacy – including reliance on phase II instead of the regular phase III trials and non-established surrogate endpoints. Since 1993, Article 13(2) of Regulation 2309/93 has allowed the EMEA to approve new drugs on to the market that lack comprehensive data on quality, safety and efficacy in 'exceptional circumstances' (European Council 1993). However, this provision was quite different from the FDA's accelerated approval rule in the sense that it was intended to be used only in cases where collection of full efficacy and safety data was judged to be impossible on ethical or scientific grounds (European Council 1975, Annex, Part 4 G).

When first introduced in the 1990s, the EU's 'exceptional circumstances' provision was well-received, particularly by groups representing patients with rare diseases because there can be genuine difficulties in obtaining adequate safety and efficacy data in the context of very small patient populations (CMS Cameron McKenna and Andersen Consulting, 2000: 45). In fact, figures on the number of products approved under the EU's exceptional circumstances regulation suggest that, in practice, it was being used more broadly than the restrictive criteria implied (Garattini and Bertele, 2001, p. 66). According to the EMEA, in the ten-year period before 2006, approximately 18 per cent of all products approved by the European agency were approved under exceptional circumstances (EMEA 2000, p. 7; FDA 2011b, p. 252).

Some former EU regulators told us that pharmaceutical firms were designing drug development programmes aimed at meeting the requirements for accelerated approval under Subparts E and H in the US and then submitting the results of those studies to the EMEA for approval under the exceptional circumstances provision. The EMEA and CPMP were then under pressure to adapt the EU's exceptional circumstances provision to companies' imperative for access to the US market. One former member of the EU's CHMP suggested that representations of patients' needs also sometimes influenced the way in which EU regulators applied the exceptional circumstances provision:

> After all, there are emotional aspects involved. And when you say: 'this is an unmet medical need, we must do something for these patients, there is nothing at this moment'. It is an argument that many people are taken in by. I don't buy that because I believe that if there is an unmet need, you need to make sure that you *meet* the need, not that you establish another way of continuing the unmet need by showing that hypothetically you have solved the problem even if you have done nothing [interviewee's emphasis].[12]

Yet despite the fact that EU regulators were clearly prepared to go beyond the limits imposed by the exceptional circumstances provision, in 2004 the new pharmaceuticals legislation included a requirement that the Commission draft regulations that would allow for conditional marketing approval. As mentioned above, the legislation also required that the EMEA establish an accelerated assessment procedure which, like the FDA's priority review, would guarantee quicker assessment times of companies' marketing applications. Specifically, the preamble to Regulation (EC) No 726/2004 stated:

> In order to meet, in particular, the legitimate expectations of patients and to take account of the increasingly rapid progress of science and therapies, accelerated assessment procedures should be set up, reserved for medicinal products of major therapeutic interest, and procedures for obtaining temporary authorization subject to certain annually reviewable conditions. (European Parliament and Council 2004, para 33)

While the official justification for introducing new mechanisms to accelerate review and grant early drug approval was the 'legitimate expectations of patients', the evidence suggests that the main impetus for more accelerated approval mechanisms in Europe came from the pharmaceutical industry and from within the Commission itself.

In a consulting report for the Commission, CMS Cameron McKenna and Andersen Consulting (2000) documented that the industry was pressing for an extension of the exceptional circumstances provision to cover products intended to treat serious or life-threatening conditions where no effective therapies were available, *and* a formal accelerated evaluation procedure for medicinal products offering a significant benefit over current therapies. Evidently, the industry was advocating for regulations in the EU that mirrored the accelerated approval and priority review procedures in the US. In their survey of pharmaceutical firms marketing drugs in the EU, Cameron McKenna and Andersen found that 63 per cent agreed or strongly agreed that the exceptional circumstances procedures should be more widely available, and 94 percent agreed or strongly agreed that a formal procedure for fast-tracking drugs should be established (CMS Cameron McKenna and Andersen Consulting 2000, pp. 172–6).

Industry was dissatisfied with the exceptional circumstances provision because companies believed it was too restrictive in scope and used too infrequently. Notably, the European Biopharmaceuticals Enterprises (EBE), a group within EFPIA representing biopharmaceutical companies, contended:

> The regulatory evaluation of certain innovative medicinal products to fulfil urgent, unmet, medical needs – and, consequently, patients' access to such therapies in Europe – is unnecessarily lengthy. We believe this is partly due to the absence of a coherent series of mechanisms in the European Union to permit accelerated market access for innovative and much needed new medicines. (EBE-EFPIA 2004, p. 1)

Such sentiments were echoed by the EU Commissioner for Enterprise, who claimed that the EU system needed to take account of the 'new global environment' by adopting conditional marketing approval to ensure that regulatory assessment of major new drugs could be 'as fast, if not faster than the FDA' (quoted in Anon. 2001b). Notably, this pronouncement also echoed arguments made by the G10 Medicines Group – a collaboration of European industrialists, Health Ministers, and patient groups, created by the European Commission to advise it on EU policy for the pharmaceutical industry (Anon. 2002d).

By contrast, a significant majority, 70 per cent, of European regulators (outside the European Commission), who were interviewed by Cameron McKenna and Andersen *opposed* making the exceptional circumstances provision more widely available. Our interviews with European regulators on this point revealed that they wanted to use, and did use, the exceptional circumstances provision for cancer drugs and drugs for serious diseases where there was little or no available treatment.[13] However, they were concerned that drug companies would abuse (post-marketing) follow-up commitments to provide more comprehensive evidence of therapeutic benefit, which also explains why these regulators were found to support stronger powers to withdraw marketing approval after one year if companies failed to comply with their post-marketing commitments (CMS Cameron McKenna and Andersen Consulting, 2000, pp. 173–4). Indeed, in 1999, some EU regulators had warned that a 'large number' of companies were failing to meet their post-marketing commitments, including safety and efficacy studies (Anon. 1999a).

Similarly, 87 per cent of European regulators surveyed were opposed to introduction of a formal fast-track procedure (like the 'priority review' system at the FDA) on the grounds that the existing systems at that time were sufficiently flexible for regulators to prioritize and accelerate new drug applications which were likely to have important public health benefits (CMS Cameron McKenna and Andersen Consulting 2000, pp. 45 and 175). As one former member of the CHMP explained, an application with sound evidence supporting the breakthrough status of a drug can be approved very quickly without special procedures.[12]

Of the patient representatives consulted in the same survey, only 49 percent supported the extension of the exceptional circumstances provision (CMS Cameron McKenna and Andersen Consulting 2000, pp. 174–6). It is unclear how many of the patient groups that responded were funded by pharmaceutical companies, and if so, to what degree. According to the consulting firm, 'national consumers organizations, together with 134 associations representing a broad spectrum of disease areas and countries were sent written questionnaires', of which 31 associations responded (CMS Cameron McKenna and Andersen Consulting 2000, pp. 25–6). There is, however, no evidence that patient associations or public health advocacy groups initially sought or demanded these changes to the exceptional circumstances provision (Anon. 2002e).

Notably, when we interviewed Commission officials, demand from patients did not feature at all in their explanations of the emergence of 'widening exceptionality'. On their account, the reasons behind the proposals related to: harmonization of European and US regulation, which the Commission saw as 'exceptionally important – both for stimulating innovation, for reducing burdens on industry, and for improving public health protection'; addressing industry demands; and levering companies to complete post-marketing studies.[14] The point about levering companies to complete post-marketing studies refers to the fact that, before 2004, the EMEA's only regulatory option if a pharmaceutical firm failed to comply with its post approval obligations made to the agency at the time of marketing approval (e.g. the completion of some post-marketing drug trial) was to remove the drug product from the market – a measure that regulators were reluctant to take when the drug benefited some patients. With the introduction of conditional marketing regulations in 2004, the EMEA was empowered to impose financial penalties on non-compliant companies, but such enforcement capabilities could have been achieved separately without conditional marketing (European Commission 2002; European Parliament and Council 2004).

While there is no evidence that EU patients were demanding earlier or faster access to innovative pharmaceuticals in the EU, there *is* evidence that some patient and consumer health advocacy groups were concerned that the draft proposals released by the Commission in 2001 prioritized industry interests over the interests of EU citizens and public health (La Revue Prescrire, 2004). In March 2002, a lobby formed in Paris – comprised of patient groups, family and consumer advocacy organizations, health insurance associations and health professional bodies – whose purpose was to 'ensure that European pharmaceutical policy serves the public interest' (European Public Health Alliance, 2003). And between 2002 and 2004, this coalition – the Medicines in Europe Forum (MiEF) – was the most visible and active civil society group lobbying the EU Parliament, Council and Commission for changes to the draft legislation during the period of the legislative review. Significantly, with respect to Commission proposals to accelerate the assessment process, the MiEF fought to *safeguard* the amount of time rapporteurs had (at least 80 days) to analyse the scientific data contained in companies' marketing approval applications. The coalition also fought for, and secured: increased transparency of both the decentralized and centralized procedures; a continued ban on direct-to-consumer advertising; and better labelling and patient information leaflets. In relation to the needs of patients with no effective treatment options, the coalition pressed, not for earlier *marketing*, but for patient access prior to marketing via compassionate use programmes (La Revue Prescrire, 2004).

Despite the absence of pressure from patient and consumer advocacy groups within the EU for earlier marketing of innovative medicines, in November 2004 the Commission published a draft regulation on conditional marketing approval, which was formally implemented as Regulation 507/2006 in April 2006 (Anon. 2006c). Under the regulation, pharmaceutical firms could request conditional marketing approval for new drugs to treat emergencies, chronically/seriously debilitating conditions or life-threatening disease. Companies could also request conditional marketing in the EU for orphan drugs. Conditional approval could be granted by the EMEA if: the drug was of public health interest;

a positive benefit–risk ratio could be demonstrated based on scientific evidence and pending completion of further studies; safety data were complete unless under very exceptional circumstances; and the company was required to finalize on-going studies or conduct new studies to verify the presumed positive benefit–risk ratio.

The purpose of the 'conditional marketing' regulation was to permit pharmaceutical companies to get innovative drugs in these categories approved on to the market faster with less clinical (and possibly less preclinical) test data than normally required for regular approval. 'Conditional marketing' was granted only if (or on the condition that) the company committed to undertake post-marketing studies to provide additional datasets that would answer any outstanding questions and data-deficiencies about the safety and efficacy of the drug in question. In this respect, the EU's 'conditional marketing' regulation was very similar to the FDA's 1992 'accelerated approval' rule in the US. Conditional marketing status in the EU was to be reviewed every year by the EMEA to assess whether any new data provided were sufficient to justify conversion from conditional to regular approval (Anon. 2006c).

Conditional marketing approval met with a mixed response from patient groups and public health advocates. The European patients' organization for rare diseases, EURODIS asserted that it was particularly important for rare diseases, but expressed concern for a more 'flexible' approach should a company fail to fulfil post-marketing conditions in time for annual renewal:

> Not renewed means that the product is withdrawn from the market. When the product is in fact useful, but only the incompetence or negligence by the applicant [company] are in question, then maybe the price [meaning loss to health] to pay for the applicant to be incompetent is too high for the patients. When the applicant does not respect the condition because for example lack of financial resources to conduct appropriate studies, it is a problem to withdraw from the market a potentially useful product that maybe has a positive benefit-risk ratio. (EURODIS 2005)

By contrast, the MiEF argued that the approval criteria were too vague and the (post-marketing) follow-up requirements too flimsy. According to the MiEF, conditional marketing could be justified only in very exceptional circumstances of urgent public health need because, by allowing inadequately assessed products to enter the European market, the procedure was inherently risky (MiEF 2004).

Probably due to the creation of the 'conditional marketing' regulation, in 2007 the European Commission introduced financial penalties for drug firms that failed to meet their post-marketing commitments for products approved through the centralized procedure. It is remarkable that 12 years elapsed before companies faced significant penalties for neglecting post-marketing obligations to regulators and public health. Initially, the Commission proposed that the fine should be ten per cent of the offending company's turnover in the previous year, but that was reduced to five per cent after the pharmaceutical industry protested that ten per cent was too punitive (Anon. 2007a).

Even after the introduction of 'conditional marketing approval' and 'accelerated review' within the EU's supranational centralized procedure, the pharmaceutical industry and the European Commission's DG Enterprise continued to press for faster marketing approval of innovative drugs. A report by a group of experts advising the Commission on how to improve the EU's performance in research and innovation asserted that, in 1992, six of the top ten selling pharmaceuticals were produced by European firms, but by 2002 this had fallen to just two of the top ten. That situation apparently prompted the inference that the Commission should do even more to accelerate patients' access to innovative drugs (Anon. 2006d). Yet the *sales* of pharmaceuticals are not necessarily an indication of innovation, let alone therapeutic advance or patient need.

In June 2005, the EMEA set up a think-tank comprising EMEA staff and the agency's scientific bodies. Aware of the declining number of pharmaceutical product innovations in the EU, the think-tank sought to identify 'bottlenecks' in the development of innovative drugs. It met with drug companies and sent 200 firms a questionnaire. In response to industry concerns, two billion euros were made available to boost drug innovation in Europe via the Innovative Medicines Initiative (IMI), a pan-European public-private collaboration involving pharmaceutical companies, regulatory authorities, patient organizations, universities and hospitals (Anon. 2006e; 2008d). Although a public-private collaboration, with one billion euros coming from EFPIA and the other billion from the European Commission, the IMI was an industry initiative. Following the priorities identified by drug companies, it was particularly focused on reducing the time and cost of drug development, especially by searching for biomarkers and surrogate endpoints that could predict clinical efficacy and minimize the failure (or so-called attrition) rate of late-stage, expensive clinical trials (Anon. 2007b). In this respect, the IMI was similar to the Lasagna Committee in the US nearly 20 years earlier, though the overall brief of the IMI was broader.

In fact, in 2006, the FDA had established a similar project, known as the Critical Path Initiative (CPI) focusing on biomarkers of drug efficacy and toxicity, whose importance to the FDA was reiterated by its commissioner at the end of the decade (Anon. 2008e; 2010b). The 'conditional marketing approval' regulation in the EU opened the door for industry to seek more extensive use of biomarkers and surrogate endpoints just as the 1992 'accelerated approval' rule had done in the US. The application of biomarkers and surrogate endpoints to drive regulatory approval decisions was much more amenable to industry and regulators in a context of early conditional marketing approval based on incomplete clinical data. The IMI was launched in 2008. Industry spokespeople, such as the director of the ABPI, were quick to represent it as 'an initiative that will bring medicines to patients more quickly and effectively' (quoted in Anon. 2008d, p. 23).

The 'Global Dossier': international harmonization of regulatory science in the EU and US

As we discussed in relation to European harmonization, according to the industry, inconsistencies between national regulatory standards produced wasteful duplication in drug testing, which drove development costs and created barriers to trade. That concern applied not only to national systems within Europe before 1995, but also more internationally (Abraham and Reed 2001).

Reflecting with the neo-liberal outlook of the Reagan Administration, in the mid-to-late 1980s, American-led bilateral initiatives between the US and Japanese governments were taken, including a determined objective on the part of the US to open up Japanese markets (Ferris 1992, 197–8). Japan represented about 20 per cent of the world pharmaceutical market at that time (Reed-Maurer 1994, 38). In response, in 1988, the first 'mission' of government regulators and industry representatives from the pharmaceutical sector in Europe was sent to Japan to discuss bilateral harmonization of regulation between Japan and the EU, so that Japanese markets might become more accessible to the European drug industry (Wyatt-Walter 1995). However, given the importance of the US market (about half the world market), the European pharmaceutical industry was unenthusiastic about solely bilateral harmonization with Japan, so the International Federation of Pharmaceutical Manufacturers Association (IFPMA) took responsibility for organizing trilateral meetings between the industry and government regulators in the pharmaceutical sectors of the EU, Japan, and the US from 1990 (Abraham and Reed 2001). These meetings became established as the International Conference on Harmonization of Technical Requirements for Registration of Pharmaceuticals for Human Use (ICH).

During the 1990s, the ICH focused on harmonizing the techno-regulatory standards for new drug approval across the three regions. The idea was that pharmaceutical firms would be able to submit the same pre-clinical toxicological, clinical trial, and post-marketing data in order to meet the regulatory requirements of the three regions – a single dossier of data instead of three (Abraham and Reed 2001).

As with advocates of accelerated regulatory review, proponents of ICH declared that, in addition to expanding industry's access to markets more quickly, it was also in the interests of patients and public health. At the opening session of the first ICH conference in Belgium in 1991, it was asserted that the savings made by companies from harmonized regulations would further the delivery of innovative research yielding therapeutic benefits to patients (Bangemann 1992, p. 4). By the end of the decade, even more emphatic claims were made by IFPMA, which had become the ICH's secretariat:

> ICH clearly enhances the competitive position of those companies that choose to operate using its standards, as well as significantly bene-fiting both the regulators and the patients, who, most importantly, receive crucial new treatments sooner. .... In summary, harmoniza-tion through ICH brings important, life-saving treatments to patients faster, while releasing the pharmaceutical companies' development funds to projects that will produce the ground-breaking treatments of the future. (IFPMA 2000, p. 1)

By 2000, the ICH had indeed harmonized the techno-regulatory stand-ards for safety and efficacy evaluation, as well as other types of new drug assessment, across the three regions. Certainly it was in the commer-cial interests of pharmaceutical firms because it cut their costs and improved their access to markets for whatever products they developed. In the context of budgetary pressures and relentless demands for more rapid regulatory review, it is easy to see why streamlining of technical data submissions could be attractive to regulatory agencies if it did not undermine their credibility as protectors of public health. Supporters of ICH from industry and regulatory agencies insisted that it did not compromise drug safety. However, independent research has shown that, in many areas of safety evaluation, harmonization actually entailed a lowering or loosening of protective standards for some countries, often the US (Abraham and Reed 2001; 2002; 2003). Whatever, the drawbacks of ICH for drug safety standards, it was yet another neo-liberal inspired process of reform supposedly to increase the number of innovative pharmaceuticals delivering therapeutic advance for patients faster, by cutting development costs and time.

Pharmaceutical outcomes and the neo-liberal reforms

Evidently, since the 1980s, the FDA and the EMEA have developed mechanisms for accelerating both regulatory review and development of innovative pharmaceuticals, including reductions in the amount of information regulators require from companies before clinical trials and/or marketing approval, and time management goals regarding formal meetings between regulators and firms encouraging regulators to provide companies with scientific advice about drug development. As we have seen those measures within the neo-liberal reform programme delivered much faster new drug review times in the EU and the US than previously. In 1994, the FDA's goal was to finish 55 per cent of its new drug reviews on time; it achieved 95 per cent. In 1995, the goal was 70 percent, but the FDA achieved 98 per cent. The goal rose again to 80 per cent in 1996 when the FDA achieved 100 per cent. In both 1997 and 1998, the goal was 90 per cent and the FDA achieved 100 per cent (Willman 2000a). To reinforce this point, FDA Commissioner Mark McClellan claimed in the agency's 2003 performance report to Congress that 'over the eleven years of PDUFA, the agency had met or exceeded nearly all of the PDUFA goals' (FDA 2003a).

There is also evidence that these measures, perhaps combined with ICH activities, had an impact on accelerating drug development. An analysis of new drug development times between 1990 and 1999 by the Tufts Center found a marked downward trend in average clinical development times in the US. For example, the mean clinical development time for priority drugs was 48 per cent lower in 1998–1999 when compared with 1990–1991, even though the previous three decades had seen steady increases in drug development times (Kaitin and DiMasi 2000). In a subsequent study, the Tufts Center reported that the fast-track designations (such as Subparts E and H) had had the intended effect of shortening drug development times – between 1998 and 2003 the average time required to develop a fast-track drug and gain approval fell by over two years. Moreover, the number of diseases attracting fast-track designation had expanded (Tufts Center for the Study of Drug Development 2004, p. 3).

Increased pharmaceutical product innovation was one of the key purposes of those decreases in drug development and regulatory review times, according to the pharmaceutical industry and governments who sought them, and the managers of regulatory agencies who adopted and implemented them in line with the demands of their political masters. For example, a report by the Charles River Associates (CRA) on behalf of the European Commission's DG Enterprise stated that the main purpose of accelerated evaluation/marketing provisions within the EU's supranational regulatory system and the increased the role of the EMEA in providing scientific advice to companies was to 'bring forward new products and increase the returns from truly innovative products' and 'act as a spur to innovation' (Charles River Associates 2004, p. viii).

This was confirmed in an interview with a European Commission official at the Pharmaceuticals Unit within DG Enterprise. As well as aiming to 'ensure a high level of health protection', the Pharmaceuticals Unit sought 'to support pharmaceutical innovation in the European Union and foster competition and transparency in the Community pharmaceutical market' (European Commission Enterprise Directorate-General 2000, pp. 4 and 5). As our interviewee explained, the view of DG Enterprise was that regulation should create an environment in which industrial innovation can flourish in order to yield important therapies for public health:

> You cannot disassociate public health protection from innovation...If you think of public health promotion not just as *protection* but as *promotion,* then there are two sides to the coin. ... You have, on the one hand, to stimulate innovation in order to find the treatments for those incurable diseases *but* you have to balance it with tough regulation on the other hand, which has demanding data requirements in terms of proving safety, quality and efficacy. We don't see it as a conflict. And we clearly do see innovation as being a crucial role of regulation.[13]

In the US, encouragement of product innovation was also central to the reasons given for the introduction of accelerated regulation. The Senate report by the Committee on Labour and Human Resources, accompanying the 1997 FDAMA legislation, asserted:

> Increases in the time, complexity and cost of bringing new products to market are borne directly by the public, in delayed access to

important new products – including life-saving medical therapies – and in higher costs. They are a growing disincentive to continued investment in the development of innovative new products and a growing incentive for American companies to move research, development, and production abroad, threatening our Nation's continued world leadership in new product development, costing American jobs, and further delaying the public's access to important new products (US Senate 1997, p. 7)

Yet, despite decreasing regulatory review times, provisions aimed at shortening development times, and increasing harmonization of regulatory science requirements in the EU and US (Kaitin and DiMasi 2000), there has actually been a *decline* in pharmaceutical product innovation since and during those neo-liberal reforms. We can see this by considering the number of innovative drugs (NMEs) that have been submitted to the FDA over time. Figure 2.2 shows that the number of NMEs submitted to the FDA for regulatory review fell from 45 in 1996 to 23 in 2010.

The falling numbers of innovative drug applications to the regulatory agencies have been mirrored by a decreasing number of approvals

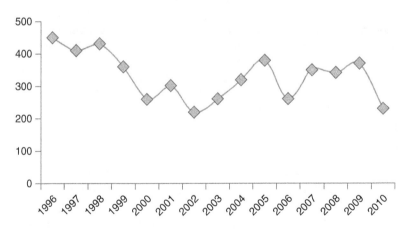

Figure 2.2   NME applications to FDA 1996–2010*

*2004–2010 represents application for new molecular entities filed under New Drug Applications (NDAs) and therapeutic biologics filed under Original Biologic Licence Applications (BLAs). 2001–2003 represents NMEs but not therapeutic biologics.

Source: FDA, 'Comparison of NMEs approved in 2010 to previous years'. Available at: http://www.fda.gov/Drugs/DevelopmentApprovalProcess/HowDrugsareDevelopedandApproved/DrugandBiologicApprovalReports/ucm242674.htm (Accessed 15 October 2011).

and market launches of pharmaceutical product innovations in the EU and US (Charles River Associates 2004; Turner 2004). In 2007, the FDA approved just 17 NMEs. This rose to 21 in 2008, but fell back to 19 in 2009. These figures were slightly higher than those for the mid-2000s, but still far behind the late 1990s and slightly less than the 22 approved in 2006 (Anon. 2008f). Similarly, EFPIA reported a decline in the number of NASs marketed in the region of Europe, falling from 89 between 1995–1999, to 57 in the period 2000–2004 (Anon. 2005d).

Indeed, by the mid-2000s, many commentators wondered whether the industry was facing a worldwide crisis in innovation (Centre for Medicines Research International 2002; 2005; Charles River Associates 2004; FDA 2004). Just 23 NMEs were launched on to the world market in 2004 – fewer than at any time in the previous 20 years (Centre for Medicines Research International 2005). The number of NDAs submitted to the centralized procedure saw a rise in 2004 of about 20 per cent over the previous year, but fell back to 2003 levels in 2005 (Anon. 2005e; 2006f). It rose again in 2006, but fell back again by 19 per cent in 2007 (Anon. 2008c). Overall pharmaceutical product innovation increased slightly in the late 2000s from its relatively very low level of the early/ mid 2000s, but never returned to levels of the 1990s. Hence, the argument put forward by industry, neo-liberal governments and their allies in regulatory agencies – that accelerated regulatory review times and a more predictable regulatory environment was needed to stimulate pharmaceutical product innovation – finds no support in trends of product innovation since the mid-1990s.

As indicated by the quotations above from the US Senate (1997) and our informant at the Commission's Pharmaceuticals Unit, governments on both sides of the Atlantic attested not only that the neo-liberal measures to accelerate regulatory review, increase predictability in the drug development process, and require less test data from drug companies would stimulate pharmaceutical product innovation, they also proclaimed that by so doing, such measures would benefit patients and public health (Milne 2000). That view was made with increasing vigour by FDA management as the reforms progressed. Commenting on the effects of the 1992 accelerated approval regulations, PDUFA, and the priority review system for allocating resources, the FDA claimed that:

> To date, the agency has cut new drug approval times nearly in half, while the number of new drugs approved in a year has doubled...U.S. drug approval times have decreased dramatically and are now among the fastest in the world. Americans have access to new therapies

faster, and, as a result, suffer less, recover more rapidly, are often cured completely, and live longer lives, or enjoy an improved quality of life. (FDA 1997a)

Similarly, at a public meeting on PDUFA in 2000, Janet Woodcock, the FDA's former director of CDER, declared:

> The public has received benefits from [PDUFA]. There is faster access to new therapies.... And ... there is increased industry incentive to direct attention into new therapeutic areas by the timeliness and the predictability of this program. (FDA 2000a)

As we explained in Chapter 1, there is no necessary correspondence between drug product innovation and therapeutic advance. The two phenomena are not entirely disconnected because without any drug product innovation, there would be fewer pharmaceutical therapeutic advances. However, neither should the two be conceptually fused because a drug may be an innovation without offering any therapeutic advance – as is clear from the FDA's own regulatory accounting. Accelerated review of 'me-too' drugs allocated standard review by the FDA might benefit industry but is likely to offer little or no benefit to patients. Conversely, in theory, it is possible that, while the overall number of NASs launched each year declined after the neo-liberal reforms, the number of NASs offering therapeutic advance could have remained constant or even increased. Hence, a partial decline in drug product innovation might not necessarily have any negative impact on public health. It follows that, to assess the validity of the claim that the neo-liberal reforms have led to an increase in the number of new drugs offering therapeutic advances and breakthroughs for patients, one must disaggregate the data on drug product innovation.

To do this, we must turn to the FDA because the EMEA does not publish data distinguishing between new drug products that offer modest or significant therapeutic advance ('priority' drug approvals) and those that offer little or no therapeutic advance ('standard' approvals). Figure 2.3 shows that from 1998 to 2010 the proportion of NME applications to FDA accorded priority review status has been fairly small. Significantly, Figure 2.3 reveals that *the number of new drugs offering therapeutic advance has also been in decline since 1998.*

In other words, the claims made by governments and regulators on both sides of the Atlantic that their neo-liberal reforms would lead to patients gaining faster access to more new drugs that they need is not

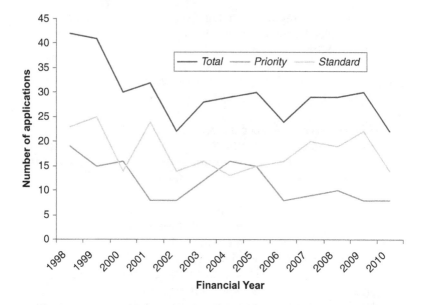

Figure 2.3   NME applications filed with FDA 1998–2010

Source: Compiled from FDA's annual PDUFA Performance Reports. Available at: http://www.fda.gov/AboutFDA/ReportsManualsForms/Reports/UserFeeReports/PerformanceReports/PDUFA/default.htm (Accessed 6 November 2011).

supported by the FDA's own quantitative evidence on the number of therapeutically valuable drugs going through the regulatory system. Moreover, the National Institute for Health Care Management (NIHCM) examined the number of therapeutic advances approved out of all the new branded medicines that entered the U.S. market between 1989 and 2000. The analysis included NMEs and 'incrementally modified drugs' (IMDs) – combinations or new formulations of existing pharmaceutical products on the market, but excluded biotechnology products. It was found that, between 1989 and 1994, out of a total of 350 NMEs and IMDs, the FDA approved 106 as priority drugs (30 per cent), but between 1995 and 2000, out of a total of 569, the agency approved only 133 as priority drugs – a decline to just 23 per cent (National Institute for Health Care Management Foundation 2002, p. 10). Subsequently, the report by Charles River Associates signalled similar conclusions, finding that about 20 per cent of NDAs approved between 1990 and 1994 were accorded priority reviews, compared with about 15 per cent between 2000 and 2003 (Charles River Associates 2004, p. 36).

Although equivalent data are not available from the EMEA, it is likely that similar trends will have occurred in Europe as there is a considerable overlap between drugs approved by the FDA and drugs approved via the EU's supranational centralized procedure (Charles River Associates 2004, pp. 29–32). Independent evaluations of the therapeutic value of drugs approved via the centralized procedure suggest that, as in the US, the proportion of new pharmaceutical products that represent significant therapeutic advance is small. Garattini and Bertele (2001, p. 65) judged that of the 126 products approved by the CPMP in its first year, only 60 could be regarded as offering any therapeutic advance. With respect to some specific disease categories these authors found that, of the nine mind-altering drugs approved through the EU's centralized procedure between 1995 and 2002 only one offered a therapeutic advance, and the 11 cardiovascular drugs approved during the same period 'contributed little to recent progress in the cardiovascular area' (Garattini and Bertele 2003a; 2003b). In direct contradiction with the assertions made by the architects and managers of neo-liberal regulatory reform in Europe, Garattini and Bertele (2003b, p. 706) reported:

> These agents seem to follow the logic of obtaining a share of large market areas rather than attempting to cover any unmet patient needs. Interestingly...the new drugs usually cost more than similar drugs already available, even if they are only equivalent in terms of efficacy and safety.

In addition, the French-based medical/healthcare professional organization, Prescrire, judged that, of the 18 innovative drug products approved through the EU centralized procedure in 2000, none represented major therapeutic advance, three offered an advantage but not in a way that would fundamentally change therapeutic practice, and three had minimal additional therapeutic value, 11 offered no therapeutic advance, and one could not be evaluated (Anon. 2001c).

Overall, the evidence suggests that the number of new drugs offering modest or significant therapeutic advance in the EU and the US represents a minority of pharmaceutical product innovations, and evidence from the US indicates that this percentage has fallen during the period characterized by neo-liberal regulatory reforms. It is not that the pharmaceutical industry's research activity had declined because between 1980 and 2005, the number of pharmaceuticals in R&D increased three-fold (Anon. 2005f). Rather, it was that R&D did not yield many new drugs offering significant therapeutic advance.

The research-based pharmaceutical industry argues that organizations like Prescrire place too much emphasis on innovative drugs that offer modest or significant therapeutic advantage and neglect the positive contributions of the other pharmaceutical product innovations that promise little or no therapeutic advance – 'standard drugs' in FDA terms, though often referred to as 'me-too' drugs because they are so therapeutically similar to medicines already on the market. In particular, the industry contends that me-too drugs create market competition, which leads to cost benefits and lower prices for patients and healthcare systems (EFPIA 2000, p. 19). Furthermore, the industry is joined by some regulators in claiming that, even if a new drug cannot be shown to offer any specific advantage over existing therapies in clinical trial populations, it may still offer an advantage to some individual patients (EFPIA 2000, p. 20).[15,16] Whatever, the merits of such arguments in favour of me-too drugs, we note that the evidence we have presented in this section demonstrates that pharmaceutical product innovation, as a whole, including me-too drugs, declined after neo-liberal reforms, presumably shifting any benefits associated with me-too drugs into decline also.

Nonetheless, in order to fully appreciate the public health implications of the fact that, despite over 20 years of regulatory reforms ostensibly aimed at accelerating pharmaceutical innovations needed by patients, only a minority of pharmaceutical product innovations offer even modest therapeutic advance, and fewer still promise significant therapeutic advance, it is important to consider these claims about the *indirect* benefits from me-too drugs. The assertion that me-too drugs provide a sensible mechanism for reducing pharmaceutical prices is not plausible. In Europe, some regulators told us that pharmaceutical companies use approval of an innovative drug through the centralized procedure as a justification for charging a very high price, even if the product offers little or no therapeutic advance.[17,18] Regarding cancer drug innovations approved during the first six years of the centralized procedure, Garattini and Bertele (2002) found that those drugs were much more expensive than existing therapies despite an absence of proven therapeutic advantages. Meanwhile, in the US, the NIHCM reported that the average price per prescription in 2000 for a standard NME (me-too) was twice that of a drug approved before 1995, implying that having innovative pharmaceutical products made up of largely me-toos has done little, if anything, to bring down American drug prices (NIHCM 2002, p. 13).

Regarding the possibility that a new drug showing no therapeutic advance in clinical trials may nevertheless provide such benefits to individual patients, one cannot entirely discount such hope, but it is rarely evidence-based. There is hardly ever data from clinical trials identifying subsets of patients with particular characteristics that might derive additional therapeutic benefit when a drug shows no overall therapeutic advance. Consequently, physicians must somehow work that out for themselves in clinical practice, selecting optimal treatments out of a range of therapeutic options for patients on the basis of individual observation. Yet such a process renders almost impossible verification of any therapeutic advantage me-too drugs may or may not provide and, unlike controlled trials, cannot be a basis for generalization beyond the individual patient to evidence-based medicine or regulatory decision-making.[19,20]

Such hope and wishful thinking about the therapeutic value of me-too drugs can, however, be the basis for pharmaceutical promotion, which may act as a substitute for scientific evidence. According to Kessler *et al.* (1994), the competitive sales behaviour of drug companies in medical fields crowded with me-too drugs has led to misleading promotion, increased healthcare costs and inappropriate prescribing, rather than patient benefits. It is claimed that pharmaceutical companies may use 'seeding trials' to make unsubstantiated claims about the superiority of their me-too drugs and run 'switch campaigns' to persuade doctors to change their patients' medication to the new me-too drug. Such promotional practices may be aided by, and reinforce, an irrational ideology that 'new' must be better (Kessler *et al.* 1994).[21]

So far in this section, we have discussed the macroscopic effects (or lack of effects) of neo-liberal regulatory reform on the stimulation of pharmaceutical product innovation and the provision of innovative pharmaceuticals that offer therapeutic advance for patients and healthcare. Scholars typically consider the impact of regulation on pharmaceutical innovation, but often neglect the effects of innovation rates on drug regulation. Since the neo-liberal reforms, such neglect can no longer be countenanced because, as we have seen, regulatory agencies in European nations, the supranational EU, and the US depend so heavily for their funding on industry fees. Consequently, the declining number of innovative drug applications has had a serious impact on the resources of regulatory agencies.

Earlier in this chapter, we explained how PDUFA caused a redistribution of resources away from some public health protection activities because of the renewed emphasis on the new drug review process. Declining numbers of NDAs and associated fees to pay for staff can only have accentuated resource problems throughout the agency. It has also had a serious impact on EMEA's regulatory activities. The crisis of 2002 is just one example. Reporting on planned new EU directives, IMS Health (2003) observed:

> The EC moves come at a time of increased concern on both sides of the Atlantic about perceived dips in output from pharmaceutical industry pipelines. Indeed, in Europe the EMEA is facing a financial crisis due to a dramatic drop in new drug applications. The number of applications fell from 54 in 2000 and 58 in 2001 to just 31 in 2002.

A commentary in the pharmaceutical trade press, *Scrip*, captured well the nature and implications of the crisis:

> On 3 October, the EMEA announced that only 25 new drug applications had been received by the end of September, down from 58 for 2001 as a whole. If orphan drug applications are stripped out of the figures, the number was 14, compared with 46 for the entire previous year. The EMEA has expected 50 new drug applications (excluding orphan products) this year. As the agency receives 80 per cent of its income from industry fees and the fees are highest for new chemical entities [NMEs], the application decline has forced the EMEA to make emergency savings such as delaying recruitment and cancelling working groups. (Anon. 2002f)

Consequently, the CPMP had to cancel expert group meetings on drug efficacy, vaccines, paediatrics, and the quality of regulatory documents (EMEA 2002a).

Risk management

In the late 1990s, senior FDA and US government officials began to articulate and practice an explicit risk management strategy for pharmaceutical regulation (FDA 1999a). The official rationale behind this strategy was that the management of pharmaceutical risks should be optimized by monitoring drug products throughout their life-cycle so that risks could be better foreseen and tailored to more nuanced regulatory interventions and patient sub-populations, thereby minimizing shocks to the medical system caused by withdrawing drugs from the market. The FDA's embrace of risk management policies was officially in response to a large number of high-profile drug products being withdrawn from the US market on safety grounds in the mid and late 1990s.

Insofar as that is an accurate explanation for the FDA's introduction of risk management policies, it also provides a superficial understanding because one must ask why that significant rise in drug safety withdrawals in the US occurred. It was not by chance (US GAO 2002). It was structurally related to the deregulatory reforms put in place during the 1980s and 1990s, which not only accelerated the pre-market review of new drugs, but also permitted more questions about safety (as well as efficacy) to be answered in the post-marketing phase, due to less demanding standards of pre-approval evidence. Olson (2002) has provided some of the most compelling evidence linking the 1992 deregulatory reforms of PDUFA and the 'accelerated approval' rules to drug safety problems. By analysing the ADRs recorded in the FDA's Spontaneous Reporting System (SRS) for 141 innovative pharmaceuticals (NCEs) approved on to the market by the agency between 1990 and 1995, she found that the reductions in new drug review times, which occurred in that period, were significantly associated with increases in both ADRs requiring hospitalization and ADRs resulting in death.

Hence, the emergence of risk management as a response to drug safety withdrawals was also largely an effect of neo-liberal deregulatory reforms. Indeed, there is evidence that risk management was also itself an expression of pro-business, neo-liberal governance of drug safety problems. Concerned about public confidence in the pharmaceutical industry and drug safety regulation, the pressure for risk management policy was also driven by the pharmaceutical industry and a strongly neo-liberal Republican-controlled Congress, as well as the FDA. As several of our FDA informants explained:

> The first time I can recall hearing about risk management [within the FDA] was 1997. It [risk management] was a reaction to a fairly large number of withdrawals in a short period of time and the agency [FDA] getting pressure from Congress – they represent industry if a drug company is in their district. There was also direct pressure from industry.[4]

We saw it [risk management] in the negotiations for the re-authorization of PDUFA. Companies were getting nervous about the number of drugs coming off the market and they wanted the standards raised [i.e. more difficult] to call an adverse drug reaction 'an adverse drug reaction'. And they wanted something to [re]assure the public – I think the FDA maybe also did – that everything was being done. The agency had taken a lot of criticism and was getting sensitive to drug withdrawals, so they tried to raise the profile of looking at drug risk.[6]

The drug industry was behind other industries in bringing the concepts of risk management into practices. So that was part of it, and highly publicized examples of problems with drugs that people felt in retrospect could have been avoided if we'd thought more about what could happen in advance.[22]

Hence, the idea of pharmaceutical risk management policy originated in the US, but it migrated to Europe. According to Demortain (2008) the 'crisis' over the safety of the lipid-lowering drug, cerivastatin, in the late 1990s and early 2000s in Europe led to an official consensus at least about the idea of pharmaceutical risk management across most western countries, culminating in an internationally agreed guideline on 'pharmacovigilance planning' at the ICH. By 'pharmacovigilance planning' was meant a plan for each new drug product detailing how possible adverse effects could be detected, understood, monitored, assessed, and perhaps even prevented. Although that was the official representation of risk management policy, some experts took an altogether more sceptical view of it, suggesting rather less consensus among regulatory agencies. In 2005, one European regulator told us: 'Risk management is an invention of industry to say that you can keep a drug even if it is toxic, if you know how to manage risk'.[12]

In the same year, Horst Reichenbach, the European Commission's director-general for enterprise, stated that the Commission was looking into whether the EU's 2004 pharmaceutical legislation was sufficient to reduce the 'risk' of product withdrawals over safety issues in the future. The implication being that the 'risk' to be managed was the risk of product withdrawal, with the risk of the product itself receiving secondary consideration. Reichenbach also linked the EMEA's conditional marketing provisions with the introduction of risk management plans, noting that, at the time of marketing approval, one cannot know the full safety profile of a new drug (Anon. 2005g). In other words, risk management plans were particularly prevalent among innovative pharmaceuticals accelerated on to the market by various deregulatory measures in the EU (Anon. 2005h). Indeed, since the mid-2000s, risk management and pharmacovigilance planning are often connected with the goal of faster drug development times within pharmaceutical policy discussions (Anon. 2006g).

In 2005, the EMEA formalized the EU's risk management plans as the 'European Risk Management Strategy (ERMS)' (Anon. 2005i). Within this framework, as in the US, pharmaceutical manufacturers may be required by regulators to submit a risk management plan to track the use of the new product on the market, along with the new drug application, though, in the US, this is known as 'Risk Evaluation and Mitigation Strategy (REMS) (Anon. 2006h; Anon. 2007c). While risk management in the US owes its origins to deregulatory reforms and associated drug safety withdrawals in the late 1990s, its development was also affected by the very public drug disaster caused by the arthritis medication, *Vioxx*, in 2004, which was estimated by FDA safety officers to have caused over 80,000 heart attacks or strokes in the US alone, with a mortality rate of 30–40 per cent (Light 2010, p. 12).

After Vioxx, arguably the worst drug disaster in the history of the US, the FDA commissioned the National Academy of Science's Institute of Medicine (IoM) to provide recommendations on how to improve drug safety, which, in turn, influenced a Congress increasingly animated by drug safety concerns in the aftermath of Vioxx. The IoM (2007) report recommended that the FDA's regulatory authority over drug safety needed clarification. In the same year, an FDA report found that drug companies had failed to even start 65 per cent of the 1200 post-marketing safety studies requested by the agency (Anon. 2008f, p.26). In response, Congress passed the 2007 FDA Administration Amendments Act (FDAAA), which authorized the agency to proactively require pharmaceutical firms to conduct studies of drug risks with the power to impose fines up to US$10 million dollars on companies that failed to comply. In truth, the FDA always had the power to request such studies, but had not enforced it. The FDAAA sent a clear message that the FDA should give more attention to drug safety pursuits and enforcement, including a re-authorization of PDUFA enabling the agency to use industry fees to improve the drug safety system (Anon. 2008f, p.26). Such developments accentuated the role of risk management and REMS in particular because the focus of the IoM and FDAAA was post-marketing safety studies and evaluation (Anon. 2008g).

Consumerism and the patient–industry complex

No organizational analysis of the neo-liberal period of pharmaceutical innovation and regulation since 1980 would be complete without some consideration of the growth of 'consumerism' and the 'patient–industry complex'. By 'consumerism' here, we are not referring to consumer advocacy to protect public health, but rather the ideology and movement regarding patients as 'consumers' in a marketplace actively seeking pharmaceutical treatments. As disease-politics theorists, such as Carpenter (2004) and Daemmrich (2004), have noticed, AIDS patient activism in the US fuelled an increase in such consumerism. However, American pharmaceutical consumerism can also be traced back to deregulatory reforms at the FDA, whose conceptualization pre-dated AIDS.

In 1982, FDA Commissioner Hayes predicted a rise in advertising of prescription drugs directly to patients, as well as the established advertising to physicians. Advertising prescription drugs to physicians, rather than patients, had long occurred precisely because such drugs could only be taken by patients on prescription, so the doctor, not the patient, was regarded as responsible for selecting the appropriate medication. The idea of advertising prescription drugs directly to patients, bypassing doctors, in effect, treated patients as consumers. Indeed, such advertising became known as 'direct-to-consumer advertising' (DTCA) of prescription drugs. Hayes attributed potential growth in DTCA to patients demanding a greater role in selection of their health care products, though advertising could only occur if it was profitable for would-be advertisers.

At that time, the US pharmaceutical industry was not particularly enthusiastic about DTCA of prescription drugs, so it seems unlikely that the FDA was responding to industry pressure on this issue in the early 1980s (US House of Representatives 1984). Indeed, historically, the claim by the research-based firms that they produced solely medical drugs for the medical profession and did not flirt with the fancies of the general public was how that part of the industry defined itself as 'ethical' (Abraham 1995a, p. 39). Political movements emphasizing citizens' rights from the civil rights campaigns of the 1960s to the environmentalist and feminist organizations of the 1970s certainly ignited an 'active citizenship' more inclined to reflect critically on its relationship with government and other powerful bodies in society, including the medical profession and the pharmaceutical industry (Abraham and Lewis 2002). Witness the creation of the environmentalist and public health advocacy organization, known as 'Public Citizen', in the US in 1970. Some scholars have considered those developments to be so fundamental that they have ascribed them the status of a social transformation, known as 'reflexive modernization' (Beck 1992).

Yet, although those movements asserted their rights to be more informed and pro-active in matters affecting their lives and their bodies, it is much less clear that they were demanding more 'information' in the form of advertising. Even if one accepts the broadly plausible theory that, as one entered the 1980s, modern society was transforming into one with many more reflexive, rather than passive, consumers, there is no necessary connection between such reflexivity and a demand for advertising. The link was made by FDA Commissioner Hayes because he chose to make it. He fused active citizenship and consumer reflexivity with a neo-liberal ideology of consumerism commensurate with demand for advertising that was consistent with the neo-liberal political aspirations of the incumbent Reagan Administration.

Given the lack of enthusiasm among the US pharmaceutical industry for DTCA of prescription drugs, in 1982, the FDA requested a voluntary moratorium, calling for a period of cautious restraint by would-be advertisers. However, in 1985, the agency withdrew its moratorium on the grounds that the existing regulations on advertising were sufficient to protect consumers. Some commentators argue that this lax approach to regulation was not intended to open the floodgates for DTCA in the US, but rather was recognition by the FDA of a new trend in society (Pines 1999). Yet that analysis begs the question why the societal trend of reflexive modernization was interpreted as a demand for advertising in this context. We contend that that is best explained by the influence of deregulatory neo-liberal political ideology gaining ascendancy at that time. Significantly, the FDA lifted its moratorium on DTCA at the request of the DHHS and Reagan's White House Office of Management and Budget (Anon. 1993b).

With the lifting of the moratorium, the industry could not resist the lure of sales and profits resulting in a considerable increase in American pharmaceuticals firms' spend on print advertising, reaching US$12 million on DTCA in 1989. However, the US regulations on 'fair balance' and 'brief summary' made DTCA cumbersome for the broadcast media. In 1997, by which time the pharmaceutical industry had become firm supporters of DTCA, even those restrictions were relaxed so that broadcasting product advertisements merely had to provide consumers with access to the drug's official labelling via a telephone number, a webpage, a concurrent advertisement, or additional information from pharmacists, physicians or other healthcare providers (Conrad and Leiter 2008). The consequence of these deregulatory measures in 1985 and 1997 was that expenditure on broadcast DTCA, a form of prescription-drug promotion which bypassed doctors, grew almost 80-fold in the US, from US$55 million in 1991 to US$4.2 billion in 2005 (UG GAO 2006).

From the early twentieth century, pharmaceutical firms have sought to develop relationships with doctors in order to promote their products. However, in the US, during the late 1980s and early 1990s, the apparent success of AIDS patient activism in influencing some FDA decision-making combined with the growth in DTCA, persuaded drug companies that it could be profitable to forge much more extensive collaborations directly with patients. Organizationally, the most amenable tactic was to develop working relationships with patient groups, often funding them. Such funding has become an increasing trend in the last decade (O'Donovan 2007). For example, the pre-eminent American advocacy group for people with ADHD is 'Children and Adults with Attention Deficit/Hyperactivity Disorder (CHADD), 22 per cent of whose revenue in 2004–2005 came from the pharmaceutical industry (Phillips 2006, p. 434). While the precise effects of pharmaceutical firms' financial support on patient groups is difficult to gauge, such close associations are clearly important to the industry as an additional pathway, beyond doctors, for creating 'consumer demand' for their products (Herxheimer 2003). In a survey of US executives from 14 pharmaceutical companies, 75 per cent of respondents cited 'patient education' as the top-ranked marketing activity necessary to bring a brand to 'the number one spot' (UK House of Common Health Committee 2005, pp. 74–6). Collaborations of this kind are emerging as, what we call, a 'patient–industry complex'.

While in the US, the 'patient–industry complex' has centred around FDA decision-making, in European countries much of the focus has been on access to drugs within national healthcare systems after marketing approval. Nonetheless, the principal modus operandi of industry–patient group collaboration, learned by American pharmaceutical firms, and enabled by a neo-liberal ideology of consumerism in the EU and the US, has been basically the same. For example, in the UK, of most significance has been patient access to new drugs on the NHS, which pays the full cost of drug treatment provided that the appropriate NHS authorities approve funding. The National Institute for Health and Clinical Excellence (NICE), which assesses the cost-effectiveness of many new drugs for use in the NHS after they have been granted marketing approval by UK drug regulatory authorities, makes key recommendations about whether many new drugs should be made available on the NHS. The significance of the patient-industry complex in the European setting is well illustrated by NICE's experience with recent drugs developed to treat Alzheimer's disease.

In March 2005, NICE recommended that four drug treatments approved to treat Alzheimer's (Aricept, Exelon, Reminyl and Ebixa) should not be funded by the NHS because they were not cost-effective. However, following a high profile campaign in the media and a formal appeal involving patient groups, such as the Alzheimer's Society, NICE revised its guidance to allow NHS funding of the drugs for people with moderate stages of the disease, but still not those with early-stage Alzheimer's. The Alzheimer's Society then took NICE to the courts, which ultimately insisted that NICE should investigate ways of making the drugs available to all those with the disease. Notably, the manufacturers of those Alzheimer's drugs were the lead claimants in the court case and centrally involved in the formal appeal to NICE (BBC News24 2007).

As we have discussed, neo-liberal ideology and its associated deregulatory politics emerged powerfully from the early 1980s in many European countries, just as in the US. However, DTCA of prescription drugs remained banned in Europe throughout the 1980s and 1990s. In the supranational EU, the socio-political roots of efforts to legalize DTCA are much more clear-cut than in the US probably because, by the early 2000s, when those efforts got underway in earnest, neo-liberal ideology was more established and the pharmaceutical industry had become a strong supporter of such advertising. Indeed, in the early 2000s, the pharmaceutical industry, with support from the European Commission's DG Enterprise, campaigned vigorously for the legalization of DTCA in the EU.

Advertising per se is not the concern of this book. However, the nature of the DTCA campaign in the EU is highly instructive in revealing the relationships between deregulatory politics and the neo-liberal ideology of consumerism, which are part of the context of drug innovation and regulation more generally. In that respect, it shares much in common with the campaign to introduce FDAMA in the US in 1997.

Although patient groups were not initially prominent in the campaign, its promoters from industry and DG Enterprise characterized patients as consumers able to decide which drugs were best for them without doctors' supervision. The campaign utilized a discourse of 'the informed patient' and the 'expert patient'. To be sure, doctors' failure to adequately inform patients about prescription medicines can be a significant problem (Britten 2008), but it was an unsubstantiated leap of faith to assume that pharmaceutical companies would fill the gap left by doctors in that respect. In addition to the use of such discourse as an ideological lever with which to achieve deregulatory goals, the industry viewed its organizational links with patient groups as a material resource with which to advance the dismantlement of DTCA bans.

The UK research-based pharmaceutical industry led the way in the European campaign, probably because London is home to the EMEA. Quoting from a speech by the Director-General of the Association of the British Pharmaceutical Industry (ABPI), Medawar and Hardon (2004, p. 121) report that the 1998 'Informed Patient Initiative' was the first part of the industry's 'battle plan'. The second part was the ABPI's publication, 'The Expert Patient', which according to the Director-General, was 'part of a softening-up assault to be mounted through those interested parties and opinion leaders by stimulating debate'. Evidently, the purpose of the campaign was to promote a consumerist ideology of patient self-care and self-medication in order to create a basis for arguing that patients were sufficiently knowledgeable to evaluate advertising claims about powerful prescription drugs. As the following passage from an article published in *Pharmaceutical Marketing* suggests, the industry hoped that the creation of such consumerist ideology would be sufficient to compel European regulators and governments to legalize DTCA throughout the EU:

> The ABPI battle plan is to employ *ground troops in the form of patient support groups*, sympathetic medical opinion and healthcare professionals which will lead the debate on the informed patient issue. This will have the effect of weakening political, ideological and professional defences. ... Then the ABPI will follow through with high-level precision strikes on specific regulatory enclaves in both Whitehall and Brussels. (Jeffries 2000, quoted in Medawar and Hardon 2004, p. 121, emphasis added)

In fact, that campaign by the industry and DG Enterprise was unsuccessful. Many EU public health organizations, medical professionals, and some national government health agencies opposed relaxation of the ban, underlining the crucial role of health professionals in the provision of tailored information to patients, and pointing to the practical difficulties of regulating and enforcing the distinction between 'information' and 'advertisement' (Association Internationale de al Mutualite et al. 2009). Consequently, to date, the European Parliament has refused complete legalization of DTCA, concluding that it would not be in the interests of patients' health. However, there are signs that the European parliament may permit the pharmaceutical industry to provide some restricted and circumscribed 'information' about their products directly to consumers via some media.

While the pharmaceutical industry and the European Commission have keenly emphasized the importance of product information provided by companies to consumers, drug regulation in Europe has been slow to develop and implement citizens' legal rights to information more broadly, compared with the situation in the US. After a drug has gained marketing approval in the US, citizens may gain extensive access to regulatory documents underpinning the approval decision under the 1967 Freedom of Information Act. In addition, FDA expert advisory committees are generally held in public (Jasanoff 1990, p. 247).

By contrast, as recently as 1995, the EMEA established only discretionary transparency in the form of European Public Assessment Reports (EPARs) for each drug approved via the centralized procedure. In the EPARs, EU regulators provided a summary basis of the approval decision, but citizens had no right to demand information beyond that. It was not until EU Regulation 726/2004 that, in principle, EU citizens could request all documents underpinning EMEA decisions, subject to restrictions regarding commercial confidentiality, which were defined broadly as any information that would harm the interests of pharmaceutical companies (Anon. 2006b). However, in practice, the EMEA has been tardy to respond to requests for information and overly protective of companies' commercial interests (La Revue Prescrire, 2009). In June 2010, the EU Ombudsman publicly accused the EMEA of maladministration after finding that the agency had refused to release information to academics on grounds of commercial confidentiality when the documents requested did not contain commercially sensitive information. Subsequently, the EMEA finally released the information – *four years after the initial request* (Gotzche and Jorgensen 2011).

Thus, not only has the campaign for advertising information directly to consumers reflected industry interests, but so too has the European Commission's focus on selective provision of information by companies, rather than concentrating on more rapidly widening citizens rights of access to information about pharmaceutical products in the EU.

Conclusion

AIDS patient activism in the US has frequently been credited in publications, conferences, and regulatory thinking with transforming the orientation and philosophy of the FDA during the late 1980s and early 1990s. In part, this may be due to an unintended effect and misunderstanding of works, such as that by Epstein (1996), who himself never made such claims. For some regulators it may be partly explained by a rationalization that is more comfortable than the alternative, namely that the FDA's mission to protect public health was compromised by various pro-industry interests in government and business.

Whatever the reasons for the popularity of that mistaken view, the forgoing analysis demonstrates beyond doubt that AIDS activists were but one part of a wider convergence of pressures affecting FDA policy and regulatory science at that time. While AIDS activism *and* the exigencies of the AIDS crisis itself undoubtedly caused the FDA to accept a greater degree of uncertainty in its risk–benefit assessments of the first AIDS treatments, it is highly unlikely that such lowering and loosening of standards for drug approval would have been formalized into new regulations, or extended to non-life-threatening diseases, had it not been for the interventions of the pharmaceutical industry and its allies within the Reagan and Bush Administrations. The AIDS treatment activists advocated 'a very limited program accessible only in life-threatening situations' and were significant in bringing about regulatory acceptance of CD4(T) cell counts as a surrogate marker of HIV/AIDS in 1991 (Epstein 1997, p. 702). Yet, that 'limited' goal was extended by the 'Subpart E' regulations in 1988 to 'severely debilitating' diseases, and then by the 'Subpart H' regulations in 1992 to 'serious' diseases (FDA 1998a, p. 4).

In particular, the President's Task Force on Regulatory Relief, followed by Quayle's White House Council of Competitiveness exerted continual pressure on the FDA throughout the 1980s and into the early 1990s to 'streamline' the drug approval process for the benefit of industry. As well as proposing accelerated approvals, Quayle's Council proposed a new 'flexible efficacy standard' whereby FDA would make a deliberate effort to 'interpret the statutory requirement of efficacy in a manner that maximizes a drug's potential for approval' and a hiring strategy with a commitment that all new staff hired should be dedicated to the drug approval process until the goals for approval times were met (Anon. 1991j) As we have seen, the primary goal of the Quayle Council, and the Task Force on Regulatory Relief before it, was to lever the FDA into implementing measures that would remove regulatory barriers to the pharmaceutical industry's access to markets for its products.

The demands of the AIDS activists for weaker regulatory standards to expedite approval of AIDS drugs in the late 1980s and early 1990s provided public legitimation and FDA rationalization for that goal, as patients themselves appeared to be asserting a coincidence of interests between industry and patients' health. As we have shown, the nature of the regulatory relationship between the FDA and industry had already shifted considerably before the AIDS crisis. Nonetheless, AIDS activists' demands strengthened the arguments of those who advocated a more co-operative relationship between industry and the FDA – if all parties shared an interest in expedited drug development and review, then conjuring a logic that all parties should work together as 'partners' became easier.

In contextualizing and explaining the role of AIDS activism, it is also important to note that the demands of the AIDS treatment activists for specific FDA 'reforms' were themselves shaped and circumscribed by external realities that cannot be reduced to a simple reflection of 'patient interest'. Early activists' demands for conditional approvals were in part determined by a situation in which pharmaceutical companies controlled all information about, and patient access to, investigational drug products. Arrangements designed to allow patient access to investigational drugs, known as 'expanded access programmes', existed before FDA reform, but depended on companies' willingness to set up such programmes. Arguably that asymmetry in power and control was at the root of what needed to be reformed. Expanded access programmes offered only weak commercial incentives to manufacturers. Aware of that, the activists sought earlier *marketing* approval (conditional approval) for AIDS drugs. As Epstein (1997, p. 702) puts it:

> Conditional approval, by contrast, was designed with the explicit goal of enlisting the pharmaceutical companies by giving them a chance to do what they liked best: earn profits.

Furthermore, it is possible that AIDS activists miscalculated the real interests of public health at that time. In subsequent years, with hindsight, some of the prominent activists from that period have expressed the opinion that they were wrong to demand earlier access to drugs whose benefit and safety had not been established (Hilts 2003, pp. 304–5). For instance, Gregg Gonsalves, of the AIDS Treatment Action Group, stated:

> We have arrived in hell. AIDS activists and government regulators have worked together, with the best intentions, over the years to speed access to drugs. What we have done, however, is to unleash drugs with well-documented toxicities onto the market, without obtaining rigorous data on their clinical efficacy. (cited in Hilts 2003, p. 251)

One may confidently conclude that it would be inaccurate to suggest that changes in the philosophy and practice of pharmaceuticals regulation in the US, which occurred throughout the 1980s and the 1990s, resulted solely, or even mainly, from the demands and activities of the early AIDS activists or patients' demands more generally. Expedited development and review was just one example of a number of measures intended to speed the marketing of *all* new drugs, regardless of whether they offered therapeutic advantage to patients. Moreover, analysis of events leading up to the 1997 FDAMA indicates that pressure for those legislative reforms came not from patients, but from industry and neo-liberal ideology in Congress. In fact, many patient groups, including AIDS activists, opposed the bulk of FDAMA reforms. In these respects, disease-politics theory has been built on flawed foundations, and its proponents have been mistaken to think that it forms the central dynamic of regulatory politics in the pharmaceutical sector, even in the US.

In Europe, the central role of neo-liberal political influence in government together with the interests of the pharmaceutical industry in shaping drug regulatory reform since 1980 is much more clear-cut than in the US. Patient activism demanding accelerated approval of new drugs in Europe was rare during the neo-liberal reforms of the 1980s and 1990s. Although it has grown in the 2000s within Europe, it has been primarily aimed at gaining access to drugs within the healthcare systems of *individual European countries after marketing approval* by EMEA or a national regulatory agency. Hence, insofar as patient activism had become significant within Europe in the 2000s, it cannot account for the neo-liberal framework that shaped the emergence of supranational EU pharmaceutical regulation during the 1980s and 1990s. Nor can it explain the deregulatory measures adopted during the last decade to accelerate drug development and approval at the *supranational EU* level. Rather, those developments were driven almost entirely by a pro-industry neo-liberal political agenda. One would be hard-pressed to find any patient or consumer group in the EU or the US that supported or demanded increased dependence of drug regulatory agencies on industry fees.

Quite distinct from the historical question of what role patient activism actually played, there is the sociological matter of how patients' interests have been *represented by others* in relation to political change. Regulatory reforms in the US and, subsequently in the EU, to accelerate drug development and review were justified by government agencies on the basis of patients' 'expectations' and the public health benefits claimed to accrue from faster approval of innovative drugs. For instance, the Prescription Drug User Fee Amendments of 2002 (PDUFA III) stated:

> The Congress finds that prompt approval of safe and effective new drugs and other therapies is critical to the improvement of the public health so that patients may enjoy the benefits provided by these

therapies to treat and prevent illness and disease. (PDUFA 2002, section 502)

The achievement of such ideological representation was also bolstered by the organizational strategies of the pharmaceutical industry, especially in forging the emergence of the patient-industry complex, and the discourses of the 'expert patient' and the 'informed patient'. All of these served as levers with which to create the impression that the commercial interests of industry coincided with advancement of patients' health, and that weakening regulatory standards was a liberatory development in the best interests of patients. Yet, the evidence suggests overwhelmingly that both PDUFA and FDAMA were designed to, and have increased, the FDA's responsiveness to industry interests, while the number of innovative pharmaceuticals offering modest or significant therapeutic advance to patients has gone into decline.

Similarly, promoting the interests of the drug industry was a predominant concern of the European Commission's DG Enterprise with responsibility for pharmaceuticals regulation in the EU, and also apparently the management of the EMEA.

By contrast, tracking the extent to which new drugs offered therapeutic advance for patients together with provision of publicly accessible information about whether the development of such drugs was increasing or going into decline proved to be a much lower priority for EMEA and the European Commission. Rather, the 2004 legislative review of EU pharmaceutical regulation took place in the context of increasing concern at the Commission that the European drug industry was losing competitive ground to the US (Gambardella *et al.* 2000, pp. 83–4). Concerns about enhancing the environment for pharmaceutical innovation, manufacture and sale were perceived as the primary goals underlying the draft legislation.

In this chapter, we have shown that, at the level of social and political organization, the vast majority of the deregulatory reforms of the neo-liberal era to accelerate drug development and approval were neither demanded by patients nor primarily motivated by patients' interests. Despite misleading representations to the contrary, the goal to stimulate pharmaceutical innovation was not necessarily in patients' interests because most innovative drugs offered little or no significant therapeutic advance. Furthermore, we have shown that, irrespective of the motivations and ideological representations behind the deregulatory reforms, the number of innovative pharmaceuticals offering modest or significant therapeutic advance actually went into decline after those reforms. Such findings certainly provide no evidence to support the claims of neo-liberal theory that the deregulatory reforms were, across the pharmaceutical sector as a whole, in patients' health interests. Indeed, our macro-political findings suggest that the neo-liberal reforms have undermined the capacity of pharmaceutical regulation to promote and protect the best interests of patients and public health.

Rather, the pharmaceutical industry has gained privileged access to the state in the neo-liberal era and worked in collaboration with its allies in the executive and legislative branches of government to bring about regulatory reforms in its commercial interests. This has made it possible for the industry and government to work together on a pro-business deregulatory agenda, including reforms of drug regulatory agencies themselves, such as appointments of more industry-friendly heads of the drug regulatory agencies, increased dependence of the agencies on industry fees, extension of informal consultation between regulators and firms, and responsiveness to commercial, rather than health priorities in terms of how quickly regulatory review of new drugs is completed. Such neo-liberal corporate bias has often operated via legislation beyond the regulatory agencies. However, throughout the neo-liberal reforms, including even the alteration of the FDA's mission statement in the FDAMA legislation, the regulatory agencies in both the EU and the US have maintained their mandate to promote and protect public health. In that context, many of the reforms to the drug regulatory agencies put in place as a result of the corporate bias during the neo-liberal period have made the agencies more vulnerable to capture by the interests of industry.

Although there are certainly differences between the evolution of pharmaceutical regulation in the EU and the US in the neo-liberal era, they are heavily outweighed by convergences between the two regulatory systems. Two phases of convergence can be identified. Initially, during the 1980s, corporate bias was much more developed in Europe than in the US, as reflected by the extent to which European drug regulatory agencies were funded by the industry, regulator–industry consultation was tolerated, drug approvals were rapid and plentiful, and senior managers in regulatory agencies had assimilated the interests of industry. Throughout the 1980s and early 1990s, the corporate bias of pharmaceutical regulation in the US converged with that of the European model in those respects. Thus, the US was 'catching up' with EU in adopting neo-liberal corporate bias as a regulatory reform programme. The second phase of convergence relates to the deregulatory reforms themselves adopted by (or imposed on) the regulatory agency, such as accelerated approval rules, use of surrogate measures of clinical outcomes, and risk management strategies. All of those specific deregulatory reforms were introduced into the US first. By the late 1990s and 2000s, the supranational EU drug regulatory system was emulating the US by introducing new legislation and regulations, such as conditional marketing approval and the European Risk Management Strategy, as well as beginning to promote further use of surrogate biomarkers.

Our macro- and meso-level investigations in this chapter have identified specific key changes to regulatory standards for innovative pharmaceuticals resulting from neo-liberal reforms in the US and the EU. These include: the broadening of drug innovations that should receive priority review; the widening of the types of conditions for which drugs intended to treat them may attain 'accelerated approval' or 'conditional marketing'; the use of non-established surrogate markers of drug efficacy together with companies' post-marketing commitments to establish efficacy; and the implementation of risk management in response to safety problems emerging with innovative drugs after marketing.

Senior officials in both the pharmaceutical industry and regulatory agencies have continually insisted that such reforms have increased the efficiency of the drug development and review process without lowering drug review standards. That new drug review times have fallen in the EU and the US following neo-liberal reforms is undeniable but, as we have indicated, whether that has delivered more 'efficiency' is dramatically called into question by the fact that the reforms were also followed by declines in innovation and new drugs that offer therapeutic advance. Regarding the issue of whether standards of safety and efficacy to promote and protect public health have been compromised by specific deregulatory measures, this question cannot be fully addressed without a micro-level examination of the regulatory science surrounding innovative pharmaceutical products themselves. It is to that issue that we now turn for the remainder of this book, before drawing out our final overall conclusions.

3
Designs on Diabetes Drugs

In this chapter, we examine how the neo-liberal deregulatory agenda has accentuated the mis-direction of innovative pharmaceutical testing and evaluation. In particular, we show how the neo-liberal hegemony within drug regulatory policy of defining efficiency of regulatory agencies almost entirely in term of the acceleration of review and marketing approval times has inhibited and neglected the need to improve the nature and design of drug testing and evaluation in the interests of health. This chapter is concerned with innovative pharmaceuticals to treat diabetes, which came to the market in both the EU and the US via the regular approval routes.

Diabetes is an inability to produce insulin effectively, or in sufficient quantities, to regulate levels of glucose in the blood properly. Consequently, the blood-glucose levels of diabetics become inappropriately elevated resulting immediately in a state known as hyperglycaemia. There are two types of diabetes. Type I typically occurs in young people before the age of 40 involving an almost total inability to produce insulin. However, most diabetics have Type II diabetes, in which the body retains the capacity to produce some insulin. This second form of the condition develops later in life with progressive deterioration over time. In this chapter we will be discussing solely Type II diabetes (hereafter simply referred to as 'diabetes').

Such diabetes is characterized by multiple defects in the pancreatic beta-cell (which secretes insulin), the liver (which produces glucose), and skeletal muscle and adipose (fatty) tissue (which take up glucose). Three major metabolic abnormalities contribute to hyperglycaemia in diabetes: (1) impaired insulin secretion in response to glucose; (2) increased production of glucose in the liver; and (3) decreased insulin-dependent glucose uptake in the peripheral tissues, such as skeletal muscle and adipose. As diabetes progresses, it is associated with long-term problems related to blood vessels and the body's circulatory system. These are known as vascular complications, which are normally divided into two categories: micro- and macrovascular. Microvascular complications include blindness, kidney disease, limb amputation, and a disease of the peripheral nervous system called neuropathy. For example, in the US, diabetes is thought to be the single largest cause of blindness and an increasingly prevalent factor in kidney disease. Diabetics have a 20-fold higher risk of eye disease than non-diabetics. Adverse macrovascular effects include cardiovascular disease, heart attack, and stroke. It is generally agreed that about two-thirds of deaths among diabetics are due to those adverse macrovascular effects. Moreover, cardiovascular disease is the principal cause of hospitalization of patients with diabetes (FDA 2008b, pp. 31–3 and 114–25).

About eight per cent of the US population (over 20 million Americans) and four per cent of Europeans are diabetics, with over 140 million sufferers worldwide – figures expected to increase with rising rates of obesity and ageing populations (Anon. 2000b; Czoski-Murray et al. 2004, p.5; EMEA 2001b, p.2; FDA 2007, pp. 22–9; US Congress 2007, p. 35). Hence, diabetes is a major and growing public health problem. By the same token, the pharmaceutical market for diabetes treatment is large and potentially expanding – worth billions of (US) dollars

The ultimate goal of diabetes treatment is to reduce, retard or prevent its adverse micro- and macrovascular effects. It is particularly important to counteract macrovascular complications, such as coronary heart disease, because they are so serious and comprise the greater burden of the disease. Within nine years of diagnosis, nine per cent of patients have microvascular problems, but 20 per cent have a macrovascular complication (UK Prospective Diabetes Study Group 1998a, p. 838). Early-stage diabetes can often be managed by dietary control and/or exercise, but as the disease progresses oral medications and/or insulin injections may be needed. Typically diabetics may take medications for 15–30 years. Most diabetics prefer oral medication to insulin injection. The latter poses difficulties for many patients, especially the elderly.

Common to all therapeutic regimens is an attempt to keep the blood-glucose levels of diabetics within a normal range, thus avoiding hyper-glycaemia (overly high blood-glucose levels) and hypoglycaemia (when blood-glucose levels are too low). This is the blood-glucose control approach to treating diabetes based on the view that the vascular complications of diabetes are directly and causally related to the extent of strict blood-glucose control. However, others have contended that the development of adverse vascular effects is a largely independent manifestation of the disease process minimally affected by blood-glucose control (Marks 1997, pp. 201–2).

The first diabetic drug was insulin, discovered in the 1920s, and administered by injection, usually by the patients themselves. The pain, inconvenience, and even social stigma of the injection led medical and healthcare professionals, as well as pharmaceutical scientists, to search for an oral diabetic medication. Insulin treatment often also led to hypoglycaemia and/or weight gain. By the late 1950s, the first oral diabetes drug, tolbutamide, was in widespread use on both sides of the Atlantic. Tolbutamide was the first of a class of drugs, known as sulfonylureas. It was probably first discovered in the 1940s in Germany. Subsequently, it was developed and marketed in the 1950s by the giant pharmaceutical companies, Hoechst and Upjohn, in Europe and the US, respectively (Greene 2007, pp. 86–92).

Although tolbutamide fell out of favour after 1969 when it was discovered that it increased cardiovascular mortality, it gave rise to a whole generation of sulfonylureas still used today, such as glibenclamide in Europe and glyburide in the US (FDA 2008b, pp. 27–8; Marks 1997, pp. 204–8). The sulfonylureas acted by increasing insulin secretion, thereby depressing blood-glucose levels. Being pills, rather than injections, they were an improvement on insulin, but for a significant number of patients the sulfonylureas also produced the adverse effects of hypoglycaemia and weight gain.

In the late 1950s, another class of oral blood-glucose lowering drugs, known as the biguanides began to be developed. These drugs acted by increasing intestinal glucose utilization, reducing production of glucose in the liver, and increasing insulin sensitivity, so that the insulin available acted more effectively. Unlike the sulfonylureas, the biguanides did not work by increasing insulin secretion. Nor did the biguanides cause hypoglycaemia in diabetic patients.

In the US, the first of the biguanides was phenformin, which was discovered by an American vitamin company in 1957. It was developed and marketed in many countries by the transnational pharmaceutical firm, Ciba-Geigy. However, by the late 1980s, it was withdrawn from most markets because, though effective in lowering blood-glucose, it caused unacceptable adverse gastrointestinal effects, most notably severe lactic acidosis, which was fatal in a large proportion of cases. Meanwhile, in Europe, another biguanide, metformin, was developed by Aron Laboratories in Paris between 1954 and 1957. The drug reached several European markets beyond France. By 1958, a subsidiary of

Aron was selling it on the UK market. Metformin also caused adverse gastrointestinal effects, including lactic acidosis, but it was much safer than phenformin because the risk of the lactic acidosis was 20–30 times less with metformin and the gastrointestinal toxicity tended to be less severe when it occurred. With the withdrawal of phenformin from the American market in the late 1970s, interest in metformin increased in the US where it was produced by Bristol-Myers-Squibb and received marketing approval from the FDA in 1994. Metformin is still used widely today as an oral diabetes medication in both Europe and the US (FDA 2008b, pp. 58–60).

The arrival of the glitazones and their innovative mechanism

Given the adverse hypoglycaemic effects of the sulfonylureas and the gastrointestinal problems associated with metformin, there was certainly room for improvement on existing diabetes therapies during the 1990s when another new class of oral diabetes medications were developed, namely, the glitazones, also known as thiazolidinediones (Czoski-Murray *et al.* 2004, pp. 5–7). All the glitazones were drug innovations in the sense that they were new molecular entities (NMEs). In addition, they were also claimed to have an innovative mechanism of action. That mechanism is based on the idea that diabetics exhibit not only insufficient production of insulin, but also reduced action of insulin to regulate blood-glucose. This is sometimes called 'insulin resistance'. The mechanism with which the glitazones were to reduce blood-glucose levels was by improving insulin sensitivity, thereby reducing insulin resistance (Diamant and Heine 2003).

Before the introduction of the glitazones in the late 1990s, metformin was the only drug able to sensitize skeletal muscle, adipose tissue and the liver to insulin. Increasing insulin sensitivity was, however, the key mechanism of the glitazones and they apparently exhibited it more specifically and powerfully than metformin in laboratory settings. The hypothesized mechanism entailed that insulin resistance was reversed by enhancing the action of insulin, which promoted the use of glucose in peripheral tissues and suppressed the production of glucose in the liver (EMEA 2001b; 2003; Krentz et al. 2000).

The first of this class was ciglitazone. Development of ciglitazone was abandoned when it was discovered that the drug was toxic to the liver (Gale 2001, p. 1870). The second drug, troglitazone, was developed by Warner-Lambert in the US. In July 1996, the firm submitted a new drug application to the FDA for troglitazone's marketing approval. Not only was troglitazone an NME, Warner-Lambert also persuaded the FDA to give it priority review status indicating that it promised modest or significant therapeutic advance. That was the first time the FDA had granted a six-month fast-track review to a diabetes pill. It was approved on to the US market by the FDA on 29 January 1997. It was also on the UK market for a few weeks in 1997 where it was distributed by GlaxoWellcome. However, after reports that troglitazone was associated with severe and unpredictable liver toxicity in Japan and the US, the drug was withdrawn from the UK market later in 1997. It was not withdrawn from the American market until March 2000 when the FDA received reports implicating the drug in 61 deaths from liver failure and the need for seven liver transplants (Krentz et al. 2000).

Meanwhile, the British company, SmithKline Beecham, which was later to merge with GlaxoWellcome to become the pharmaceutical giant, GlaxoSmithKline (GSK), was developing rosiglitazone as an oral blood-glucose lowering drug to treat diabetes. As an innovative pharmaceutical, the firm submitted an application for its marketing approval throughout the EU to the EMEA's regular supranational centralized procedure on 3 December 1998. The EMEA's expert Committee for Proprietary Medicinal Products (CPMP) delivered a positive opinion regarding rosiglitazone on 16 March 2000 and the drug was approved by the European Commission for marketing throughout the EU on 11 July 2000 (EMEA 2003). Like most transnational pharmaceutical companies, SmithKline Beecham wanted to market its products in the US. On 25 November 1998, the firm submitted a new drug application for rosiglitazone's marketing approval in the US to the FDA. The application went through the FDA's regular approval process, but was granted priority review status by the agency marking out this NME as promising modest or significant therapeutic advance that should be reviewed within the six-month fast-track timeline. Rosiglitazone was approved for marketing in the US on 25 May 1999 (Anon. 2000c). It was marketed under the brand name 'Avandia' in both regions.

At about the same time, the Japanese pharmaceutical firm, Takeda, was bringing a third glitazone to fruition, namely, pioglitazone. As an innovative pharmaceutical, like rosiglitazone, pioglitazone qualified to be submitted to the EMEA's regular supranational centralized procedure for marketing approval throughout the EU. The application, which was made on 30 March 1999, received a positive opinion from the CPMP on 29 June 2000, and marketing approval from the European Commission on 13 October 2000 (EMEA 2001b). Takeda also applied to the FDA on 15 January 1999 for approval to market pioglitazone in the US. That also went through the regular approval process, but received priority review status from the agency. It was approved by the FDA on 15 July 1999 (Anon. 2000c). Pioglitazone was marketed under the brand name, 'Actos' in both Europe and the US.

Pioglitazone and rosiglitazone remained on the American and European markets under the brand names Actos and Avandia respectively, throughout the 2000s. Both drugs achieved blockbuster status, by which is meant their global sales exceeded a billion (US) dollars (Anon. 2007d). In 2005, these two glitazones accounted for 21 per cent of the oral diabetic drug market in the US and five per cent of that market in Europe (Yki-Jarvinen 2005). For the remainder of this book, when we refer collectively to 'the glitazones' we will be referring only to pioglitazone and rosiglitazone.

The carcinogenic potential of pioglitazone

As well as being tested for clinical safety and efficacy, new drugs must also undergo a battery of non-clinical laboratory studies. Perhaps the most important non-clinical studies are those that test whether new drugs have cancer-inducing potential (carcinogenicity). All new drugs, except for some intended to treat life-threatening conditions, need to be tested to assess their carcinogenicity before marketing approval. If a drug were to cause cancer in people, then clearly it could not be appropriate to prescribe it to diabetics or any other patients unless they were already suffering from a severely life-threatening condition, which was itself expected to be terminal. Hence, carcinogenicity testing of pharmaceuticals is a standard part of nearly all drug development before submission of a new drug application to regulatory agencies for marketing approval.

Such testing does not take the form of clinical trials for both practical and ethical reasons. The onset of most types of cancer may develop over the lifespan, taking many years (20–50) after first exposure to the cancer-inducing compound (carcinogen) and with increasing likelihood if exposure to the carcinogen is continual over a long period. That makes it imperative that pharmaceuticals taken chronically, such as diabetes drugs, are carefully assessed for carcinogenic risk, but it also makes it impossible for carcinogenicity testing to be done directly on people because if a new drug were proven to be a carcinogen by that method, then that finding would have been achieved by giving many people cancer on a 20- or 30-year trial. Consequently, carcinogenicity testing is done on disembodied human cells and whole live animals (usually rodents) in laboratories. The key tests are the animal tests because they monitor any carcinogenic effects of the new drug over the lifespan of the rodents (about 18–30 months). As such, they attempt to model the intended reality of a patient taking the drug as a medication over some considerable period of time. Given that new drugs are not, and effectively cannot be, tested directly on humans for carcinogenicity, the animal tests have a particular significance because, aside from short-term studies of cells, they provide the bulk of information about pharmaceutical carcinogenicity. There is, of course, the problematic uncertainty of extrapolating from experimental results in rodents to real risks for patients. If a drug is carcinogenic to rats, it might nevertheless be harmless to humans, but in the absence of any direct knowledge about its carcinogenicity to humans, regulators must make a judgement about the risk posed to patients.

Both glitazones were tested for carcinogenicity in rats and mice. Rosiglitazone was not found to be carcinogenic in mice or rats, though the drug produced a statistically significant trend in non-cancerous fatty tumours (lipomas) in rats during a two-year carcinogenicity study (EMEA 2003, p. 5; Moh-Jee Ng 1999, p. 7). Although pioglitazone was not found to induce cancer in mice, in June 1999, FDA pharmacologists found that the drug produced cancerous bladder tumours in rats (Steigerwalt 1999, p. 44). Notably, pioglitazone induced the bladder cancers in rats at the same or at a similar dose as would be taken by diabetic patients.

EU regulators were also aware of pioglitazone's carcinogenicity in rats when the drug was approved on to the European market, but the CPMP, while considering the finding, seemingly registered little concern (EMEA 2001b, pp. 5–6). FDA scientists regarded pioglitazone's carcinogenicity in rats as significant enough to mention on the drug's label and to cite as an important difference in its safety profile compared with rosiglitazone. Nonetheless, the extent of risk posed to patients by this finding during a two-year carcinogenicity study of pioglitazone was never resolved throughout the 2000s after the drug's approval on to the markets in Europe and the US. This is clear from a statement made to a 2007 FDA advisory committee meeting by Dr Meyer, Director of the FDA's Office of New Drug Evaluation, seven years after the drug had been on the market:

It is worth noting that important differences in the safety profile of these [glitazones] exist. For instance, pioglitazone's labelling mentions positive carcinogenicity studies in animals, specifically bladder cancers in rats given the drug at the same level of exposure as those used clinically. While we don't know, and I would stress that we don't know that there is a clinical correlate to this animal finding, at least some evidence from clinical trials also cited and available in the labelling raised the possibility that there might be a human risk, and that continues to be evaluated in postmarketing studies. (FDA 2007, p. 246)

A post-marketing three-year clinical study, known as PROactive, and published by Takeda in 2005 did not detect a significant presence of bladder cancers in patients taking pioglitazone (Yki-Jarvinen 2005). However, as we explained above, three years of human exposure is far too short a time to detect a carcinogenic effect unless it was highly potent. Indeed, despite such apparent post-marketing monitoring, the issue remained entirely unresolved years later. In a briefing document to a 2010 FDA advisory committee, agency scientists reminded the committee that pioglitazone had been found to induce bladder cancer in rats and commented:

The bladder tumours occurred at a dose that approximates human exposure. ... The weight of evidence suggests that rats are more susceptible to [this type of] bladder tumours than other species (including humans), but the risk of patients developing bladder tumours with chronic exposure cannot be entirely excluded. (Bourcier 2010)

The FDA advised that pioglitazone should not be used by patients with 'prior bladder cancer', but everyone else could take their chances (Woodcock 2010, p. 2).

On 9 June 2011, the drug approvals committee of the French national health insurance agency (CNAMTS) reported the result of their gigantic retrospective cohort study aimed at assessing the risk of developing bladder cancer after exposure to pioglitazone. The biggest study of its kind, it ran from 2006 to 2009, involving 1.5 million diabetes patients aged between 40 and 79, of whom 155,535 were exposed to the drug. The study found that there was a statistically significant link between exposure to pioglitazone and incidence of bladder cancer. The hazard ratio was 1.22, meaning that the patients taking pioglitazone had a 22 per cent higher risk of contracting bladder cancer than the other diabetic patients not taking the drug. Just one day later, the French and German drug regulatory agencies announced that they were suspending its marketing approval in their countries (Bruce 2011). The FDA and European Medicines Agency (EMA – the EMEA's successor since 2010) did not respond by immediately withdrawing the drug from their markets (Young 2011a).

Clinical testing for blood-glucose control

Evidently, none of the laboratory tests of either pioglitazone or rosiglitazone were deemed to raise concerns sufficient to prevent development moving forward into the clinical trials phase. The controlled clinical trials examining the drugs' ability to lower and control blood-glucose formed the central basis for their marketing approving. In particular, such trials were pivotal to regulators' assessment of their clinical efficacy. These trials, which typically lasted between 12 and 26 weeks, except for one of a year's duration, are summarized in Table 3.1.

Table 3.1   Key double-blind clinical (efficacy) trials with glitazones

Study no.	Treatment groups	Duration (weeks)	in NDA/EPAR	N
(a) Pioglitazone (pio)				
PNFP-012	pio vs pl	24	NDA + EPAR	260
PNFP-026	pio vs pl	16	NDA + EPAR	197
EC204	pio vs su	26	EPAR	270
PNFP-010	pio + su vs su +pl	16	NDA + EPAR	560
CCT-003	pio + su vs su +pl	12	EPAR	237
CCT-012	pio + su vs su + pl	12	EPAR	119
PNFP-027	pio + met vs met + pl	16	NDA + EPAR	328
PNFP-014	pio + ins vs ins + pl	16	NDA + EPAR	566
(b) Rosiglitazone (rsg)				
011	rsg vs pl	26	NDA + EPAR	493
024	rsg vs pl	26	NDA + EPAR	908
020	rsg vs su	52	NDA + EPAR	587
093	rsg + met vs rsg + pl vs met + pl	26	NDA + EPAR	306
094	rsg + met vs met + pl	26	NDA + EPAR	339
015	rsg + su vs rsg + pl	26	EPAR	574
079	rsg + su vs rsg + pl vs su + pl	26	EPAR	296
096	rsg + su vs su + pl	26	EPAR	346

Sources: EMEA European Public Assessment Reports (EPARs) and FDA New Drug Applications (NDAs) for pioglitazone and rosiglitazone.

(a) Pioglitazone (actos) controlled trials

A Japanese company with transnational ambitions, Takeda sought to market pioglitazone in Europe and the US. The new drug application for pioglitazone submitted to the FDA by Takeda in January 1999 contained five placebo-controlled trials, of which two compared pioglitazone alone (known as monotherapy) with placebo (trials PNFP-012 and -026). The other three placebo-controlled trials involved comparisons of combination therapies, that is, pioglitazone combined with an established diabetes therapy compared with placebo combined with the same established therapy. The established therapies used in these three trials were sulfonylurea, metformin or insulin (PNFP-010, -027 and -014). Notably, Takeda's submission to the FDA did not include any active-controlled trials comparing pioglitazone directly with existing drug therapies (see Table 3.1a).

All FDA regulators agreed that the placebo-controlled trials indicated that pioglitazone was statistically significantly better at blood-glucose control than placebo. However, as explained in Chapter 1, such placebo trials were not designed to provide knowledge about whether pioglitazone was a useful addition to the existing diabetes drug therapies at that time, that is, whether it offered a therapeutic advance (Gale 2001). The only indication of how pioglitazone *might* compare to other existing oral medications, such as the sulphonylureas or metformin, was among the subset of patients who were on existing diabetic medications at screening before the monotherapy placebo trials began. Those patients had their blood-glucose levels measured just before coming off their existing medication, then a 'wash-out' period without diabetic medication (on placebo) to establish 'baseline' blood-glucose levels before going on to pioglitazone therapy in the trial proper. Among those patients, it was found that, after taking pioglitazone, blood-glucose levels improved relative to the baseline values, but *worsened* relative to the levels that the patients had attained on their prior diabetic medication. The FDA's Medical Reviewer for the application put it as follows:

> Pioglitazone appears to be inferior to patients' previous anti-diabetic medication. Since patients' hyperglycaemia deteriorates when they are switched to pioglitazone from other medications, it is hard to see how these data can be used to support an indication [use] of initial [pioglitazone] monotherapy. (Misbin 1999a, p. 4)

The pioglitazone application to the EMEA in March 1999 included the same five placebo-controlled trials that had been submitted to the

FDA. The EU regulators on the CPMP concurred with the FDA that the drug controlled blood-glucose levels more effectively than placebo. In addition, the European application contained an active-controlled trial comparing pioglitazone monotherapy with sulfonylurea (EC204), as well as two further placebo-controlled combination therapy trials with sulfonylureas, listed under the codenames CCT-03 and 012 (see Table 3.1a). When the CPMP reviewed the active-controlled trial (EC204), they considered that the daily dose of sulfonylurea used as a comparator against pioglitazone was sub-optimal. Takeda had given patients a daily dose of 2.5–5.0 mg of the sulfonylurea, but the CPMP concluded that a dose of 10–15 mg would have been more appropriate. Despite this, the sulfonylurea still produced *greater* blood-glucose control than pioglitazone, though the difference was not statistically significant (EMEA 2001b, pp. 12–13).

The combination studies of pioglitazone with sulphonylureas (PNFP-010, CCT-003, CCT-012), metformin (PNFP-027) and insulin (PNFP-014) demonstrated a beneficial add-on effect of pioglitazone for patients whose blood-glucose was inadequately controlled by sulphonylurea, metformin or insulin monotherapy. However, there were no direct comparisons between pioglitazone and what was arguably the most important and widely used existing therapy at that time, namely, *metformin-plus-sulphonylurea* (EMEA 2001b, p. 21; 2003, p. 11).

(b)   Rosiglitazone (avandia) trials

Just two months earlier than Takeda, SmithKline Beecham submitted its new drug application for rosiglitazone to the FDA. The application contained three placebo-controlled trials, of which two were monotherapy (011 and 024), and the third, a combination therapy with metformin (094). It also included two active-controlled trials, one comparing rosiglitazone monotherapy with sulphonylurea (020), and the other, a three-arm design, comparing rosiglitazone monotherapy, metformin monotherapy and metformin–rosiglitazone combination-therapy (093). In addition to these trials, SmithKline Beecham's rosiglitazone application to the EMEA contained three further placebo-controlled trials involving combination therapy with sulfonylurea, listed as 015, 079 and 096 (see Table 3.1b).

As with pioglitazone, regulators on both sides of the Atlantic agreed that the placebo-controlled trials demonstrated that rosiglitazone significantly outperformed placebo at blood-glucose control. Regarding the active-controlled trial (020) comparing rosiglitazone with sulfonylurea, submitted to both agencies, the FDA found rosiglitazone barely equivalent to sulfonylurea, while the CPMP suspected that there had been sub-optimal use of sulfonylurea here too (EMEA 2003b, p. 11; Misbin 1999b, pp. 13–15). The other active-controlled trial (093) lasted 26 weeks in total. According to FDA scientists, it showed that patients had a rapid loss of blood-glucose control when they switched from metformin to rosiglitazone, though the loss of control subsided and stabilized by week 18 of the trial. However, by week 26, blood-glucose levels had still not returned to baseline. Specifically, on reviewing this trial, American and European regulators noticed that when patients switched from metformin to rosiglitazone, blood-glucose control *decreased*, with patients on rosiglitazone monotherapy still having 1.3 per cent higher blood-glucose at the end of the trial than they had on metformin (Misbin 1999b, pp. 21–2).

Combination studies of rosiglitazone with sulphonylureas (015, 079 and 096) and metformin (093, 094) demonstrated a beneficial add-on effect of rosiglitazone for patients whose blood-glucose was inadequately controlled by sulphonylurea or metformin monotherapy. However, as with pioglitazone, there were no direct comparisons with the most important and widely used existing therapy, metformin-plus-sulphonylurea (EMEA 2001b, p. 21; 2003, p. 11).

The approval decisions in the EU and the US

A number of implications follow immediately from these trials. The glitazones were more effective at blood-glucose control of diabetics than placebo. However, the predominance of placebo-controlled trials generated considerable uncertainty about whether the glitazones were more effective at blood-glucose control than existing oral medications. When an active-control design was employed by the drug firms, they selected sub-optimal or inappropriate comparator drug treatments, rather than a gold standard (for instance, the most effective dose of metformin-plus-sulphonylurea). That practice, therefore, similarly failed to address the most important question for patients and clinicians, namely if the glitazones offered any therapeutic advance in blood-glucose control for diabetics. Yet, insofar as the trials did provide some information about how the glitazones compared with sulfonylurea or metformin monotherapy, that limited evidence suggested that the glitazones were less effective in reducing and controlling blood-glucose than those existing therapies.

Despite this, and in the case of pioglitazone despite some carcinogenic risk, the FDA approved the glitazones on to the US market as monotherapies and combination-therapies (pioglitazone with sulphony-lurea, metformin, or insulin, rosiglitazone with metformin) because the drugs had proved superior efficacy to placebo in blood-glucose control (SmithKline Beecham 1999; Takeda 1999). After marketing approval, the glitazones became pharmaceutical product innovations in the official statistics informing pharmaceutical policy analysis. Moreover, because they were given priority review they came to be counted in FDA statistics as drug approvals that offered modest or significant therapeutic advance, even though the details of the clinical evidence supported no such characterization.

Partly, as a consequence of concerns over the insufficient evidence of the efficacy of rosiglitazone monotherapy relative to existing treatments, and the lack of any data on rosiglitazone combination therapy compared with the metformin plus sulphonylurea combination, the CPMP initially turned down, by a majority, SmithKline Beecham's application to market rosiglitazone throughout the EU (EMEA 2003). Following an appeal by the company, the Committee recommended marketing approval, but with a much more restricted indication than the FDA. The CPMP decided that an unmet medical need existed for diabetics who could not tolerate the widely used combination of sulpho-nylurea plus metformin because the only alternative option would be to start taking insulin. The Committee considered that insulin therapy, with the need for frequent blood measurements and the risk of hypogly-caemia, made the therapy less appropriate in some patients, particularly the elderly and groups with severe insulin resistance. The CPMP, there-fore, requested a post-hoc analysis of efficacy and safety results in the subset of diabetic patients where there might be such an unmet medical need. The same approach was applied to pioglitazone several months later, though without the need for an appeal.

According to the EPARs for rosiglitazone and pioglitazone, data submitted by SmithKline Beecham and Takeda revealed no difference in efficacy among patients for whom the sulphonylurea plus metformin combination was inappropriate. This suggested that glitazone therapy could be an effective treatment for that group of diabetics, but the CPMP remained resolute that glitazone monotherapy could not even be a second line indication, so the glitazones were approved as second-line combi-nation therapies only. Specifically, rosiglitazone and pioglitazone were approved on to the EU market as: second-line combination-therapies with sulphonylureas in patients unable to take metformin; and second-line combination-therapy with metformin in obese patients – a group thought likely to benefit from the glitazones' insulin sensitizing effect because obesity is associated with insulin resistance (EMEA 2001b; 2003). A minority of CPMP members opposed approval even for that restricted indication, though due to secrecy we do not know their reasons (EMEA 2003). It may be that they were not convinced that diabetics who could not be treated effectively by metformin plus sulphonylureas combination therapy should be offered glitazone combination therapy instead of insulin. This was a point made by an eminent UK specialist in diabetes drugs, Edwin Gale, within one year of the glitazones coming on to the European market:

> There is no evidence at all that a change to rosiglitazone might work when the standard combination [sulfonylurea plus metformin] has failed, and...there is no reason to believe that there are any clinical benefits in postponing insulin treatment in this situation. (Gale 2001, p. 1873)

The FDA approved the glitazones for monotherapy, despite all the uncertainties that that entailed for prescribing practice and patients about whether the glitazones were (a) more effective monotherapies than sulfonylurea or metformin, or (b) more effective combination-therapies than metformin-plus-sulfonylurea (Gale 2001). By contrast, due to insufficient evidence of the glitazones' mono-therapeutic efficacy relative to existing treatments, and lack of data comparing glitazone combination-therapies with metformin-plus-sulphonylureas, EU regulators approved the glitazones only for diabetics who could not tolerate the combination sulphonylureas-plus-metformin, because (without glitazones) the only alternative for those patients would be insulin therapy.

Political culture and the battle for standards

Let us remind ourselves that the FDA and the EMEA are two of the world's leading and best-resourced drug regulatory agencies. Yet, despite marketing approval by both those agencies, expert clinicians soon declared that the trial evidence upon which the drugs were approved provided little or no knowledge informing doctors and patients about whether the glitazones were useful additions to existing diabetes drug therapy (Gale 2001). The clear implication of our analysis is that more and better data about the efficacy of the glitazones needed to be collected and analysed before marketing approval in order to determine if they provided any therapeutic advantage. This raises the crucial question of why these regulatory agencies, especially the FDA, did not demand higher standards of approval.

A superficial answer might be to say that the legislation and regulations in Europe and the US permitted proof of clinical efficacy based on superiority to placebo. That is true, though as explained in Chapter 1, for decades, regulators on both sides of the Atlantic have required non-inferiority trials, instead of placebo trials, for antibiotics, betraying the idea that regulators' hands were tied by the law. It is also true that legislation and regulations permitting proof of clinical efficacy based on superiority to placebo pre-dates the neo-liberal era in both regions. In the case of the US, that approval standard for efficacy dates back to the 1962 Kefauver-Harris Drug Amendments to the 1938 Food, Drug & Cosmetic Act (Abraham 1995a; FDA 1991). In Europe, the transposition of a similar standard into the supranational EU regulatory system did not occur until 1995, but the standard had existed previously in many individual European countries since the late 1960s, and even in a primitive form as early as the 1930s in some Scandinavian countries (Abraham and Lewis 2000; Temple and Ellenberg 2000). We do not suggest, therefore, that neo-liberalism is solely responsible for the application of the placebo-controlled trial standard of clinical efficacy (hereafter 'the placebo standard'). However, as we saw in Chapter 2, drug legislation and regulations could be, and were, significantly altered during the neo-liberal period. This raises the deeper question of why the placebo standard was not also altered in regulations to require the collection and analysis of some data comparing new drugs with prevailing gold standard therapies (hereafter the gold standard).

It is not that there were no voices calling for gold standard regulation in Europe and the US at the time. For instance, the FDA's acceptance of placebo-controlled trials of new diabetes drugs, where patients in the control arm of the study were 'washed out' from active treatment, was contrary to the ethics of some of its scientific reviewing staff because it left those patients' hyperglycaemia untreated (Misbin 1999c). According to Gale (2001, pp. 1870–1), in the glitazone trials, this design meant that hundreds of patients were potentially exposed to blood-glucose levels higher than that recommended by the American Diabetes Association. As the FDA's medical reviewing officer for the glitazones put it:

> If it is unacceptable to allow patients with this degree of hyperglycemia to go untreated, how does one justify intentionally causing hyperglycemia by withdrawal of active treatment? It is unethical to conduct a study that is designed to demonstrate the benefits of a particular treatment in comparison to harm done to patients who do not receive that treatment. (Misbin 1999c)

Indeed, the very dilemmas that doctors treating diabetes faced after the approval of the glitazones was anticipated by the US public health advocacy group, Public Citizen, who exhorted at the time:

The problem of the use of placebos when an effective treatment for the condition exists is not only a problem of the violation of accepted ethical guidelines. These trials often do not provide the information that is most useful clinically. A drug treatment professional, for example, is not interested in whether a new treatment is better than nothing. To optimize therapy for a patient, the physician needs to know how the new treatment compares to the older, known effective treatment. These treatments need not be exactly equal in efficacy to be useful; depending on side effect profile, patient characteristics and even cost, the physician may even select a somewhat less effective medication. But trials that compare new treatments to placebo, with predictable results, do not aid physicians in making these decisions. (Public Citizen 1998)

In Europe, an EU Council Directive 75/318/EEC, which was in existence since before the neo-liberal period, stated that 'all clinical trials shall be carried out in accordance with the ethical principles laid down in the current revision of the Declaration of Helsinki', the revised version of which declares:

The benefits, risks, burdens and effectiveness of a new method should be tested against those of the best current prophylactic, diagnostic, and therapeutic methods. This does not preclude the use of placebo, or no treatment in studies where no proven prophylactic, diagnostic or therapeutic method exists. (World Medical Association 2000, para. 29)

As an EMEA/CPMP position statement on the use of placebo controls observes, the Declaration of Helsinki may be read as implying that all clinical trials intended to inform new drug evaluation should test the new drug against gold standard therapy if one exists (EMEA 2001a). Furthermore, in 1999, Patrick Deboyser, Head of the European Commission's Pharmaceuticals Unit, stated that some evaluation of comparative effectiveness of new drugs during the approval process would become 'unavoidable', while EU Health Committees proposed that the approval process should define new drugs as 'genuinely innovative' if they brought efficacy/safety benefits compared with existing therapies (Anon. 1999b; 2000d; 2000e).

Yet no substantive reform of the placebo standard towards the gold standard came. While neo-liberal governments were busy reforming drug regulatory agencies and legislation to accelerate new drug approvals and foster greater collaboration between industry and regulators, determination to change the placebo standard, which remains the legal standard, was piecemeal at best.

The EMEA (2001a) took the view that a strict interpretation of the Declaration of Helsinki was not appropriate and that there may be situations where the use of placebo remains essential to demonstrate the value of a new product even where alternative therapies exist. This meant that, in practice, little changed – even though a few years later the EMEA (2004a) acknowledged that new drugs *should* be tested against best available therapies. Similarly, the FDA was free to develop 'technical' objections to altering the placebo standard, which were somewhat of a diversion. Some senior scientists at the FDA argued, with some merit, that in a trial comparing a new drug with an active control – the typical design of non-inferiority trials – one cannot be sure if either of the drugs are superior to placebo in that particular trial. In technical terms, one cannot be sure if the trial has assay sensitivity without a placebo (Temple and Ellenberg 2000).[1] As one highly experienced FDA scientist put it, one needs to know:

> That had there been a placebo group present, the active therapies could have been distinguished from placebo *in that trial* ... Obviously, equivalence of two active treatments in a trial that could not distinguish active drug from placebo is wholly meaningless. (Temple 1997, p. 617)

We are not, of course, suggesting that such technical arguments by senior FDA scientists should have been suppressed in any way by US governments. Rather, we merely wish to draw attention to the contrast between successive American Administrations' silence on such matters compared with their relentless efforts to introduce, for example, accelerated approval regulations and many other reforms discussed in Chapter 2.

So far as the technical arguments themselves are concerned, a fairly obvious solution is to require a clinical efficacy standard that includes some *three*-arm trials, each simultaneously involving the new drug, a gold standard therapy, and a placebo. FDA scientists have suggested that such trials must be very large, unless they are 'non-inferiority' trials, which allow up to a 20 per cent chance that the test drug is less efficacious than the active control (EMEA 2004b; 2004c, p.2; Li Bassi *et al.* 2003, p. 249). From the perspective of providing valuable public health information, such three-arm 'non-inferiority' trials would seem to be preferable to placebo trials provided that the active control was gold standard therapy. Moreover, pharmaceutical firms often conduct many clinical trials on their drugs when only two pivotal studies are required. Evidently, there is scope for a smaller number of larger, more informative, trials instead of large numbers of less informative small trials. By moving in that direction, it becomes more feasible to conduct three-arm superiority trials or at least reduce the margin of difference ('delta') between the new drug and the gold standard in 'non-inferiority' trials, as has been suggested by some expert science advisers to regulatory agencies and public health advocacy groups (Public Citizen 1998; Garattini and Bertele 2001; Garattini 2004). It is for these reasons that we say that the technical arguments presented by some senior FDA scientists were rather diversionary and certainly not decisive as a justification for resistance to reforming the placebo standard.

Exactly how persuasive those technical interventions by FDA scientists were in helping to maintain the placebo standard in the US and Europe is hard to say, but there were certainly other significant factors. The pharmaceutical industry opposed, and continues to oppose, the introduction of approval standards requiring trials comparing new drugs with the best existing therapies (EFPIA 2000; Shimmings 2011). Companies perceive comparative active-controlled trials as commercially risky because it is easier to demonstrate a new drug's superior efficacy to placebo than superiority or even 'non-inferiority' to effective treatments.[2] Moreover, active-controlled trials are usually more expensive than placebo-controlled trials because, as discussed above, they typically require more patients (European Platform for Patients' Organisations, Science and Industry 2004; Public Citizen 1999).[3]

In addition, the industry has argued that an approval standard of comparative active-controlled trials would be a disincentive to pharmaceutical innovation, thus denying patients access to potentially valuable drugs (EFPIA 2000). The underlying basis for that argument is the oft-repeated claim that the pharmaceutical industry is a victim of over-regulation. For instance, at a workshop on 'the Value of Innovation', jointly organized by the European Commission's DG Enterprise and the European Platform for Patients' Organisations, Science and Industry (EPPOSI) – an alliance of patient groups, industry and academic scientists founded in 1994 – industry representatives argued that industry was 'massively' over-burdened by regulation. As an example of such over-regulation, they cited the existing requirements for demonstrating the non-inferiority of new antibiotics:

> In anti-bacterials, the application of the '10 per cent delta' rule – designed to elicit a comparison in effectiveness with existing clinical practice – has led to large trial numbers. In one example, 8,326 patients were required to be enrolled in a trial for a new anti-bacterial agent, more than double the number required for 15 per cent delta. (EPPOSI 2004)

Objections to placing economic burdens on the industry by raising regulatory approval standards were influential on the FDA, the EMEA, and the European Commission, all of whom came to accept during the neo-liberal era that the R&D costs borne by the pharmaceutical industry were too high (European Commission 2003; Kaplan and Laing, 2004). Such acceptance was, however, based on the industry's own estimates of its R&D costs, whose validity has been repeatedly challenged by analysts outside the industry (Angell 2004; Goozner 2004; Lexchin 2006; Public Citizen 2001). Thus, while the placebo standard was not created by neo-liberal regulatory reforms, the context of such reforms, which focused on encouraging industrial innovation by reducing data demands on drug companies for proof of efficacy, helped maintain it by making it difficult for the regulators and others to challenge the standard successfully.

This is illustrated by the fact that, even when American or European regulators expressed a wish for active-controlled trials to reduce potential uncertainties about how to prescribe a new drug, pharmaceutical manufacturers have not necessarily undertaken the desired comparator trials (Anon., 2000d).[4] Indeed, even when companies do conduct active-controlled trials they may be inadequately designed for valuable therapeutic information,[3] perhaps as some argue, driven instead by the priorities of marketing consultants and strategists (Dehue 2010). As we have seen in this chapter, manufacturers may select a weak comparator drug, rather than the gold standard, and/or select a sub-optimal dose of the gold standard, as occurred with the comparisons of the glitazones with sulfonylurea by both Takeda and SmithKline-Beecham. Those types of active-controlled trials maximize commercial protection for companies, but impair regulators' ability to determine whether the new drug offers therapeutic advance, thereby transporting pharmaceutical uncertainties into clinical practice and health-care systems when the drug is approved by outperforming placebo (Avorn 2005).

The glitazones are not isolated examples in this respect. According to an experienced FDA reviewer of diabetes drugs, techno-scientifically inappropriate and uninformative study designs are a persistent problem in the field:

> Drugs are being developed in ways that are really not trying to define what they can actually do. This is a big problem that I see. I see protocols which are really very self-serving. Protocols that are constructed in such a way that it's almost impossible for the drug that's being developed not to succeed. At the moment, really, the only way of stopping a trial is because of a gross safety concern, otherwise firms really do what they want.[5]

We suggest that that state of affairs is, in part, a product of the deregulatory political environment fostered during the neo-liberal period, which arguably was at one of its high points in the US in the late 1990s when the glitazones were being reviewed and approved.

Neo-liberalism flourished in both Europe and the US during the late 1990s and early 2000s when both the FDA and the EMEA granted marketing approval to the glitazones. However, there is some evidence from the regulation of the glitazones that the CPMP were more critical of the placebo standard than the FDA and less affected by neo-liberal reforms, which may partly explain why the drugs were granted much more restricted approval in Europe.[6] EU regulators more willingly accepted the drawbacks of 'non-inferiority' trials to obtain more comparative efficacy information (Anon. 2000d). According to a former CPMP member, such thinking influenced the EU's assessment of the glitazones:

> The reality in Europe is they are asking for active-control studies more and more ... The preferred option is a three-armed study. So you have placebo-control, you have an active control, and you have the drug. So the guidelines are interpreting the legislation as saying: 'in order to prove efficacy you must compare it with what's in the marketplace at the moment', otherwise you may have something superior to placebo, but significantly inferior to active comparators.[7]

These policy differences also reflected contrasting American and European political cultures towards the ability of 'the market' (prescribers) to produce optimal medical and health outcomes. One European regulator told us that the FDA insisted on superiority over placebo and assurances about safety, but beyond that, it was willing to let the drug 'find it's own place in the market'.[8] An oversimplified characterization, perhaps, but it chimes remarkably well with some of the comments above made by FDA scientists themselves.

Thus, while the FDA required data from active-control studies as an exception, EU regulators were increasingly viewing such studies as necessary. Furthermore, there is evidence that EU regulators would have sought greater use of active controls, but for the international hegemony of the US pharmaceutical market, which, at 50 per cent of the world market, is twice the size of the EU's market. This results in firms designing drug development programmes first to meet FDA regulatory requirements, and then presenting those programmes to the EMEA (Li Bassi *et al.* 2003, p. 248).[9] Hence, the uncertainties concerning the therapeutic value of new diabetes drugs generated by the FDA's approval standard had knock-on effects in Europe, although in the case of pioglitazone EU regulators appear to have demanded a further trial against active control.

International dissimilarities in political culture, therefore, have filtered the influence of neo-liberal regulatory reforms on challenges to the placebo standard differently in the EU and the US. However, the significance of those differences should not be exaggerated either in terms of political analysis or health outcomes. It is telling that EU regulators did not demand gold standard trials before approving the glitazones. The neo-liberal context militated against it. For instance, during the European Commission's review of pharmaceutical legislation, 2001–2004, whose background we discussed in Chapter 2, some Members of the European Parliament, European patient and consumer groups, health professionals, and CPMP members proposed an amendment requiring the EMEA to consider 'added therapeutic value' in drug approval decisions. (Anon., 2002b; European Commission 2001b).[10] However, the European Commission rejected the proposal mainly due to opposition from its Directorate-General Enterprise, which was responsible for promoting the European pharmaceutical industry.

While the CPMP's restricted approval afforded many fewer uncertainties in diabetic drug treatment than the FDA's more permissive approach, significant uncertainties also faced doctors and patients in Europe due to policy interpretations of approval standards. Nonetheless, to counter the US hegemony of the placebo standard experienced with the glitazones, in 2002, the CPMP developed guidance on the clinical investigation of diabetes drugs, specifying that applications would be expected to contain placebo- and active-controlled efficacy trials (EMEA 2002b,

p. 7). By contrast, the FDA had not established such requirements for approval in the US by the late 2000s, where marketing approval applications for diabetes drugs typically continued to be dominated by placebo-controlled trials (FDA 2008b). However, specifically regarding the glitazones, in September 2003, the CPMP recommended their approval as monotherapies for diabetes patients in the EU (Anon. 2003a). Thus, the earlier differences between EU and US approval of those particular diabetes drugs became much diminished thereafter.

Expectations, promissory science and the ideology of innovation

The influence of neo-liberalism on political culture within the drug regulatory agencies went well beyond a lack of determination to insist on more comprehensive and relevant efficacy data from industry. It also pro-actively shaped what Doran *et al.* (2008, p. 39) call an 'ideology of innovation'. Neo-liberalism perpetuated the misleading ideology that innovation and public health benefit were as one and that, therefore, the goal of regulation should be to promote innovation *per se*, or at least that a drug's claim to innovative mechanisms should be given great emphasis based on expectations of health benefits. That regulatory culture is revealed by the case of the glitazones.

The FDA's willingness to approve the glitazones despite lack of evidence favourably comparing them with existing therapies, was partly because agency scientists were excited about their innovative mechanism of increasing insulin sensitivity. One senior FDA scientist commented:

> It was a very appealing mechanism. If the lesion in type 2 diabetes is insulin resistance, and we can reverse that underlying pathology, then we're attacking where the disease begins... [And then there is] the idea that there may be patients who don't respond to sulphonylureas, who *will* respond to a thiazolidinedione [glitazone], and so when we give them new mechanisms to attack a disease problem, we're basically expanding the possibilities that [physicians] are going to be able to help particular patients.[11]

The hypothesized mechanistic benefit of the glitazones was that they 'corrected' insulin resistance so might offer an advantage over existing therapies by providing more *enduring* blood-glucose control or even permanently prevent the long-term progressive deterioration of diabetes.[5] However, this was not proven because pre-approval trials for pioglitazone or rosiglitazone extended only to 26 or 52 weeks, respectively. As one FDA reviewer explained:

> If one looks at how the thing has developed, it's still not known. There are still people who say that the glitazones should be used when everything else fails, but there's data to show – based on troglitazone actually – that it prevents diabetes in a permanent way, as opposed to metformin, where the effect is lost. So there are many people who don't have an axe to grind who say we should be using the glitazones very early, that it corrects insulin resistance, that it resets the pancreas and that you could prevent the fact that people deteriorate if you use the glitazones early.[5]

Consequently, this FDA scientist believed that the agency was right to approve the glitazones so long as there was appropriate information for physicians to make up their own minds.[5] Another (former) senior FDA scientist viewed the agency's approval of the glitazones as a desirable encouragement of innovation that should be given priority over evidence about efficacy compared with existing therapies:

> I'm glad we encouraged development of the thiazolidinediones [glitazones]. We could have really retarded development in this field, but I'm glad we encouraged development in this field... I think the innovation is not lodged in comparative efficacy, but in its novel physiologic action. Comparative efficacy can always be studied post-approval. For example, the glitazones, I think they had a novel action, and that should have been enough [for approval] even though maybe metformin and the sulphonylureas can do better in a particular study. But to have an agent with a unique pharmacologic action – that is innovation...Just to block off an entire area of physiologic action based on [comparative efficacy being a requirement of approval] – I don't think that's a good idea. I think that you should encourage novel pharmacologic agents. The comparative efficacy is something that can be established by subsequent studies. Because some patients respond to one thing, and don't respond to another. There are patients that will respond to glitazones, and not to metformin. So it's good to have those options. I mean it gives a little more flexibility to practice...It's always good to have another way of outflanking the disease.[12]

Scholars, such as Brown and Webster (2004) and Hedgecoe (2004) have discussed the role of expectations of companies, research scientists, and clinicians in early-stage medical and pharmaceutical innovation. Here we see the assimilation of such expectations by FDA regulators. A similar process took place in Europe. According to Professor Edwin Gale, an expert diabetologist and adviser to the EMEA in the years before marketing approval of the glitazones, some of the diabetology profession, encouraged by pharmaceutical industry claims about their innovative potential, pressed for these new drugs to be made available. He elaborated:

> There were tremendous expectations about Avandia [rosiglitazone] – partly because scientifically it was extremely interesting. It was a whole new model of the way a drug could act. It affected the body and its energy metabolism in totally new ways that were very interesting and fascinating. (Gale cited in Cohen 2010, p. 531)

At the annual conference of the European Association for the Study of Diabetes in September 1999, one speaker declared:

> Unlike most traditional drugs for type 2 diabetes, rosiglitazone works in a novel way to reduce insulin resistance, helping the body's own insulin work more effectively and offering patients improved glycaemic control. ... The hope is that this will slow long-term deterioration. (Matthews cited in Cohen 2010, p. 531)

Some diabetologists at the conference even suggested that the glitazones should be approved as first-line treatment for diabetes in Europe (Cohen 2010, p. 531). EU regulators echoed such enthusiasm about the glitazones' hypothesized mechanistic benefits. One former CPMP member remarked:

> If you look at how the glitazones work in the test-tube or on the page, they're wonderful because they're insulin sensitizers ... There was even a suggestion that they help your pancreas repair. And so from that point of view these are ideal drugs. ... [But] they didn't have the data to support it.[7]

However, as the end of the above quote signals, an important difference between the American and European regulatory approaches to those hypothetical mechanistic benefits was that the EU regulators took much more seriously the lack of evidence from controlled clinical trials to substantiate them. That also partly explains why the CPMP approved the glitazones in a much more restricted way than the FDA. By 2007, the FDA had reached the uncomfortable conclusion that their medical optimism about the glitazones' 'wonderful' and 'unique' mechanism had been little more than wishful thinking:

> it is disappointing...I think there was a feeling out in the research community that they might provide a major reduction in cardiovascular disease by reducing insulin resistance which had been hypothesized as an underlying factor contributing to these complications. We have really seen nothing that would suggest the glitazones are going to provide that benefit. (FDA 2007, p. 413)

Not only did the expectations about the glitazones' innovative mechanism turn out to be groundless in terms of health benefit, but they were highly ideological and unscientific with respect to rational drug use even at the time of the drugs' regulatory review. This is because the hypothesized clinical benefits associated with the glitazones' innovative mechanism, on the one hand, and the comparative clinical evidence actually available, on the other, implied *diametrically opposed clinical use*. Rationally, on the clinical trial evidence that the glitazones were less effective than existing drugs at lowering blood-glucose, glitazone use would be recommended only *after* those other drugs had failed (Yki-Jarvinen 2004). By contrast, the hypothesized benefits from the glitazones' innovative mechanism implied that the drugs should be used *early* in therapy to prevent disease progression. The fundamental contradiction of the FDA's approach, and others clamouring for extensive first-line marketing approval of the glitazones at that time, is that to suggest such potential health benefits from an innovative mechanism underlines the need for good comparative gold standard efficacy data because it is all the more important that information about the glitazones' most effective use is available. Without such data, the contradictory implications for the drugs' clinical use could not be resolved.

There is certainly little evidence, and perhaps less logic, to support the suggestion that the 'market' should be relied upon to produce rational drug use. Lack of comprehensive clinical trial data does not necessarily result in cautious prescribing, as shown by the glitazones, which achieved blockbuster status without clear evidence of advantages over existing therapies. Sales for rosiglitazone reached over US$505 million during the first 15 months of marketing alone, mainly in the US (Anon. 2000c). By 2006, worldwide sales of rosiglitazone had reached a staggering US$3.2 billion (Young 2011b).

Furthermore, industry promotion encouraging inappropriate prescribing may 'fill' the information gap – a gap frequently caused by how drug firms design and/or conduct trials (Anon. 2001h). Within 15 months of rosiglitazone reaching the US market, the FDA had to instruct the drug's manufacturer to refrain from using promotional material containing misleading claims that suggested the drug was 2–3 times more effective at reducing blood-glucose levels than had been demonstrated (Anon. 2000c). By 2011, the manufacturer (which had become GlaxoSmithKline) was being investigated by the US Department of Justice regarding its marketing of rosiglitazone (Virji 2011).

Macrovascular effects: limitations of the surrogate efficacy regulatory paradigm

Although the therapeutic goal of diabetes treatment is to improve patients' clinical symptoms and reduce the risk of long-term micro- and macrovascular clinical complications, in both the EU and the US (and elsewhere), diabetes drugs are, as we have seen, approved on to the market based on their effects on a 'surrogate marker/predictor' of those clinical outcomes, namely, blood-glucose levels, rather than the clinical outcomes themselves. (FDA 1992, pp. 132–5; 2008b, pp. 19–21). Effectiveness in reducing and controlling blood-glucose can be measured relatively quickly and cheaply in randomized controlled trials taking months, as occurred with the glitazones. By contrast, trials investigating vascular clinical complications take years. As we noted in Chapter 2, trials involving surrogate measures, therefore, typically satisfy the commercial interests of pharmaceutical manufacturers much more than long-term clinical outcome studies.

During the 1960s, the US University Diabetes Group (USUDG) conducted a long-term trial to try to settle the debate among diabetes specialists about whether vascular complications were causally linked to blood-glucose control (Marks 1997, pp. 201–2). Specifically, the USUDG sought to determine if the first sulfonylurea (tolbutamide) reduced vascular complications, but the trial was inconclusive because it had to be abandoned due to excess mortality among the subjects taking the drug (Greene 2007, pp. 120–1).

Such uncertainty provided little basis for change, so the regulatory paradigm defining drug efficacy in terms of blood-glucose control persisted. In 1998, the ten-year results of the massive, long-term trial by the UK Prospective Diabetes Study Group (UKPDSG) reported that blood-glucose control with sulfonylureas decreased the risk of diabetes-related microvascular, but not macrovascular, complications, and that blood-glucose control with metformin decreased the risk of both more than sulfonylureas or insulin (UKPDSG 1998a; 1998b). This reinforced the regulatory paradigm that drugs' capacity to lower blood-glucose appropriately was correlated with improvements in clinically relevant outcomes – and with good reason regarding *micro*vascular benefits. However, the superiority of metformin regarding *macro*vascular outcomes was *not* related to the degree of blood-glucose control because insulin, metformin and the sulphonylureas all produced similar reductions of blood-glucose (FDA 2008b, pp. 145–6). Indeed, the two largest randomized controlled trials with diabetes patients, namely the USUDG and UKPDSG studies, failed to find a significant reduction in cardiovascular events despite excellent blood-glucose control (Anon., 2007e). Furthermore, at the very time the FDA was approving the glitazones, one of its most senior and experienced scientists, Robert Temple, was warning that surrogate measures of drug efficacy could lead to the adoption of useless or even harmful therapies (Temple 1999).

Despite the unconvincing correlation between blood-glucose control and macrovascular effects, the surrogate measure of diabetes drugs' efficacy in terms of blood-glucose control remained the dominant and largely unchallenged regulatory paradigm. That is, blood-glucose control continued to be regarded as an *established* surrogate measure of the vascular effects of diabetes within the American and European regulatory agencies. The alternative, namely, a shift to a regulatory policy requiring long-term trials directly investigating the macrovascular effects of new diabetes drugs would have delayed drug approvals significantly. Yet again such a policy change would have had to swim against the tide of neo-liberal reforms sweeping across the EU and the US at that time, which were concerned with *reducing* drug approval times.

As we saw in Chapter 2, in response to pressure from the pharmaceutical industry and its pro-business allies in American and European governments, the supranational EU centralized system set relatively short and decreasing time limits on regulatory review of new drugs, while, in the US, under the 1992 Prescription Drug Users Fee Act and subsequent legislation in 1997, Congress made about half of the FDA's funding dependent on its performance in reducing new drug approval times (Abraham 2008). This resulted in enormous emphasis on acceleration of new drugs on to the market, especially those deemed 'innovative' (Abraham and Lewis 2000; Hilts 2003). Consequently, we suggest that the neo-liberalism of this period created a regulatory culture in the EMEA and the FDA so focused on accelerating drug approval, that the possibility of making policy changes that would significantly *prolong* drug development was antithetical to managerial thinking in the regulatory agencies. Any attempt by expert scientists to challenge the dominance of the established surrogate efficacy measure as a paradigm and standard for drug approval faced managerial resistance propelled by the neo-liberal regulatory politics of the period. In particular, rather than demanding long-term trials before approval, neo-liberal regulatory reforms led regulators increasingly to defer aspects of drug evaluation to the post-marketing phase. As one senior FDA scientist explained:

> The balance has been tipped so that the benefits of the doubt will be given to the company [at time of approval] and there's greater reliance that we can take care of things after drug approval. So we have all these post-marketing agreements [with firms].[11]

As we have discussed, when SmithKline-Beecham and Takeda applied to market rosiglitazone and pioglitazone respectively, they submitted short-term (six-month) trials showing the drugs' capacity to reduce blood-glucose as demanded by the EMEA and FDA, but not long-term trials demonstrating beneficial vascular benefits or improvements in the mortality or morbidity outcomes for diabetics (Misbin 1999a). By necessity, the lack of long-term studies of the glitazones' capacity to reduce adverse macrovascular effects, at the time of approval, simultaneously created uncertainty about whether they might have the potential to increase macrovasular risks. As the CPMP put it: 'due to the absence of long-term data, the overall effect of pioglitazone on cardiovascular risk cannot be assessed' (EMEA 2001c, p. 21). In this respect, efficacy and risk were two sides of the same coin – not knowing about the glitazones' efficacy to improve adverse macrovascular outcomes, such as heart attacks and strokes, also meant not knowing about the drugs' risk of exacerbating those outcomes.

Before approval, both regulatory agencies identified that the glitazones could cause oedema/fluid-retention, induce significant weight gain, and exacerbate congestive heart failure (EMEA 2003, p.15; Malozowski 1999a; 1999b). Congestive heart failure is a gradual condition, distinct from what medical scientists call cardiovascular 'events', such as heart attacks. FDA scientists clearly felt that the information they had received about pioglitazone's propensity to cause fluid retention was sub-optimal. They specifically noted that Takeda could have investigated how the drug produced fluid retention in patients, but did not do so, thus 'hindering the ability to alert subjects prone to get this complication and to develop rational treatment for patients that do'

(Malozowski 1999b, p. 2). Nonetheless, the FDA did not demand any further pre-approval studies to clarify that issue.

Regarding rosiglitazone, the FDA's medical reviewer raised concerns about the drug's cardiovascular risks. In his analysis, he reasoned:

> As an 'insulin sensitizer', rosiglitazone appears to lower glucose levels by converting glucose to fat.... My concern about deleterious long-term effects on the heart should be addressed by requiring the sponsor to provide adequate information in the label about changes in weight and lipids. A postmarketing study to address these issues needs to be a condition of approval. (Misbin 1999d, p. 40)

The regulatory culture of pushing questions about efficacy and safety to the post-marketing phase had become so ingrained that we have found no evidence that anyone at the FDA suggested that SmithKline Beecham should have been required to undertake *pre*-marketing trials before approval to address Misbin's concerns about 'deleterious' cardio-vascular effects. Nonetheless, as shown by the above quote, FDA scientists did recommend a long-term post-marketing study of the drug's cardiovascular effects, as a condition of approval, to allay their concerns about the matter. Yet the FDA's approval letter in 1999 vaguely required the company to undertake a long-term safety and efficacy trial without specifying an investigation of cardiovascular effects.

In Europe, a minority of CPMP experts felt that the short-term nature of the glitazones' pre-market trials (mostly six months or less) was insufficient to grant marketing approval. They wanted long-term trials to determine the cardiovascular benefits and/or risks of both pioglitazone and rosiglitazone before approval on to the market where millions of patients were likely to be taking the drugs (EMEA 2001b, pp. 21–22; 2003, pp. 16–18). Some expert advisers to the EMEA noted that without such long-term studies it was unclear whether rosiglitazone would have any beneficial impact on cardiovascular disease (Cohen 2010, p. 8). Nevertheless, taken as a whole, the CPMP and the EMEA decided that these matters could be resolved by requiring the manufacturers of both pioglitazone and rosiglitazone to conduct post-marketing trials as conditions of their approval (EMEA 2001b, pp. 21–22; 2003, pp. 16–18). As these matters concerned safety and risk, not only efficacy, the regulators in Europe and the US had the legal power to demand such trials before approval, but declined to do so (Abraham and Lewis 2000; US Congress 2007, pp. 65–6).

Those regulatory decisions in Europe and the US highlight the dominance of the surrogate efficacy paradigm within the regulatory agencies, who did not even require the companies to begin, let alone complete, long-term trials on the glitazones' macrovascular effects before approval. The policy contributed directly to uncertainty about new diabetes drugs' macrovascular effects in clinical practice, while doctors awaited study results. Furthermore, deferring trials until after approval reduced pharmaceutical companies' incentive to conduct the studies quickly or to high standards because they already had their products on the market making commercial gains (FDA 2008b, pp. 215–16 and 381).

What neither the regulators in Europe or the US, nor probably anyone outside the pharmaceutical industry, seem to have known was that, as early as 1999, SmithKline Beecham (subsequently GSK) started a trial comparing the cardiovascular effects of rosiglitazone with pioglitazone, but they never posted it on their website or published the results. There is also no evidence of it being reviewed by American or European regulators. It may have begun after marketing approval in the US, but it must have been underway before EU approval, which was not until 2000. The existence of this trial only came to light in 2010 after a *New York Times* investigation (Harris 2010a). GSK has accepted that the trial took place. According to GSK, they did not publish the trial because its findings were insignificant (Harris 2010a). However, the *New York Times* alleged that GSK suppressed the data because rosiglitazone was found to be much worse for the heart than pioglitazone (Harris 2010a). The newspaper also released an internal company memo regarding the trial, which stated: 'these data should not see the light of day to anyone outside of GSK' (cited in Cohen 2010, p. 534). It is unclear whether those allegations form part of the US Department of Justice's ongoing investigation into GSK's development and marketing of rosiglitazone (Virji 2011).

Whatever the significance and probity of that 'secret trial', quite separately, SmithKline Beecham did conduct, on the record, a long-term post-marketing study of rosiglitazone in response to the condition of approval set by the FDA. By the time the post-marketing study was completed and published Smith Kline Beecham had become GSK. The study was called, 'A Diabetes Outcome Progression Trial (ADOPT)' and lasted for four years (Kahn *et al.* 2006). Although FDA scientists had recommended a long-term post-marketing trial of rosiglitazone to investigate its cardiovascular effects, ADOPT compared the durability of rosiglitazone's blood-glucose control with metformin or sulfonylurea in 4360 patients (Anon. 2006i; FDA 2008b, p. 188; Kahn *et al.* 2006; US Congress 2007, pp. 42–43 and 67–68).

In other words, as FDA experts subsequently acknowledged, ADOPT was primarily a surrogate efficacy trial. It was not designed to assess cardiovascular outcomes, so could not 'provide any meaningful evidence' on the matter (FDA 2007, p. 226). It was 'too small to get at heart-attack risk' – a characteristic that 'increased the uncertainty of estimates' about cardiovascular effect; and perhaps most damning, it had no pre-planned process for adjudication of cardiovascular events necessary to assess rosiglitazone's potential to increase or decrease heart attacks (FDA 2007, pp. 152–3; FDA 2008b, p. 188; US Congress, 2007, pp. 42, 67–8 and 132). The limitations of ADOPT's design were further demonstrated by the fact that it failed to adequately detect some of rosiglitazone's known adverse effects, such as coronary heart failure (FDA 2007, p. 217). Moreover, ADOPT was not published until December 2006, seven years after rosiglitazone entered the market (Anon. 2006i; US Congress 2007, p. 42).

Whether the FDA's vague approval letter allowed GSK the latitude to conduct a trial that was deficiently designed, instead of the sort of trial FDA scientists wanted is an arguable point. It is clear, however, that the FDA did not rigorously insist on the type of post-marketing trial that the agency's scientists wanted. Indeed, an FDA report to Congress in 2005 noted that only 14 per cent of post-marketing study commitments agreed by companies as conditions for approval had been completed (FDA 2008b, p. 324). While the company reaped the commercial rewards from seven years of marketing, concerns about rosiglitazone's cardiovascular effects remained unresolved. Those concerns were only heightened when two cardiovascular experts from The Cleveland Clinic, Nissen and Wolski (2007), conducted a meta-analysis of 42 completed randomized controlled trials involving rosiglitazone and found that the drug was 'associated with a significant increase in the risk of myocardial infarction [heart attack] and with an increase in the risk of death from cardiovascular cause that had borderline [statistical] significance' (p. 2457).

Nissen and Wolski's (2007) findings sparked a Congressional investigation of how the FDA was regulating the cardiovascular risks of rosiglitazone, and prompted the EMEA to require GSK to insert a warning about increased risk of heart attack on the drug's label in the EU (US Congress 2007, p. 127). An FDA expert advisory committee meeting followed just one month later to consider whether rosiglitazone increased cardiovascular risk, and whether it should be removed from the US market. David Graham, a pharmacoepidemiologist and Associate Director for Science and Medicine at the FDA, who analysed all the key rosiglitazone trial data available at that time, presented the meeting with a stark message:

What it says is that there are about 80,000 excess cases of cardiac death and myocardial infarction [heart attack] attributable to rosiglitazone use over the seven and a half year period that this analysis covers. For that estimate the numbers range from 30,000 to 140,000. The number needed to harm was 114, what that tells you is that at this point for every 114 patients that we treat with rosiglitazone for a year we produce one extra case of serious coronary heart disease. (FDA 2007, p. 229)

GSK did not agree, but the FDA's expert advisers agreed by 20 votes to 3 that the drug increased 'cardiac ischaemic risk' (the risk of cardiovascular disease). However, the committee voted 22 to 1 that rosiglitazone should remain on the market apparently because of the considerable uncertainty generated by the lack of good quality trials addressing the long-term macrovascular effects of the drug and a lingering hope that the drug's effectiveness in blood-glucose control might be delivering microvascular benefits to diabetics – though even that had not been demonstrated in rosiglitazone trials.

Indeed cardiovascular risk was only one element of a much larger picture about the adequacy of pre-approval and post-approval testing requirements for diabetes drugs. Regulators' willingness to accept short-term trials demonstrating blood-glucose control as a surrogate for effectiveness in reducing the macrovascular outcomes of diabetes, despite the evidence from the UK Prospective Diabetes Study Group in 1998, was beginning to unravel. In 2006, a large three-year randomized control trial comparing 2635 patients taking rosiglitazone with 2634 patients receiving placebo was published. It was called 'Diabetes REduction Assessment with ramipiril and rosiglitazone Medication (DREAM)' and was carried out by academics, not GSK (DREAM Investigators 2006). The patients in DREAM were 'pre-diabetic' (that is, at high risk of developing diabetes), rather than diabetics, because the purpose of the trial was to discover if use of early treatment with rosiglitazone could forestall the development of overt diabetes. The findings from DREAM implied that rosiglitazone was effective in delaying the onset of diabetes in terms of loss of blood-glucose control, but it simultaneously showed a 40 per cent *increase* in cardiovascular risk in a pre-diabetic patient population, who would normally be expected to be at lower cardiovascular risk than diabetics (FDA 2007, pp. 201–15). In other words, rosiglitazone was more effective than placebo in controlling blood-glucose (the measure used to approve new diabetes drugs as efficacious), but increased patients' risk of suffering an adverse macrovascular event.

Meanwhile, in 2005, Takeda's long-term three-year, post-marketing placebo-controlled trial of pioglitazone, which had been mandated as a condition of approval by the EMEA, was published (Dormandy *et al.* 2005). It was entitled: 'PROspective pioglitAzone Clinical Trial in macroVascular Events (PROactive). The trial was designed to examine cardiovascular outcomes in 5238 patients, so it did at least address the concerns of the EU regulators. As we have seen PROactive also monitored pioglitazone's carcinogenic risk to a limited extent. The trial investigators reported that pioglitazone was associated with a non-statistically significant decreased risk of macrovascular events (heart attacks and strokes), but an increased risk of heart failure, which was larger than the decreased risk of macrovascular events. While the FDA judged that PROactive provided 'no conclusive evidence of macrovacular risk reduction', the *Lancet* noted that the placebo-control design told doctors little about how it compared with other diabetes drugs (FDA 2008b, pp. 26–27, 147–8 and 278; Yki-Jarvinen 2005). At best, it seems that PROactive implied that pioglitazone was marginally better than placebo at improving macrovascular events for diabetics, as against risks of heart failure and carcinogenicity associated with the drug.

Also underway was GSK's six-year trial of rosiglitazone's cardiovascular effects on 4447 diabetics mandated by the EMEA, which became known as RECORD (Home *et al.* 2009). In this trial, which was un-blinded (open-label), patients were treated with rosiglitazone as add-on therapy to either metformin or sulphonylurea and, unlike PROactive, compared with patients taking metformin-plus-sulphonylurea, rather than placebo. In addition, unlike ADOPT, there was a predefined adjudication process for cardiovascular events (Unger 2010; US Congress 2007, p. 44). As such, RECORD was potentially more informative to clinical practice than PROactive or ADOPT, but being a 'head-to-head' trial, RECORD needed to be much larger than placebo-controlled trials. By 2007 many expert clinicians and statisticians believed that RECORD had insufficient 'statistical power' to reliably detect differences between treatments' effects on macrovacular events, such as heart attacks and strokes (US Congress 2007, pp. 46 and 131–2; FDA 2007, pp. 235–38 and 411–24).

Thus, even before RECORD was published in 2009, the FDA instructed GSK to conduct a further, adequately powered, seven-year trial comparing the cardiovascular risks of rosiglitazone with pioglitazone, which would not be completed until 2014 or 2015 (FDA 2007, pp. 411–12; 2008b, pp. 333–6; US Congress 2007, p. 151). Frustration at the inadequate databases exploded publicly in FDA advisory committee meetings in

2007 and 2008 (FDA 2007, pp. 351–2; 2008b, p. 130). Steve Nissen, a world-leading specialist in cardiovascular medicine, who had undertaken a meta-analysis of rosiglitazone trials declared:

> So here we are 50 yrs after the initial introduction of anti-diabetic agents and although cardiovascular disease is the cause of death in 75 per cent of diabetics, there exists no well-designed, adequately powered, comparative effectiveness trials evaluating macrovascular outcomes for diabetes drugs.... I think the absence of information on macrovascular effects is unfortunate. It is, however, a consequence of a regulatory policy that emphasizes glucose-lowering as a therapeutic goal.... There are major consequences of using blood-glucose control as the primary driver of drug approval. Pre-approval studies focus on demonstrating maximal glucose lowering effects, so patients are selected with relatively high blood-glucose levels because that enhances apparent efficacy. Patients at high cardiovascular risk are deliberately avoided. Firms say, 'why take a chance of an adverse safety signal?'... What we really need to know are what agents can we develop beyond current therapies that will improve health outcomes. We have lots of ways to reduce blood-sugar. What we really want to do is find the right way to lower blood-sugar. ... We need more robust outcome data to inform physicians on how to use these drugs safely and effectively. ... If a large, well-powered outcomes trial had been mandated in 1999, we would not have to wait until 2014, 15 yrs after approval to determine if this drug [rosiglitazone] is safe or not. (FDA 2008b, pp. 310–24 and 333–6)

In August 2007, the Chair of the FDA's expert advisory committee on diabetes drugs declared that reducing macrovascular complications of diabetes should become the primary focus of drug trials, rather than surrogate measures of blood-glucose control (Anon., 2007e). The following year, the committee discussed whether and when pharmaceutical manufacturers should be required to conduct trials investigating such macrovascular effects, especially cardiovascular risks (FDA, 2008b). It recommended that such trials should be required for new diabetes drugs. However, the crucial question of whether the trials should be at least underway (if not completed) pre-approval or merely mandated as post-marketing studies was evaded and unresolved.

When RECORD was finally published, it suggested that rosiglitazone was associated with an increased heart attack risk – although not to a statistically significant degree – and yet also associated with decreased overall mortality, compared with the metformin-sulfonylurea combination (Home *et al.* 2009). The *Lancet* commented that, 'owing to [RECORD's] study limitations, ... uncertainty remains regarding the effect of rosiglitazone on cardiovascular disease' (Retnakaran and Zinman 2009, p. 2089). FDA scientists concurred less diplomatically that: 'the open-label design made it susceptible to bias' and it 'was inadequately designed and conducted to provide any reassurance about the cardiovascular safety of rosiglitazone' (Marciniak 2010, p. 1; Unger 2010, pp. 14–16).

In particular, when FDA scientists reviewed GSK's RECORD trial, they sampled 549 (12 per cent) of the 4447 individual patient case-report-forms. They found 70 instances of problematic trial conduct, such as 'unacceptable case handling' and 'failures to refer events for adjudication [heart attack or stroke]' (Marciniak 2010, pp. 3–4). The former included a case where 'no information was obtained regarding a 67-day hospitalization for a severe stroke and the stroke was adjudicated as non-cardiovascular', while the latter included a heart-attack event which was 'deleted 16 months later and never referred for adjudication' (Marciniak 2010, pp. 92–5). Of those 70 cases, FDA reviewers found that the problematic conduct favoured rosiglitazone in 57, but the comparator treatment in only 13. They concluded that RECORD's results might contain four-fold biases favouring rosiglitazone (Marciniak 2010). The conduct of RECORD contributed to further uncertainty about the macrovascular effects of rosiglitazone by reinforcing the agency's belief that the trial could not provide conclusive, or even reassuring, evidence about the drug's cardiovascular risks. It provides further evidence that permitting companies to conduct crucial trials in the post-marketing phase creates problems not only in terms of whether the trials will actually be completed in a timely manner, but also whether they will be designed and conducted appropriately.

At an FDA expert advisory committee in July 2010 convened to discuss rosiglitazone, FDA scientists revealed that a new trial, not due for completion until 2015, would be necessary to determine conclusively whether the glitazones decreased or increased adverse cardiovascular effects (Parks 2010). Underlining this uncertainty, the expert advisory committee was split 20 votes to 12 to maintain rosiglitazone on the US market, having just voted 18 to six, with nine abstentions, that the drug increased the risk of heart attack (Fitzgerald 2010; Harris 2010b). The FDA decided to keep rosiglitazone on the market, but imposed severe restrictions on its use, requiring GSK to develop a risk evaluation and mitigation strategy involving a limited access programme. By contrast, after a meeting in September 2010, the European Medicines Agency (the EMEA's successor) suspended marketing approval for rosiglitazone in the EU because of concerns about its cardiovascular risks (Faigen 2010).

Conclusion

The glitazones were innovative pharmaceuticals both in the formal administrative sense that they were NMEs brought to the market, and in the additional technical sense that the mechanism by which they were hypothesized to treat diabetes was novel. Some regulators, especially in the US, thought that this implied that they offered modest or significant therapeutic advance and they were so classified by the FDA decision to fast-track them under priority review. Yet in reality clinical trial evidence to support that characterization was lacking. Indeed, the available evidence suggested that the glitazones were less effective than existing oral diabetes medications. If nothing else, that raises searching questions about the validity of official FDA judgements that 40 per cent of NMEs offer modest or significant therapeutic advance. Our case study of the glitazones suggests that figure may be much lower in reality. Evidently, not only is it a mistake to conflate innovation with therapeutic advance, it is also a mistake to conflate the official categorizations of 'modest and significant therapeutic advance' with actual evidence of therapeutic advance.

The overwhelming insight gained from our case study of the glitazones is that the goals of public health benefit, especially for diabetic patients would have been best achieved by gathering and analysing more and better data on the drugs' clinical efficacy and outcomes before marketing approval. That would have delayed drug approval and required regulators to resist the arguments by the pharmaceutical industry that requiring gold standard blood-glucose trials and long-term studies of macrovascular effects would damage innovation. However, the American and European regulatory agencies, on the whole, did not challenge this ideology of innovation, though EU regulators were more sceptical and accordingly regulated the glitazones more strictly during their first few years on the market. Nonetheless, both EU and US regulators approved the glitazones and, with some notable exceptions, were highly influenced by non-evidence-based expectations and promissory science generated about the glitazones' therapeutic potential. Not only were those expectations never realized, they were scarcely, if at all, valid at the time they were propagated. Specifically, there was no convincing evidence that the glitazones were even more effective at controlling blood-glucose than existing therapies, let alone offering advances in the reduction of micro- and macrovasular adverse effects, about which adequate collection of data had not even begun before approval. Given that neither pioglitazone nor rosiglitazone was demonstrated to offer therapeutic advance, together with the fact that the former was associated with carcinogenic potential, and the latter with cardiac risk, the regulatory agencies' marketing approval of the drugs prioritized the commercial interests of the pharmaceutical manufacturers over and above those of public health by giving the glitazones an unwarranted benefit of the doubt. Potential carcinogenic and cardiac risks were marginalized and not regarded as 'expectations' at the time of approval, while expectations of the glitazones' innovative potential for therapeutic advance was permitted to take centre-stage. *Thus, expectations were structured around interests (consciously or unconsciously).* Furthermore, maintenance of rosiglitazone on the market for over a decade, despite post-marketing studies that failed to provide reassurance about its cardiovascular risks because of inadequate design and/or conduct, also awarded the manufacturer enormous benefit of the doubt. In this sense, expectations theorists are right to refer to the significance of *promissory* science, because the promise of therapeutic benefit tends to be granted much greater importance than the science-based expectations about drug risks.

These empirical realities of diabetes drug regulation from the late 1990s to the late 2000s in Europe and the US are consistent with a corporate bias in which the closeness and interdependence between the pharmaceutical industry and government agencies encouraged by neo-liberalism led to an overly permissive approach to innovative pharmaceutical regulation. It is important to appreciate, however, that the translation of such corporate bias into permissive product approval is not necessarily characterized by the infiltration of government regulatory agencies by pro-industry personnel (though as we saw in Chapter 2, that certainly occurred). An uncritical acceptance of the neo-liberal ideology that industry's interests in innovation equated with patients' interests in therapeutic advance helped to persuade regulators that the approval of the glitazones should go ahead. Indeed, this case study suggests that even if post-marketing trials are therapeutically disappointing, *sales* can thrive for years afterwards on the back of immense promotional efforts by manufacturers and the ideological sway of the idea of innovation within neo-liberal regulation (Dehue 2010; Doran *et al.* 2008).

There are also much more direct and specific ways in which the neo-liberal regulatory reforms affected the micro-sociological context of the glitazones. For example, although the surrogate efficacy paradigm is partly rooted in pursuits endogenous to scientific and professional exploration, its resilience in regulation stems from its congruence with quick and cheap drug development, which is in industry's interests and with an ideology that mistakes technological novelty for public health benefit. In particular, since 1990, the industry has also been a principal driver of an international regulatory environment in which drug testing has been streamlined to accelerate new drugs on to the market, irrespective of their therapeutic advantage (Abraham 2008; Hilts 2003, IFPMA 2000; Kaitin and Di Masi 2000; O'Donovan 2008). This involved transferring many aspects of clinical trial evidence-gathering from pre-approval to post-marketing (Abraham and Reed 2001). Consequently, crucial information about the efficacy and safety of new drugs came to be collected after, not before, doctors and patients medicated illness, as occurred with the trials of the glitazones' macrovascular effects.

Indeed, as we have discussed in Chapter 2, the whole thrust of the drug regulatory reforms in the neo-liberal period was to expand the use of surrogate measures even to those that were 'non-established', to shorten approval times and, through increased industry consultation, to reduce the risk of non-approval in response to industry's demands. Those specific deregulatory reforms were instigated first in the US during the 1980s and 1990s and largely mimicked in the EU in the late 1990s and 2000s. The glitazones case-study shows that, in the interests of public health, the placebo standard and regulatory reliance on some 'established' surrogate measures of drug efficacy, such as blood-glucose control, needed to be challenged in ways contrary to the commercial interests of the pharmaceutical industry by prolonging approval times and increasing the risk of non-approval. We draw two conclusions from that. First, the neo-liberal deregulatory environment militated against the measures needed to regulate the glitazones optimally for public health. And second, our analysis of the glitazones regulation reveals why such deregulatory policies are not in the interests of patients or public health.

Regarding disease-politics theory, we note that the fast-tracking of the glitazones went ahead together with the creation of expectations about their innovative mechanism and concomitant claims about therapeutic advance with little or no involvement of patient activism or patients' demands on either side of the Atlantic. Yet here we see highly significant under-cutting of the type of test data needed to assess drug benefit because that was what the industry wanted and what the regulators ultimately were willing to accept in the climate of neo-liberal deregulatory reforms. Evidently, neo-liberal corporate bias proceeded unabated irrespective of whether patient activism was demanding access to a drug. The glitazones case-study is further evidence that patient activism beyond AIDS activism, is a very limited explanation for regulatory trajectory, such as whether gold standard trials and surrogate measures are employed.

The glitazone experience also highlights substantively the problems with the 'risk-risk thesis' – the idea that regulators' dilemma should be framed as the risk of approving an unsafe drug too quickly versus the risk of withholding a therapeutically valuable drug for too long. In fact, as demonstrated vividly in this chapter, the problem of approving too quickly may crucially relate to allowing a drug on the market, whose therapeutic value is poorly understood and questionable. A clear lesson from the glitazones is that fast approval should not only be associated with safety risks, but also compromise of benefit assessment. Conversely, delayed approval should not only be linked to denial of important medicines, but also provision of information necessary for effective clinical use of medicines. Yet, despite all the uncertainties that the placebo standard and the blood-glucose approach generated about the therapeutic value of the glitazones for over a decade, they remain significant regulatory approaches in the EU and the US (Shimmings 2011). The promissory science and ideology of innovation, encouraged by the industry, continues to outweigh demands for approval standards based on evidence-gathering most relevant to enhancing users' knowledge.

4

Desperate Regulation for Desperate Cancer Patients: The Unmet Needs of Accelerated Drug Approval

As we explained in Chapter 1, the *regular* process of drug development (as occurred with the glitazones in Chapter 3) includes three sequential phases of clinical trials before government regulatory agencies consider granting marketing approval for new drugs. Specifically, under the regular drug regulatory process, the EMEA and the FDA have required *at least two* 'pivotal' phase III randomized controlled clinical trials (RCTs) to demonstrate new drug efficacy before marketing approval. Often that involves trials designed to provide evidence of direct efficacy of the drug on clinical symptoms or survival. However, as we saw in Chapter 3 regarding diabetes drugs, sometimes regular marketing approval is based on a new drug's effectiveness on an 'established' surrogate measure of clinical efficacy, that is, a laboratory or physical measure, which is accepted by regulatory agencies to reasonably substitute for clinical efficacy. For example, as discussed in Chapter 3, in the case of diabetes drugs, regulatory agencies have long accepted (perhaps too readily and wrongly) that a drug's capacity to control blood-glucose is a valid predictor, and hence surrogate marker, of clinical efficacy to reduce the micro- and macrovascular effects of diabetes.

In Chapter 2, we discussed the 1988 and 1992 regulations introduced by the FDA to accelerate marketing of drugs intended to treat *serious* or *life-threatening* conditions and expected to address an 'unmet need'. In the US, those *non-regular* regulations, known as 'Subpart E' and 'Subpart H', respectively, enabled the FDA to approve such drugs with less data than normally required to support clinical efficacy (Federal Register 1988; 1992). In particular, marketing approval could be granted *without phase III trials*, with just *one* phase II trial (Subpart E), and/or using '*non-established*' surrogate measures of clinical efficacy in trials. Specifically, a 'non-established' surrogate marker was one that was '*reasonably* likely to predict clinical efficacy', but not demonstrated to be a valid substitute for clinical outcomes (Subpart H). Drugs approved under Subpart H, based on non-established surrogates, were referred to by the FDA as 'accelerated approvals'. From 1992 in the US, upon receiving accelerated approval under Subpart H, drug manufacturers were required to conduct post-marketing (phase IV) '*confirmatory trials*' to demonstrate drug efficacy according to regular approval requirements. If such studies failed to confirm clinical efficacy or were not forthcoming from manu-facturers with 'due diligence', that is, were either not conducted in a timely manner or conducted improperly, then the FDA was expected to remove the drug from the market, though not strictly required to do so by law (US GAO 2009, p. 34).

As we also explained in Chapter 2, in the EU, similar types of regulations were introduced by the EMEA and the European Commission, albeit later and initially in a much more restricted way than pertained in the US. The EU's 1993 'exceptional circumstances' provisions permitted marketing approval without regular ('comprehensive') efficacy and safety data in circumstances where it was neither ethically nor scientifically reasonable to expect such data to be provided (EC Regulation 2309/93). Ostensibly, this was much more restrictive than the FDA's accelerated approval regulations under Subparts E and H because the European 'exceptional circumstances' provision was supposed to be confined to situations in which the collection of data needed for regular approval was impossible. By contrast, the FDA's Subparts E and H allowed 'accelerated approval' by deferring to the post-marketing phase various types of data collection that would be pre-market requirements for 'regular approval'. Indeed, in the US, a product's 'accelerated approval' status could be converted to 'regular approval' if post-marketing study commitments demonstrated efficacy in accordance with regular regulatory standards.

Following the EU's 2004 review of pharmaceutical legislation, in April 2006 the European Commission implemented Regulation 507/2006, which introduced a 'conditional marketing approval' provision avail-able to pharmaceutical companies seeking to market new drugs to treat emergencies, chronically/seriously debilitating conditions or life-threatening disease. As we mentioned in Chapter 2, that provi-sion allowed a company to gain marketing approval in Europe for new drugs in those categories based on incomplete data (relative to regular approval standards) merely because such data had not been collected at the time of application, provided that the company committed to conduct post-marketing studies that would supply additional datasets addressing outstanding questions and data-deficiencies about drug safety and efficacy at the time of 'conditional marketing approval'. If such datasets were provided, then the marketing status of the drug could be upgraded from 'conditional approval' to 'regular approval'. In this respect, the EMEA's 2006 'conditional marketing' regulation is almost identical to the FDA's 1992 'accelerated approval'.

The EU's 'conditional marketing' regulation opened the door for industry to seek more extensive use of biomarkers and surrogate endpoints just as the 1992 'accelerated approval' rule had done in the US. Clinical trials using surrogate endpoints are in manufacturers' interests because they require fewer patients, have shorter duration, and are cheaper, than trials tracking direct clinical effects. The most obvious illustration of the increased search for biomarkers and surrogate measures of clinical efficacy in Europe is the two-billion-euro Innovative Medicines Initiative (IMI), which was jointly funded by the pharmaceutical industry and the European Commission and launched in 2008. Hence, there was a period between 1992 and the mid-2000s when the EU and the US had fairly different regulations regarding acceleration of approval of new drugs intended to treat, serious, debilitating, or life-threatening conditions. However, from 2006, the European Commission and the EMEA effectively adopted the FDA's procedures and trajectory towards this aspect of pharmaceutical regulation.

Micro-sociological analysis of these neo-liberal deregulatory reforms is important not only because they concern drugs intended to treat serious and life-threatening conditions, but also because they have affected a significant proportion of pharmaceutical regulation as a whole. Between 1992 and 2008, the FDA approved 64 new drugs under accelerated approval regulations based on non-established surrogate measures of clinical efficacy – about 15 per cent of new molecular entities (NMEs) approved on to the US market (US GAO 2009, p. 15). Similarly, in the EU, between 1995 and 2005, 18 per cent of marketing approvals were accelerated under 'exceptional circumstances' (Boone 2011). These figures suggest that while the criteria for accelerating approval of new drugs onto the market prior to 2006 were *formally* more restrictive in the EU compared to the US, *in practice* the difference between the two regions may not be as significant as the regulatory rules imply. Indeed, there is evidence that the EU's 'exceptional circumstances' rule was not strictly applied, at times operating like 'conditional marketing' (Garattini and Bertele 2001).

Gefitinib and lung cancer: the innovation and the need

To examine the health implications of the 'accelerated approval' neo-liberal deregulatory reform regarding new drugs intended to treat serious, debilitating, or life-threatening conditions, we consider a lung cancer drug as a case study. There are three principal reasons for that selection. First, cancer is a disease of major public health significance, the burden of which doubled globally between 1975 and 2000 (Dennis 2008, p. 1). In particular, lung cancer is the most common form of cancer in the US and Europe. Second, it is important to investigate a scenario in which it is absolutely clear that the condition is life-threatening to avoid the analysis being confounded by disputes about whether the case-study merely represents a mis-application of the deregulatory reform. Indeed, gefitinib, the drug we have chosen to study in this chapter, was not only intended to treat a life-threatening condition, it was developed partly as a third-line treatment for patients with non-small cell (NSC) lung cancer who had exhausted all approved pharmaceutical treatment options. Effective third-line treatment was their last hope from medico-pharmaceutical intervention against a highly lethal condition. Thus, if 'accelerated approval' regulations for cancer drugs has operated in the interests of patients and public health, then one would expect to find that manifested *most* in the context of gefitinib where patient need for treatment could scarcely be more urgent. And third, a large and increasing proportion of accelerated approvals involving non-established surrogate measures of clinical efficacy are cancer drugs. Forty-two per cent of accelerated approvals granted by the FDA based on surrogate markers from 1992 to 2000 were cancer drugs, rising to 59 per cent for 2001–2008 (US GAO 2009, pp. 15–17). The evidence also suggests that regulators on both sides of the Atlantic are increasingly likely to approve cancer drugs under accelerated approval mechanisms. In the US, The FDA approved 20 percent of all new indications for oncology drug products (14 out of 71 new indications) under the accelerated approval regulations between 1992 and 2002 (Johnson *et al.* 2003). Between 2005 and 2007, this proportion had risen to 26 percent (14 out of 53 new indications) (Sridhara *et al.* 2010). Meanwhile, in the EU, 27 per cent (8 out of 30) of all new oncology drug products were 'non-regular' approvals between 1995 and 2005, rising to 36 per cent between 2006 and 2010 (Boone 2011, FDA 2011b, p. 252). Hence, by investigating cancer drugs, we attain insights about a very large proportion of the pharmaceuticals accelerated on to the market by this neo-liberal reform.

Lung cancer is a worldwide epidemic primarily caused by tobacco smoking. About one million new cases of lung cancer are diagnosed each year globally, resulting in more than 900,000 deaths. Of those, approximately 175,000 new cases and 160,000 deaths occur annually in the US – with similar, but higher, figures for the EU (Argiris and Schiller 2004,

p. 499; Ranson 2002, p. 16). Accounting for 28 per cent of all American cancer deaths, lung cancer is the most common cause of cancer death in the US, more than breast, prostate, and colorectal cancers combined (FDA 2002a, p. 35). At least 80 per cent of all lung tumours are NSC lung cancer. In 2000, there were 991,089 cases of NSC lung cancer worldwide with a mortality of 882,495 (Onn *et al.* 2004). Two-thirds of patients who present with NSC lung cancer are beyond curative surgery so effective drug therapy is vital (Delbaldo *et al.* 2004). Unfortunately, the prognosis for lung cancer patients is often terminal. The five-year survival rates for lung cancer patients are only about 15 per cent. However, systematic chemotherapy is estimated to reduce the risk of death for patients with advanced NSC lung cancer by 26–32 per cent, though survival benefits remain small (Herbst and Bunn 2003, p. 5813).

In 1948, nitrogen mustard was introduced as an early lung cancer therapy for bronchogenic carcinoma (Karnofsky *et al.* 1948). Between then and 1990, only a handful of drugs were found to be clinically effective against NSC lung cancer, most notably those collectively known as the 'platinum-based agents', such as cisplatin and carboplatin. By the mid-1990s, it was well established that the platinum-based drugs, which had become the mainstay of lung cancer chemotherapy, provided a small, but statistically and clinically significant survival advantage compared with best supportive care and no therapy. Specifically, a meta-analysis demonstrated that cisplatin-based chemotherapy led to an increase of one-year survival for 10 per cent of patients and an overall increased (median) survival of two months (NSC Lung Cancer Collaborative Group 1995).

During the 1990s, five newer drugs became available, namely the taxanes (paclitaxel and docetaxel), vinorelbine, gemcitabine, and irinotecan (Argiris and Schiller 2004, p. 500). When those were used with the platinum-based drugs in two-drug combinations, the resulting 'cocktail' was found to be more effective than best supportive care or a single drug, while less toxic than combinations of three or more drugs (Herbst and Bunn 2003; Onn *et al.* 2004). Also, in a second-line setting, that is, when a patient is unresponsive to, or intolerant of, platinum-based drugs, docetaxel has been shown to provide significant survival benefit over supportive care and no therapy (Cohen 2002, p. 21). Nevertheless, a major impediment to major treatment advances involving drug combinations was the emergence of intolerable toxicities, which may themselves be life-threatening and/or significantly detract from the patient's quality of life (Argiris and Schiller 2004, p. 500; Onn *et al.* 2004). Furthermore, significant therapeutic progress remained elusive. A review in the early 2000s of 22 years of RCTs on advanced NSC lung cancer in the US showed that the improvement in (median) survival between drug treatment and controls rarely exceeded two months (Breathnach *et al.* 2001). Subsequent research reported that, during the 14-year period between 1983 and 1997, the average life expectancy for elderly lung cancer patients over 65 years of age rose by less than one month (Dennis 2007, p. 1)

Evidently, in the early 2000s (and beyond) enormous challenges remained in the treatment of lung cancer. Some scientists believed that progress could best be made by developing drugs aimed at inhibiting specific pathways and key molecules pivotal in cancer cell proliferation and tumour growth, while sparing normal cells, thereby improving drug efficacy and reducing toxicity (Argiris and Schiller 2004, p. 500; Herbst and Bunn 2003, p. 5814). One such pathway was thought to involve epidermal growth factor, a chemical 'signal' that instructs cells bearing epidermal growth factor receptors (EGFRs) to proliferate. Epidermal growth factor was discovered about 50 years ago by Dr Stan Cohen, who also identified its receptor (EGFR) 20 years later – discoveries for which he was awarded the Nobel Prize in 1986 (Cohen 1962; Cohen *et al.* 1980; Mendelson 2000). Since then research has indicated that the 'activation' of the EGFR pathway generates cell proliferation and contributes to the malignant progression of solid tumours (Ranson 2002; Stephenson 2000). This is because laboratory tests have found EGFRs to be 'expressed' in human tumours, including tumours of NSC lung cancer (Onn *et al.* 2004). Hence, targeting drug development at EGFR with a view to blocking or inhibiting its activation came to be regarded among scientists and clinicians as a promising molecular approach in NSC lung cancer treatment (Ciardiello *et al.* 2004; Kris *et al.* 2003).

In 1990, research scientists at AstraZeneca began to investigate molecular targets in common solid tumours. They focused on the EGFR pathway and drugs that could inhibit it. In 1994, they discovered a compound, which they called ZD1839 – later to become known as gefitinib. In laboratory studies of human cell cultures and mice, this drug reportedly selectively inhibited the EGFR pathway responsible for driving proliferation, invasion and survival of cancer cells (Sirotnak 2003). According to the clinical vice-president of oncology at AstraZeneca, Dr Blackledge, the drug showed 'excellent tolerability' in pre-clinical safety studies (FDA 2002a, p. 58). Consequently, the firm proceeded with clinical development of the drug to treat advanced NSC lung cancer. Phase I trials in healthy subjects began in 1998 (FDA 2005c, p. 19). At that time, first-line therapy was platinum-based drugs and second-line treatment was docetaxel, but there was no (third-line) drug treatment for patients refractory (unresponsive or intolerant) to both platinum-based (first-line) and docetaxel (second-line) therapy. That was the unmet need ostensibly to be addressed by AstraZeneca's gefitinib.

Clinical development: promissory science and the construction of expectations

In 2001, AstraZeneca began a phase II trial with gefitinib. By this time the drug was also known by its potential brand name, 'Iressa'. The phase II trial was entitled, 'Iressa Dose Evaluation in Advanced Lung cancer' (IDEAL 1) and took place in Europe and Japan. It investigated the potential of Iressa to treat patients with advanced NSC lung cancer, who had failed one or two chemotherapy agents, at least one of which was platinum-based. That is, Iressa was being tested as a second or third line monotherapy. An abstract from a meeting of the American Association for Cancer Research (AACR), the US National Cancer Institute (NCI) and the European Organization for Research and Treatment of Cancer (EORTC) in October 2001 outlined interim results of IDEAL 1. The researchers claimed that Iressa provided clinically meaningful symptom relief as a second and third-line treatment for NSC lung cancer patients previously treated with platinum-based therapy. Financial analysts at Morgan Stanley believed that the drug had a large market. They forecast it had potential blockbuster annual sales of US$1.6 billion by 2007(Anon. 2001d).

In May 2002, at a meeting of the American Society for Clinical Oncology (ASCO) in Orlando, researchers reported results from a second phase II IDEAL trial, which took place in the US, known as IDEAL 2. Patients enrolled in IDEAL 2 had more advanced NSC lung cancer than in IDEAL 1 because they had received at least two prior treatments – one docetaxel and the other a platinum-based drug. Hence, Iressa was tested as a third-line therapy. Measurement of the drug's efficacy relied on the non-established surrogate marker, tumour shrinkage/response. Tumour response (with 50 per cent shrinkage or greater) to Iressa was reported as 11.8 per cent of patients in IDEAL 2 compared with 18.4 per cent in IDEAL 1. At a press briefing sponsored by AstraZeneca during the ASCO meeting, Dr Kris, one of the researchers, characterized Iressa's performance as 'nothing short of amazing' and remarked that 'people were shocked that any lung cancer could be shrunk' (Anon. 2002g). Other researchers elsewhere described Iressa's clinical benefit for NSC lung cancer patients in the IDEAL trials as 'unprecedented' (Cortes-Funes and Soto Parra 2003).

At about that time, AstraZeneca also ran what is known as an 'expanded access programme'. The programme involved over 12,000 patients with NSC lung cancer. In the neo-liberal era, the FDA encouraged, and continues, to permit, patients who have exhausted all treatment options and are ineligible for clinical trials because of poor fitness or illnesses, to receive new drugs that show therapeutic promise in Phase II or III trials (Stahel 2003). Before that those patients could only receive experimental drugs on a case-by-case basis. The decision to offer cancer drugs to patients through expanded access programmes is entirely up to the pharmaceutical company, though the firm is required to submit its protocol to the FDA (Baldwin 2002). Such programmes are un-blinded, non-controlled, and not designed to evaluate efficacy (FDA 2002a, p. 19; Williams 2002, p. 5). Although they have financial costs for companies, industry executives acknowledge that firms 'benefit from a lot of goodwill from patients' by offering them (Baldwin 2002, p. 1669).

Just five months after filing a marketing application with the Japanese drug regulatory authorities, AstraZeneca was granted marketing approval for Iressa in Japan in July 2002 as a monotherapy to treat inoperable and recurrent NSC lung cancer. The Japanese regulatory authorities were the first in the world to approve the drug. There are about 50,000 NSC lung cancer patients in Japan. The company predicted sales of 'several tens of billions of yen', while financial analysts forecast sales of about $US500 million by 2006 (Anon. 2002h).

Regulatory review of Iressa in the US

Meanwhile AstraZeneca had been making a 'rolling' new drug application to the FDA, the last section of which was submitted on 5 August 2002 (Cohen 2002). Under FDA regulations, companies seeking *regular* approval for drugs to treat NSC lung cancer were required to demonstrate that a drug provided a meaningful therapeutic benefit for patients such as increased survival or symptom improvement. However, FDA also stipulated that tumour response/shrinkage could qualify as a non-established surrogate measure for accelerated approval under Subpart H 'in refractory cancer settings, where therapies with meaningful benefit are unavailable' (i.e. 'an unmet medical need'), and where evidence to confirm true clinical benefit could be obtained through post-approval phase IV studies (FDA 1998b, p. 3).

On this latter stipulation, AstraZeneca sought accelerated approval of Iressa for the 'unmet medical need' of third-line treatment of NSC lung cancer – that is, in patients 'refractory' to (unresponsive to or intolerant of) both platinum-based (first-line) and docetaxel (second-line) therapy. The company's case for the clinical efficacy of the drug to deliver such treatment was based on IDEAL 2, that is, one small uncontrolled, un-blinded, phase II trial, involving 216 patients (Cohen 2002, p. 10). When the FDA came to review the trial, they determined that only 139 of the 216 patients involved actually met the eligibility criteria of the study, so it was really a trial of 139 patients given Iressa, rather than 216. This was because only 139 of the patients were refractory to both platinum-based drugs and docetaxel, and thus had an unmet medical need (FDA 2002a, p. 116). Although researchers collaborating with AstraZeneca had declared Iressa's efficacy to be 'amazing' and 'unprecedented', FDA scientists found that, of the two primary measures of clinical efficacy in the trial, one, namely, symptom improvement, could not be accurately evaluated because of the way the trial had been designed and conducted. The other efficacy measure was tumour response (shrinkage), which could be evaluated, but it was only meaningfully achieved in ten per cent of patients on the trial (FDA 2002a, p. 117).

We may note from Chapter 2 that the Lasagna Committee, which was established in 1989 as part of the Bush Administration's deregulatory drive to expedite drug approval, set a standard that accelerated approval of cancer drugs should be based on tumour regression (shrinkage) in at least 20–30 per cent of patients (Anon. 1989f;1989g; 1989h; 1990c). Iressa's performance (of ten per cent) fell well below even that standard – a standard that was endorsed by Vice-President Quayle's Council on Competitiveness from the early 1990s against the advice of the FDA at that time that it was *too low* a standard. Not surprisingly, therefore, FDA scientists reviewing the Iressa data questioned whether a ten per cent response rate was 'likely to predict clinical benefit' (FDA 2002a, p. 110). That a ten per cent tumour response rate was a very low standard was confirmed by an FDA specialist in oncology, who told us:

I think if you ask anybody on the unit here [FDA Oncology Division] people would say that [tumour] response rate is not a reliable surrogate for clinical benefit in non-small-cell lung cancer. ... When I started out in oncology, which was the mid-sixties, if a drug didn't produce a 20 per cent [tumour] response rate, nobody paid any attention to it. So we're going low and, you know, with a 90 per cent confidence interval for a ten per cent response rate, you get to a lower bound of five per cent [tumour response rate].[1]

A former FDA pharmacologist, who had worked on Iressa further commented:

> The actual cells responsible for tumour growth are very small in number within a tumour and it's the other cells that are most susceptible to drugs so that you're getting rid of the cells that are not involved in tumour growth. So tumour shrinkage is most likely not a valid endpoint in most tumours. The thing with Iressa was pretty sad. I mean, even using tumour shrinkage, only ten per cent responded and without overall mortality data that [information alone] could be completely useless [for predicting clinical benefit].[2]

Moreover, even if one were willing to accept ten per cent response rate *per se* as a standard, in this case, FDA scientists were concerned that the ten per cent finding overestimated the actual number of patients who might respond to the drug if marketed because the patients in the trial were atypical of the general lung cancer patient population. Seventy per cent of the patients in the trial had 'less aggressive, slow-growing tumours' than the American population with NSC lung cancer (Cohen 2002, pp. 13–15). As Dr Martin Cohen, the FDA Medical Reviewing Officer for the drug elaborated:

> The predominant histology was adenocarcinoma [which] generally have the slowest doubling time of all lung cancers, and this is reflected in the long interval between date of initial diagnosis of lung cancer and the date of trial randomization. The median time was 20 months. ... Since the median survival of newly diagnosed lung cancer patients with metastatic disease is in the range of six to nine months, this, again, suggests that the study population was enriched with slow growing, less aggressive tumours. (FDA 2002a, p. 114)

Bearing this in mind, FDA statistician, Dr Rajeshwari Sridhara, estimated that the true tumour response rate might be as low as 5.6 per cent (FDA 2002a, p. 128).

Nonetheless, confident that their drug was worthy of accelerated approval, AstraZeneca began what were intended to be confirmatory trials designed to demonstrate direct clinical efficacy of Iressa on survival, so that the drug's 'accelerated approval' status (if granted) could be converted into regular approval. Those confirmatory studies were entitled the 'Iressa Non-small-cell lung cancer Trial Assessing Combination Treatment' (INTACT). However, the results of the INTACT trials became available in early September 2002, before the FDA had granted any marketing approval. Consequently, the INTACT trials, initially intended as post-marketing confirmatory studies, became pre-market phase III studies by default. As with IDEAL, there were two INTACT trials, but otherwise IDEAL and INTACT were very different. Measurement of direct clinical efficacy (survival), rather than the surrogate marker (tumour response), was not the only difference that distinguished the two INTACT trials from the IDEAL trials. The INTACT trials were large phase III controlled clinical trials, each involving more than a thousand patients. INTACT 1 was conducted mostly outside the US, while INTACT 2 was mainly an American trial. Most importantly, the INTACT trials were designed to test whether Iressa could be used as a first-line therapy in combination with either platinum-based drugs or other established medications. Hence, the INTACT trials were first-line combination placebo trials in which patients were given either placebo or Iressa along with standard therapies – cisplatin or gemcitabine in INTACT 1 and carboplatin or paclitaxel in INTACT 2 (Anon. 2002i). Drug firms often switched from third- to first-line trials when designing confirmatory studies because, if successful, the ensuing approval would grant the company access to the much larger market of first-line patients.

Later that summer, FDA scientists confirmed that INTACT comprised two well-conducted, double-blind, placebo-controlled, randomized trials evaluating the effect of standard chemotherapy combined with Iressa on overall survival in more than 2,000 first-line patients with NSC lung cancer (FDA 2002a, p. 129). If Iressa had significantly improved survival in these trials, then its application status could have been upgraded from accelerated approval to regular approval because survival is a key *clinical* endpoint (Williams 2002, p. 4). However, INTACT failed to demonstrate that Iressa prolonged survival (FDA 2002a, p. 132). Indeed, patients survived a month *longer* on placebo, though the superiority of placebo over Iressa was not statistically significant (FDA 2002a, p. 129; Williams 2002, p. 4). The chief executive of AstraZeneca acknowledged that first-line use of Iressa in this way was not appropriate for patients with NSC lung cancer (Anon. 2002i)

Moreover, there was 'no difference between Iressa and placebo with respect to tumour shrinkage', even though this ought to have been easier to demonstrate in first-line patients because their cancers should be more responsive to therapy than third-line patients (FDA 2002a, p. 132). Thus, rather than confirming the efficacy of Iressa, the INTACT trials cast further doubt over it. One FDA scientist, who specialized in cancer drugs, characterized the scientific evidence from these phase II IDEAL studies and phase III INTACT trials as: 'It's 2000 patients saying Iressa doesn't work versus 139 saying it works marginally'.[1] While the chief executive of AstraZeneca accepted Iressa's failure in INTACT, he noted that the future market for Iressa did not depend on its approval as a first-line therapy, implying that the company would proceed with its application for Iressa's accelerated approval as a third-line monotherapy based on the IDEAL trials (Anon. 2002i). The firm did just that.

Expectations, expert advisers and the patient–industry complex

For its part, the FDA deferred making an approval decision on Iressa as a third-line monotherapy and sought the advice of its expert Oncology Drugs Advisory Committee (ODAC), which met on 24 September 2002. That morning, the *Wall Street Journal* carried an editorial arguing that inadequate trial data 'wasn't a good reason, and certainly not an ethical one, for delaying approval'. According to the editorial, 'particularly in cases of terminal disease, any safe drug with even a hint of effectiveness should be brought to market as quickly as possible' (Wall Street Journal 2002). The *Wall Street Journal* was clearly campaigning for FDA approval of Iressa, which became even more evident from some of its later publications (Wall Street Journal 2003). Even if one were to accept the *Journal's* unscientific 'standard' of 'a hint of effectiveness', Iressa did not meet that 'standard' because tumour shrinkage reported in the IDEAL trials was not confirmed in the much larger and rigorously designed INTACT trials. The INTACT trials had shown not only that Iressa provided no first-line survival benefit, but also no sign of efficacy based on the non-established surrogate measure of tumour shrinkage. As Dr Robert Temple, director of the FDA's Office of Drug Evaluation, put it:

> Usually in refractory disease, we have a history of accepting tumour response data as reasonably likely to correspond. The problem here is that in the first-line therapy where that question was tested – nothing; no hint; no nothing. (FDA 2002a, p. 194)

Before discussing and voting on whether Iressa should be approved as a third-line therapy, ODAC heard analyses of the drug's trials from AstraZeneca and the FDA. Prior to that, however, the advisory committee heard testimonies from cancer patient organizations and eight individual patients out of the 12,000 who had been on AstraZeneca's access programme for Iressa. Specifically, the first part of the meeting was devoted to accounts by members and/or representatives of the Kidney Cancer Association, the National Organization for Rare Diseases (NORD), the Cancer Research Foundation of America, the National Patient Advocate Foundation, and The Wellness Community, plus eight individual patients (FDA 2002a, p. 3). The representative of the Kidney Cancer Association made no comments particularly about Iressa, but the contribution from NORD was more substantial.

NORD ran the access programme with AstraZeneca. The representative of NORD spoke supportively about both the company and Iressa, though noted, albeit implicitly, the distinction between scientific validity and experience:

> I can't make any judgement about the scientific viability of Iressa, but we have heard stories about some people doing very well and some people who didn't do so well. But to people who were on their death bed and are now alive, a lot of them with no tumour at all, it has been an extraordinary experience for us as well as for them. (FDA 2002a, p. 20)

The president of the Cancer Research Foundation of America also welcomed Iressa and praised greater partnership between the pharmaceutical industry and the FDA, while disclosing that the Foundation received 'unrestricted educational grants' from AstraZeneca (FDA 2002a, p. 23). The director of patient education and outreach at the Wellness Community also declared that her organization received 'unrestricted educational grants' from the company. She urged the expert advisory committee to 'seriously consider the need that patients with lung cancer have for a new and broader range of treatment options' (FDA 2002a, p. 55).

The other accounts, including that by the representative of the National Patient Advocate Foundation, were of personal experiences of being on Iressa. One patient told how within a month of taking the drug, her tumour shrunk by half, and had disappeared after two months. That patient commented: 'If Iressa works for other people like it did for us, it is the best thing that has ever happened' (FDA 2002a, pp. 17–18). Another patient from AstraZeneca's expanded access programme recounted:

> In my case, Iressa began eliminating cancer symptoms in precisely seven days. In just five weeks, my scans showed a significant decrease in my tumours. Tumour shrinkage is now up to 90 per cent and I am

completely symptom-free with a normal breathing capacity. ... I speak for many by asking that you approve Iressa as simply another choice for the cancer patient. Give others the gift that we have received, which is the gift of comfort and time. (FDA 2002a, p. 28)

Similarly, yet more patients among the eight related how Iressa caused their tumours to be '90 per cent gone in three months', was 'a wonderful drug', 'light-years better than previous treatment', and 'will save lives' (FDA 2002a, pp. 22–52). Perhaps the most forceful testimony came from a patient who attributed her access to Iressa as life-saving:

> The very fact that I am standing here today is nothing short of a miracle. In February of this year, I was at the end of the line. I had received the best standard care available, surgery, radiation, numerous chemotherapy regimens. ... On February 16th, I began taking Iressa. Within days, and I mean days, I felt significantly better. ... With Iressa, I'm not merely living, I'm thriving. ... Iressa is a blessing. When this drug works, it works fast and it works well. My physician apologized to me. He said that if he had had any idea Iressa would work so well, I could have been spared much suffering. But that decision was not in his hands. He could not have prescribed it to me and I had to fail several chemotherapy regimens before I could receive it through the expanded access programme. (FDA 2002a, pp. 40–1)

According to Goldberg (2005, p. 15), some of the patients, who testified, were funded to attend the meeting with AstraZeneca money distributed through a patient group, though that does not invalidate their accounts or diminish their sincerity.

At the ODAC meeting, AstraZeneca repeated the claim that Iressa had demonstrated 'unprecedented clinical effects' in NSC lung cancer patients refractory to first-line and second-line treatments (FDA 2002a, pp. 57 and 65).[3] Afterwards, FDA scientists reiterated their analysis that the IDEAL 2 trial provided no convincing evidence of symptom improvement, while the two INTACT trials failed to confirm any survival benefit. Hence, as FDA scientists explained to the Committee, there was no case demonstrating direct clinical efficacy of Iressa from scientific trials. The only scientific evidence supporting Iressa's efficacy was in terms of the surrogate measure, tumour shrinkage, but as revealed by the FDA reviewers, even that could have been as low as 5.6 per cent in the IDEAL trial, and was not confirmed at all in the INTACT trials (FDA 2002a, pp. 103–34).

In addition, the FDA's Dr Grant Williams reminded the Committee of the ironical situation that the two pre-approval INTACT trials, which initially had been intended to be post-approval confirmatory trials, had not confirmed Iressa's clinical efficacy, as hypothesized from tumour response (FDA 2002a, pp. 110–11). Or as Professor Tom Fleming, from the University of Washington and consultant to ODAC, impressed upon the Committee:

> It appears that the strategy that was in place several months ago, prior to the release of the INTACT trials, was that the INTACT trials were going to give us the truth. (FDA 2002a, p. 206)

Had the INTACT trials actually been post-approval confirmatory trials, then their failure to demonstrate direct clinical efficacy would have implied that Iressa should be withdrawn from the market. Results from the INTACT trials coming to the FDA before approval seemingly provided regulators with 'advance information' that could prevent approval followed by withdrawal, by not approving in the first place. Of course, as the senior FDA scientist, Robert Temple made clear, there was never any question of withdrawing Iressa from the small number of patients who were already benefiting from it (FDA 2005c, p. 71). The issue was whether the drug should be approved for general market release as a third-line therapy.

Despite the negative INTACT trial results, and the FDA's analysis of the IDEAL trial, the advisory committee voted 11–3 in favour of accelerated approval for Iressa. The ODAC is only advisory and final approval decisions rest with the FDA. Nonetheless, advisory committees can have a major influence on FDA decisions. For that reason, it is important to understand ODAC's recommendation regarding Iressa, which was not consistent with the prevailing techno-scientific or regulatory standards – an inconsistency emphasized by one FDA scientist, who told us 'the negative data in the INTACT trials should have caused Iressa not to be approved'.[1]

Carpenter (2010a, p. 743) has hypothesized that the editorials of the *Wall Street Journal* were a major factor in ODAC's considerations, and that major national newspapers, such as the *Washington Post* and the *Wall Street Journal* served to accelerate new drug approval in the US through pressure on the FDA more generally by arguing that the agency's regulatory demands prevented patients access to important new therapies (Carpenter 2010a, pp. 498–9). Our research into the Iressa case does not support that hypothesis. Interviews with FDA staff, experts, and patient groups suggested that the influence of the *Wall Street Journal* was, at most, minimal, not least because FDA scientists dismissed the *Wall Street Journal* commentaries as ignorant – and with considerable justification.[4] Indeed, some senior FDA scientists directly denied such influence. One commented:

> They [*Wall Street Journal*] are so poorly informed and wrong in what they say. I mean in oncology there's a million ways for people to get an experimental drug. Sometimes the companies don't want to. But it really isn't us, we're perfectly happy to do that. To call what we ask for over-demanding is, you know – I can't imagine what they're smoking when they say that.[5]

Moreover, during interviews with us, FDA scientists revealed that in refractory situations (second-line or third-line therapy), the agency had, under the 1992 accelerated approval regulations, previously accepted tumour response rates lower than 20 per cent (though never before together with trials, like INTACT, also showing no clinical benefit or tumour response).[1,5] As one FDA oncologist explained:

> We [FDA] have a history of accepting [tumour] response rate as a surrogate reasonably likely to predict clinical benefit, so they [AstraZeneca] just had to show that they had a meaningful response rate. We're going low [below 20 per cent] because all the drugs we're seeing are in that ballpark. Unfortunately it's what's available. ... The problem is that the drugs don't work that well. What people don't understand is that the drugs really aren't that great. You're not depriving them of all that much. And if the drug is great, we'll get it out. ... If we had good choices, then a lot of these marginal drugs wouldn't get out on the market. The problem is that, for whatever reason, there is nothing that's coming along that looks outstanding. And so all we deal with is marginal drugs.[1]

While we found scant evidence that the press had influenced the ODAC, our interviews with representatives of patient organizations, public health advocacy groups, and FDA scientists who attended the meeting, strongly indicated that some advisory committee members were swayed by the testimonies of patients on the access programme. One informant, for instance, recounted how two committee members acknowledged that their votes in favour of accelerated approval of Iressa were influenced by patient testimonies despite recognizing that such testimonies did not constitute sound, scientific evidence of drug efficacy.[6] An FDA oncology specialist, who was at the meeting was left with the same impression, asserting:

> I think they [patient testimonials] definitely have an influence over advisory committees. That's what Iressa proves.[1]

Furthermore, most of the committee members were from the oncology profession, who tend to be keen to have as many drugs available to them as possible to offer to their patients.[1,2,7]

FDA's accelerated approval of Iressa: disease-politics or corporate bias in action?

Some months after the ODAC meeting, the FDA granted Iressa accelerated marketing approval as a third-line monotherapy by using the 1992 Accelerated Approval regulations. It is clear that patients' testimonials at the ODAC meeting prevailed upon the advisory committee to recommend approval of Iressa. Furthermore, FDA scientists believed that the advisory committee's recommendation was a key factor in the agency's decision to grant accelerated approval to the drug. Indeed, one FDA specialist in oncology told us that the internal recommendation by the agency's oncology division to recommend approval of Iressa was 'based on the advisory committee'.[1] He elaborated further:

> Iressa was a tough drug. With Iressa there was no evidence of clinical benefit. We [FDA] don't have to accept the advisory committee recommendations, but we had no good reason not to accept them with Iressa. You know sometimes the advisory committee goes off where you can scientifically argue that the decision was ridiculous, and that you should reverse it. But you know, they heard the same ten per cent response rate that everybody else heard. The INTACT data did come up in the advisory committee, and they heard the negative INTACT data, and they decided that the third-line setting might be different than first line.[1]

Thus, on the face of it, it seems that patients influenced ODAC's recommendation to approve, which in turn pressured the FDA decision to approve. This finding appears to support the thesis of a new disease-politics of the kind asserted by Carpenter (2004; 2010a) in which the FDA had become primarily concerned to maintain its reputation for being responsive to patients and the medical profession, rather than industry. Or even, perhaps, a disease-politics of the kind suggested by Daemmrich, Krucken, Edgar and Rothman, in which the agency had become primarily responsive to patients' interests, with the interests of pharmaceutical firms merely a secondary consideration? (Daemmrich 2004; Daemmrich and Krucken 2000; Edgar and Rothman 1990).

Yet those formulations, while containing an element of truth, are highly questionable and superficial even in the Iressa case where patient advocacy clearly played some role. In fact, the interests of the pharmaceutical industry and its connections with neo-liberalism were never far away. The possibility of AstraZeneca gaining marketing approval for Iressa on the basis of a small percentage of tumour response rate arose in the first place because the industry's interests combined with the political power of Quayle's Council and the Lasagna Committee to force through the use of that non-established surrogate measure against the FDA's advice (see Chapter 2). Secondly, AstraZeneca largely determined *which patient voices* were heard by the advisory committee – ODAC heard from eight patients out of 12,000. Thus, 'pressure' felt by the committee from patients to approve Iressa should not be regarded as entirely separate from 'pressure' exercised by the manufacturer. While we do not doubt the sincerity of the eight individual testimonies given by those patients, the committee's overall encounter with patient experience of the drug was shaped by the company. And thirdly, the manufacturer also influenced oncologists directly, such as those on the committee.[2] As one FDA oncologist revealed:

> I'm not saying it was a lot of pressure that caused Iressa to get approved. But there *was* a lot of pressure. AstraZeneca did a fantastic job of marketing Iressa to the oncology world, and everybody was thinking it was a *wonder* drug! When it clearly isn't [speaker's emphasis].[1]

There is no doubt that lung cancer patients are desperate for treatment and that doctors want to be able to help them with more effective therapies, but it is the pharmaceutical firms that promote specific claims about the benefits of particular drugs to those expectant audiences. So here we see the creation of expectations fused with the business of marketing and commercial interests. One consumer advocate we interviewed, who was entirely independent of the pharmaceutical industry, saw this as part of a wider industry influence on American culture:

There's a culture in the country, and a lot of it is fanned by the media, you know: cancer patient on his death bed and then he gets this experimental drug and he rises again. That's what's out there in the mythology. What people don't get is that sometimes it's three months – that's all you get. That's where the media has been in collusion with the pharmaceutical industry for not making that stuff clear – that just because a tumour shrinks that doesn't mean you're going to live longer.[6]

Crucially, FDA scientists emphasized *manufacturers'* determination of knowledge about, and availability of, cancer drugs – something entirely neglected in disease-politics theory and dominant media narratives about pharmaceutical development and regulation.[1,5] The FDA's Oncology Division has asked, but cannot force, firms to conduct trials employing survival, rather than surrogate, endpoints without blocking the drug altogether – a course of action that would run counter to the spirit of the neo-liberal reforms of accelerated approval mandated by Congress. According to an FDA oncologist, the agency pressed AstraZeneca to conduct trials comparing the survival of patients taking Iressa with the second-line therapy, docetaxel, but the company refused for two years due to fear of results reflecting negatively on Iressa.[1] This suggests that industry priorities built into accelerated approval regulations frames the FDA's regulatory options significantly, long before patients' testimonials at FDA advisory committees.

Moreover, although some FDA scientists stated that it was difficult to reverse the ODAC's recommendation because the advisory committee had heard and discussed all the relevant evidence pertaining to approval, that explanation was not strictly accurate. Additional significant evidence, albeit pertaining to Iressa's safety, became available to the FDA after the ODAC's recommendation on 24 September 2002, but before the drug's approval in late Spring 2003. In mid-October, AstraZeneca reported that Iressa had been associated with 39 deaths in Japan where it had been on the market since July 2002. In particular, there were 69 cases of acute-interstitial pneumonia of which 27 were fatal (Anon. 2002j).

The FDA first became aware of the reports of Iressa-induced interstitial pneumonia when they received a submission from AstraZeneca on 17 October 2002 outlining 41 such cases post-marketing in Japan out of an estimated 10,000 users of the drug. The number of such reports coming into the agency grew steadily by the month. By 13 January 2003, the number of cases in Japan had risen more than ten-fold to

446 out of an estimated 21,990 users, so in three months the incidence had grown from 0.41 to 2.03 per cent (Sullivan 2003). In 30–40 per cent of these cases, the outcome was fatal within one or two weeks (FDA 2003b, pp. 107–20). AstraZeneca argued that those post-marketing reports of interstitial pneumonia were 'anecdotal' – meaning that they did not have the validity and reliability of a controlled clinical trial setting, but then neither did the patients' testimonials of Iressa's efficacy at the advisory committee meeting in September 2002. Notably, the company was willing to promote claims derived from the uncontrolled setting of an expanded access programme when those claims supported Iressa's approval, while casting doubt on adverse safety reports that could damage the drug's chances of approval on the grounds that those reports came from an equally uncontrolled context.

On 1 May 2003, the Public Citizen Health Research Group, a US public health advocacy organization, wrote an open letter to the FDA, which received press coverage.[7] The letter urged the FDA not to grant Iressa accelerated approval because the drug was associated with a growing number of fatal adverse reactions and there was not at that time evidence of the drug's clinical efficacy (Public Citizen 2003). At that stage, the FDA had a clear option to reject its advisory committee's recommendations of September 2002 to approve Iressa on the grounds that new and disturbing information about the toxicity of the drug had come to light and tipped the risk–benefit balance against granting accelerated approval. Specifically, insofar as one accepted that a very small proportion of new third-line patients *might* benefit from Iressa – based on patient testimonials at the advisory committee, no such patients could be identified *prospectively* either on the product label or by prescribing physicians. Consequently, the vast majority of users would be exposed to potentially life-shortening fatal reactions without deriving any therapeutic benefit from the drug. Nevertheless, on 5 May 2003, FDA approved Iressa as third-line treatment for NSC lung cancer.

This chain of events not only raises difficulties for FDA staff, who have attempted to explain their decision to approve Iressa as mere deferral to the views of the ODAC, it also points to limitations of reputational theory as a fundamental explanation for regulatory approvals. For if the FDA's regulatory decisions were driven by concerns about audience reputation for good science and service to public health, as implied by scholars like Carpenter (2010a), then one would expect the science-based analysis of Iressa's non-efficacy by agency staff combined with the demands to protect patients from fatal adverse drug reactions to have militated against approval. At the very least, one is left with an analysis which suggests that the FDA's reputational concerns not to withhold the new drug from patients and the oncology profession outweighed the agency's reputational considerations about maintaining regulatory standards consistent with its own science and protecting patients from life-shortening toxicities. Yet that begs the question: Why did the former reputational concerns outweigh the latter? That question points to the need for a deeper aetiology capable of disaggregating reputational phenomena.

Post-market 'confirmatory' studies of Iressa in the US

As a condition of Iressa's accelerated approval by the FDA, AstraZeneca agreed to conduct a post-marketing 'confirmatory' randomized control trial with *third-line* patients using survival as the clinical efficacy endpoint (FDA 2003b). As previously explained, that was standard procedure under the 1992 Accelerated Approval regulations, which permitted new drugs on the market based on non-clinical surrogate measures of efficacy. The 'confirmatory' trial was known as ISEL. It was a large placebo-controlled trial involving 1692 highly refractory, third-line NSC lung cancer patients (FDA 2005c, p. 25).

In December 2004, the company informed the FDA that the ISEL trial showed no survival benefit with Iressa compared to placebo (FDA 2005c, pp. 11–12). Although the patients in ISEL taking Iressa had a higher tumour response rate than those on placebo, that tumour shrinkage had not translated into any survival advantage (FDA 2005c, p. 29). The findings were unequivocal. As Robert Temple at the FDA put it:

> This was a very large study. You would expect it to detect an overall survival effect if there was one, and the fact that it didn't tells you something. (FDA 2005c, p. 127)

Dr Richard Pazdur, Head of the FDA's Oncology Division, echoed the view that the results of ISEL were clear cut: 'This is a placebo-controlled trial. It is about as clean as you get here' (FDA 2005c, p. 130). Yet the FDA did not withdraw Iressa from the market. As with the decision about whether to approve the drug, there was never any question of removing the drug from patients already benefiting from its use. Rather, the issue was whether any new patients (or patients taking the drug without any prior benefit) should be exposed to a toxic drug when the scientific evidence pointed to its lack of clinical efficacy. As for the regulations, ostensibly the idea behind the 1992 Accelerated Approval rules was that patients with life-threatening conditions could get faster access to new drugs based on surrogate measures of efficacy, but that 'surrogate efficacy' would then have to be confirmed as clinical efficacy in the post-marketing phase, otherwise the drug would no longer be therapeutically viable on the market.

After the results of ISEL became available, AstraZeneca suspended promotion of Iressa to doctors, but continued to make it available to patients who appeared to benefit from the drug. In addition, the company sent a letter to doctors and patient groups, and posted notices in medical journals urging physicians to consider therapeutic options other than Iressa (FDA 2005c, pp. 12–13). The company contended, however, that there might be subgroups of patients within the ISEL trial who might have benefited, and that it might be that lung cancer patients with similar characteristics to that hypothetical subgroup in ISEL could benefit from the drug. Instead of withdrawing Iressa from the market, the FDA called a meeting of its advisory committee, ODAC, to discuss the way forward with the drug, including the firm's proposals.

The ODAC did not meet until 4 March 2005, so Iressa had remained on the US market for nearly three months after a well-conducted trial had demonstrated that the drug provided no clinical benefit for the purposes it had been granted marketing approval. At the meeting, AstraZeneca argued that they needed until June 2005 to complete their post-hoc subgroup analysis of ISEL in the hope of showing that some identifiable type of lung cancer patient, perhaps of a particular genetic profile, might derive some clinical benefit from the drug (FDA 2005c, p. 23). Commercially, a narrow indication could still command some market if doctors significantly prescribed Iressa off-label, that is, to patients beyond the subgroup specified on the label – quite a likely scenario in the field of lung cancer therapy.[8] Whatever their reasons, even in the aftermath of unequivocal failure, the firm sought to create renewed expectations about the drug's potential efficacy. From a scientific perspective, that proposal received a fairly frosty response from Robert Temple at the FDA, who commented:

> This [AstraZeneca's proposal] was, shall we say, an optimistic presentation. The study [ISEL], after all, failed. You had opportunities to identify subsets before the study that would be your primary analysis, but you didn't think that they were good enough to do that. So, these [AstraZeneca's proposals] are – and it's an important distinction – these are after-the-fact subset analyses in a study that did not win.

That is different from subset analysis in a study that did win. ... You are looking at the mutation status of some of the people in the trial, about 200 patients, ... but at the moment you have no prospective data on that subgroup for survival. (FDA 2005c, pp. 71–72 and 74)

Temple's point was important because, even if one accepted the hypothesis that there might be some subgroup of NSC lung cancer patients who could benefit from Iressa, without a prospective trial that hypothesis had not been confirmed. Hence, it would have been impossible for prescribing doctors to know which lung cancer patients would benefit from the drug. All others taking Iressa (the vast majority of third-line NSC lung cancer patients) would derive no benefit and be exposed to potentially fatal toxicities.

Some expert advisers on the committee believed that patients' interests would have been better served if AstraZeneca had conducted post-marketing phase IV trials that aimed to identify and study the small group of third-line patients with Iressa-responsive genetic profiles immediately after approval, rather than waiting until ISEL failed. Frustration at the lack of such investigations was evident at the ODAC meeting, when one expert, Dr Otis Brawley from Emory University, declared:

I think the development of this drug has been mishandled. It has been mishandled by AstraZeneca and by this committee. I, myself, take some blame for that because I voted for approval of it two years ago. The fact remains that this drug has been available for seven years, and we still haven't figured out exactly how it should be used in the treatment of lung cancer. Perhaps if we had held off in getting it available to people two, three years ago, those studies would have been done. (FDA 2005c, p. 124)

As pointed out to us by some experts, the company may not have wanted to embark on such subgroup analyses before ISEL for commercial reasons because it would have reduced the number of patients to a relatively small number. The firm had hoped that ISEL would demonstrate clinical efficacy across the third-line lung cancer patient population as a whole, which was a much larger market than a subgroup specific to a particular genetic profile.[8] As one FDA scientist surmised:

The problem for the drug company is that they don't want to limit their market and they want the broadest possible indication [therapeutic application to patients]. Now, if you start rigorously defining which subgroup is the appropriate subgroup for treatment, that's good scientifically and it's good for patients, but it's not good for the drug company, so they're anxious not to do those studies. ... I'm trying to think what the phase IV studies for Iressa are and whether we're getting any of the information. I really don't think we are. ... The drug companies would fight not to do those studies.[1]

The FDA's decision not to withdraw Iressa from the market was even more remarkable because by early 2005 an alternative drug, known as Tarceva, had become, and remains, available on the US market after having demonstrated survival benefit in a randomized controlled trial in third-line NSC lung cancer patients (Anon 2004a; 2009b; 2009c; FDA 2005c, pp. 99–100). Not only was Tarceva another third-line NSC lung cancer drug, but it also worked by the same mechanism as Iressa, EGFR inhibition, and had similar tumour response rates (FDA 2005c, pp. 114–15). Thus, Tarceva's demonstrated survival benefit in trials makes it difficult to see why any physician would choose Iressa, rather than Tarceva, for a new third-line NSC lung cancer patient or for one showing no improvement on Iressa. Even AstraZeneca accepted that Tarceva was an 'alternative therapeutic option with proven survival benefits' that physicians should consider for their NSC lung cancer patients, and that for 'those patients who fit into a class where it is felt appropriate that an EGFR inhibitor should be utilized, Tarceva would in fact be the preferred agent' (FDA 2005c, pp. 39 and 80–1).

Nonetheless, the FDA's response to ISEL was to allow continued marketing of Iressa in the US provided that AstraZeneca sent doctors letters advising them against starting new patients on the drug (FDA 2005c, pp. 12–13). Thus, a quite extraordinary situation obtained in which the drug was left on the market with both the manufacturer and the FDA agreeing that physicians needed to be advised not to start patients on it, while the manufacturer searched for some types of patients for whom the drug might be clinically effective. A significant drawback of this approach from the perspective of protecting public health was that the advice was neither always taken nor effectively conveyed. According to Public Citizen, who had petitioned the FDA to remove Iressa from the market after ISEL, there were 331 new prescriptions for Iressa in the US in one week of February 2005 alone (FDA 2005c, p. 94). In June 2005, the FDA altered Iressa's labelling directing physicians to limit its distribution to patients in clinical trials or to those who were benefiting, or had benefited, from the drug (Anon. 2005m). Although that strengthened the advice to prescribing doctors, it remained a more limited regulatory approach than market withdrawal. Had Iressa been withdrawn from the market, new patients could not have been put on the drug and those benefiting from it could have received it under a compassionate-use programme. In 2007, results of a US National Cancer Institute (NCI) trial showed that Iressa led to worse survival outcomes in lung cancer patients than placebo (Anon. 2007g).

Maintaining Iressa on the market after failing to confirm clinical efficacy when an alternative drug with proven efficacy to treat the same condition was available was in AstraZeneca's commercial and institutional interests, *but inconsistent with patients' health interests and the FDA's own techno-scientific regulatory standards*. That leaving Iressa on the market under such conditions was contrary to FDA standards was highlighted by a senior FDA scientist involved in management, who had previously told us: 'We don't want drugs for life-threatening diseases that are clearly worse than other drugs'.[5] Indeed, even marketing approval of Iressa in the first place was inconsistent with those standards and hardly in the health interests of patients given that tumour response was unconvincing, trial patients survived longer on placebo, and there was about a one-per-cent chance or more of being killed by the drug's adverse effects.

This shows that the dichotomy within disease-politics theory of an over-cautious FDA acting against patients' interests by denying them timely market access to drugs *before AIDS*, on the one hand, and an FDA responsive to patients' interests by accelerating access to drugs *after AIDS*, on the other, is flawed because Iressa highlights FDA's *accelerated approval against* the health interests of patients. Given that the FDA's regulatory decisions about Iressa were neither in the best interests of public health nor consistent with the agency's own scientific and regulatory standards, two crucially important political and social scientific questions are raised. Why did the FDA not reject the testimonials from a small number of patients on company-funded access programmes, uphold its own techno-regulatory standards, and refuse Iressa approval? Second, why did the FDA allow the drug to remain on the market after ISEL failed to demonstrate any clinical benefit?

Scholars inclined to reputational theory, such as Carpenter (2004; 2010a), might argue that this occurred because the FDA, motivated by reputational considerations, did not wish to be accused of ignoring the demands of some patients and the advice of its expert committee. While that explanation may go some way to answering the first question, it is ultimately unsatisfactory. The FDA could have responded to patients' demands by implementing a compassionate-use programme making the drug available to patients already benefiting from it without permitting it on to the market. As we suggested earlier, an explanation relying on reputational factors also faces the question of why the FDA exhibited no such reputational concerns about transgressing its own technoregulatory standards in the face of public calls by health advocacy groups not to approve Iressa on to the market – transgressions that came with their own reputational costs for the agency, not least with respect to its own scientists, the international expert regulatory community, and US consumer and health groups.

Furthermore, the reputational strand of disease-politics theory struggles even more to adequately address our second question. At the 2005 ODAC meeting to discuss the ISEL results there was scarcely any actualization of patient demand for Iressa – a fraction of the palpable emotions generated at the pre-approval September 2002 meeting. In fact, in addition to a proposal from Public Citizen to remove the drug from the market, there were just three presentations from patients and/or patient groups, not all of which supported Iressa's maintenance on the market. One patient recounted that Iressa had worked for him and wanted the drug to remain available, while the President of the Lung Cancer Alliance, which had been funded by AstraZeneca, implored the company and the FDA to continue to allow doctors and patients access to the drug. It is worth reiterating here that the drug could have remained available to those already benefiting from it, as an experimental drug, without retaining the product on the market. By contrast, the presentation by a cancer survivor, who was also Chief Executive of a cancer advocacy organization, implied Iressa's continuance on the market was problematic for patients given the availability of Tarceva:

> Patients and medical consumers deserve choice, but most importantly, they need and expect full disclosure and rational explanations, and what patients especially need are adequate safeguards to protect them from erroneous choice..... How am I to respond to the man who tells me that he has read that Iressa has no survival advantage, that it is not being used in Europe, yet he will begin receiving it here? I find I have no reasonable and satisfactory answer. But what the patient is really asking is, how many patients are being harmed by not receiving the most effective and safest product for their disease. (FDA 2005c, pp. 101–3)

One could scarcely argue that the FDA permitted Iressa to remain on the market after ISEL in order to protect its reputation in the face of patient or professional pressure, both of which were markedly muted. Indeed, as we have seen, the mood of some of the oncology professionals at the 2005 ODAC meeting could be depicted as disillusionment with the drug. The explanation for the FDA's decisions about Iressa is not best captured by reputational aspects of disease-politics theory, but lies in the relationship between politics, industry interests, and regulatory culture – to which we now turn.

Neo-liberal political reforms and abdicatory regulatory culture

The key to a deep understanding of why the FDA did not uphold its own techno-regulatory standards relates to how neo-liberal deregulatory reforms led to changes in FDA management, which in turn generated a culture of resignation among many FDA scientists who reviewed new drug applications. That abdicatory culture manifested in the perspective that there was little point in making a stand to defend regulatory standards against the interests promoting drug approval. In theory, that resignation could apply to conflict with patient groups or the medical profession, but in practice its origins and persistence derived from challenges from the drug industry, as a whole, and individual pharmaceutical companies.

It is not that there was necessarily suppression of FDA scientists' views about particular drugs, though something close to that may have happened in some cases (see Chapter 5). Many FDA scientists told us that the agency's management was sensitive about allowing discussion and debate up to a point, and that reviewers' views that a drug should not be approved would be 'heard out'.[9] So it was with Iressa in the FDA's Oncology Division. The Medical Reviewing Officer, Martin Cohen, was able to provide his full critical assessment of Iressa both as part of the written public record of the drug's approval package and orally before the ODAC. What he and his colleagues at the Oncology Division did not do was fight against the ODAC's recommendation, even though some reviewers in the Division clearly believed that accelerated approval of Iressa after the INTACT studies did not meet with the established techno-regulatory standards.

The link between this culture of resignation in the regulatory approach to Iressa and the neo-liberal reforms enacted under the 1992 Accelerated

Approval rules for drugs to treat life-threatening diseases is revealed in our interviews with scientists working at the FDA's oncology division, who were involved with Iressa. For example, we recorded the following field-notes from a senior scientist in the oncology division, who preferred not to be tape-recorded:

> The main problem with the accelerated approval regulations was recognized before the regulations even came into effect – and this was that the FDA wouldn't have the courage to withdraw drugs where confirmatory studies were not being done, or where confirmatory studies failed to confirm a benefit. Despite FDA urging companies to start the confirmatory studies, it remains a constant problem that hasn't changed over time. The FDA should be requiring that confirmatory studies are already underway before granting accelerated approval, but no one has the courage to do that. There is no insistence that the studies be underway.... FDA management were very reluctant to withdraw Iressa [and never did]. The official justification for taking no action on Iressa [after ISEL] is that the FDA do not have the study reports for the latest trials! [indicating disbelief].[10]

The connection is substantiated further by our interview with another experienced oncologist at the FDA, who was intrinsic to discussions about Iressa at the agency:

> **FDA Oncologist:** Drug companies have no incentive to do phase four studies, and so they frequently delay as much as they can and it's only in the last year that we've actually tried to push companies into completing phase four studies. But by and large most phase four studies have never been completed. And there hasn't been a phase four study result that has resulted in the drug being withdrawn from the market. But theoretically it could be. A negative phase four study could theoretically result in a drug being removed from the market. It's never happened.
>
> **CD:** Would it be likely to happen?
>
> **FDA Oncologist:** It will *never* happen, I don't think. [Speaker's emphasis].
>
> **CD:** What would be more likely to happen? That they [the company] would be allowed to do different studies in a different setting?
>
> **FDA Oncologist:** Yeah right. And then they would do another study and spend another four or five years leaving *that* study, and

hopefully somebody would forget about it by then! Well, Iressa's the classic example. You know, we *already* knew the phase four study results in Iressa, and the drug is still approved ... and if *that* can happen then it would be hard to see any phase four study leading to a drug being withdrawn. [Speaker's emphasis].

CD: It would be even harder to withdraw a drug after it had been given marketing approval?

FDA Oncologist: Yeah.[1]

Thus, at the FDA's oncology division, we found a sense of resignation that the deregulatory reforms of the neo-liberal era regarding drugs for life-threatening diseases had created weak standards, and further resignation that the agency's management was unlikely to uphold even those weak standards, at least regarding cancer drugs like Iressa.

Yet the oncology division did not exist in a bubble. In fact, there is evidence from our research that the oncology division has made *more* effort than some other divisions within the FDA to uphold FDA's regulatory standards. There is also evidence of variation between divisions with respect to the extent to which FDA reviewers feel confident that their division Directors are committed to transparent and publicly defensible regulatory decision-making, and in this respect Richard Pazdur (Head of the FDA's oncology division) is held in high regard – not just by FDA reviewers but also by independent and knowledgeable cancer patient advocates and the wider oncology community. This suggests that problems of an abdicatory culture may be even more pronounced within other Divisions, and if we turn our attention to this culture of resignation more widely within FDA, then one quickly discovers how the connections between FDA management, industry interests and neo-liberal reforms strike at the heart of the regulatory review of individual drugs. A Medical Reviewer at the agency's Division of Cardio-Renal Drug Products recounted similar frustrations of dealing with post-marketing commitments resulting from reforms of the neo-liberal era:

A sponsor [pharmaceutical firm] comes in, they [the drug company] say, 'things aren't going as well as we planned', and the agency [management] has repeatedly said, 'take some more time'. And so I certainly have no faith at all in post-marketing proposals. The sponsor has not much incentive to do them, not much incentive to do timely reporting of the results. It's just not likely you're going to get much information out of that. And with subpart H there's at least the potential of threatening to withdraw a drug from the market, but as far as I can see we've never exercised it.[11]

The sense of resignation about successfully challenging the claims and conduct of pharmaceutical firms was tangible in these interviews. It extended even to the bio-statisticians, who worked across all therapeutic areas covered by reviewing divisions within the agency. One FDA bio-statistician explained how the dynamics of this abdicatory regulatory culture at the agency formed and affected new drug review:

> The burden of proof is definitely not on the company. It's on the [FDA] reviewer, if they [the reviewer] think there's a problem, to prove that. ... If you, let's say, did a poor review and missed things, you'd probably never hear much about it. But if you do an in-depth review and find problems [with the drug], then there are lots of meetings and you've got to justify everything, even before it gets to an advisory committee. So I think a lot of people took the path of: 'it's not worth the effort'. They know that companies will complain to their superiors if a drug is not approved. People who raise questions about drugs are not the ones who get promoted, and I think people were aware that I was taken off the review of certain drugs. ... Everyone understood what was going on scientifically, but it was a situation where my bosses felt they didn't want to fight the battle. That kind of thing happened all the time. ... There were certainly other cases where reviewers would have reservations about a drug that were ignored and if reviewers tried to do something about it then someone from the higher management would get involved. And I think people learned from experience. Certainly it was fairly widely known within the statistics group what was going on with me so people could easily draw the lessons from that. And there'd be discussions all the time about, 'there are problems with something, what should I do about it? It seems they want to approve it anyway'.[12]

In this account, the role of the commercial interests of pharmaceutical firms in seeking to influence FDA management to approve drugs is stark and unmistakeable. This FDA bio-statistician was convinced that the agency's senior management tended not to resist industry interventions. Upon probing why that might be, if indeed it were true, one FDA scientist from the Division of Metabolism and Endocrinology Products told us:

I think that has to do with how people are selected for management. I think there is a plan, really, to have FDA managed by people who are just vanilla, without anything more to them than that. And [people] who will go along with a general concept of how things should be done and not make any waves. It's very difficult when something is coming down the pike to stand up and say, 'no'. People at upper management though, view that as some kind of iconoclastic reaction and were much more willing to go along with the pressure that was being applied to them. The reason for all this is that they were selected on that basis. And that really is the major problem we face.[13]

Yet, insofar as senior FDA managers were 'just vanilla', that insight is only part of the story. Although FDA managers might sometimes have been 'vanilla' towards pharmaceutical firms and government pressure from Congress and the Administration, evidently a different flavour was applied to FDA reviewers below the managers in the agency's hierarchy. The interview above with the FDA bio-statistician reveals not only the mechanism with which industry interests influenced drug regulation via pressure on FDA management. Regarding agency reviewers, it also outlines what sociologists refer to as 'acculturation' – the process by which one learns the acceptable norms of a culture. Evidently, within the FDA during the neo-liberal era, agency reviewers could certainly have their say, including negative assessments of drugs, but it was important to know when to stop pushing such negative criticisms. Otherwise, management might intervene at the behest of the pharmaceutical firm. This bio-statistician declared that such interference 'happened all the time'. We do not claim that, but our research suggests that it was sufficiently common to have a significant effect on the regulatory culture of the agency.

From our investigation, it seems fairly clear that the management culture which regarded FDA scientists, who persistently raised problems about new drugs, as a nuisance was *accentuated by the neo-liberal reforms putting an emphasis on rapid drug approval*. This became particularly apparent when we asked a former FDA pharmacologist how the agency responded to such agency scientists:

Well, you certainly weren't popular. You'd work well with reviewers. But management does not want to hear problems because that makes their life difficult and then they have to face the pressure from the drug company to ignore it. *The pressure was on to approve things*, and to figure out how to deal with them in the labelling so that you *could* approve them [speaker's emphasis].[2]

The key connection to appreciate here is the confluence of industry pressure *and* the neo-liberal reforms promoting accelerated drug approvals together shaping the culture and priorities of the FDA's senior management. It explains why there was a perception, or perhaps even a reality, that FDA reviewers who found problems with drugs were not favoured by agency management. More significantly, in relation to Iressa, it accounts for the abdicatory culture towards battling to uphold the appropriate public-health protective techno-regulatory standards. The comments of one senior FDA scientist, also outside the oncology division, who had been with the agency for much of the neo-liberal period are particularly apposite:

Since PDUFA and FDAMA there's been a lot pressure to approve drugs quickly. It's driven by corporate pressure and economics and that came through Congress. The whole notion of PDUFA which was, 'we're [industry] now paying for a service, and that service is you [FDA] review our drug' – well the service is really, 'you approve our drug'. So, what happens now is that if a [FDA] medical reviewer finds a problem with a drug and it's seen that the problem is real, what they'll [senior FDA staff] frequently look for is a narrow indication upon which to base approval: 'Is there someplace, somewhere, within this application that we [FDA] can say yes'? I think 15 years ago they [FDA] would just have said, 'this is not going to make it'. So I think now the agency is very much afraid to regulate, to tell a company, 'No'. I'm sure there are instances when they do, but generally the threshold for doing that has been raised quite a bit. The FDA's Center [CDER] is highly dependent now on the group [industry] that it regulates. And I think that that changes perspective. Over the same time, this philosophy has come down which is that it's not so much 'the regulator' and 'the regulated', it's 'colleagues', and we [FDA] work with industry to get the drug approved. Within the agency, there are people who view drug companies as their customers or their client. With that perspective all of a sudden, it's 'who do we [FDA] have to please?' And 'how do we give them [industry] what they want?' Whereas before [PDUFA and FDAMA], I think maybe things were 'we're [FDA] not for you [industry], we're not against you, but you have to show us [why your drug should be on the market]'. Now I think the balance has been tipped so that the benefits of the doubt will be given to the company and I think that there's a greater reliance on the idea that we [FDA] can take care of this [regulating new drugs] after the drug's approved. So we have all these postmarketing agreements, these phase IV commitments.[14]

This senior scientist identified links between the neo-liberal reforms of accelerated approval and industry funding of new drug review, on the one hand, and the internal regulatory culture at the FDA, on the other. In particular, how the neo-liberal reforms affected the way in which agency management mediated its relationships with pharmaceutical firms and FDA reviewers. Senior management's propensity to foster an internal agency culture of collaboration with industry to accelerate approval of new drugs, in line with neo-liberal political demands, fuelled the abdicatory culture among FDA scientists that compromising on regulatory standards in order to approve a drug was a more congenial option than fighting for non-approval. Although this senior scientist was not referring to Iressa, once again, some of his remarks relate very directly to that case, particularly the comments about the FDA not being willing to say 'No' to a drug company, but rather allowing it to look for increasingly narrow indications to allow the drug to enter, or remain on, the market. In effect, that was what happened after the ISEL trial. Instead of withdrawing Iressa from the market after ISEL, the FDA permitted AstraZeneca to leave the drug on the market while the firm searched for evidence of more limited ways in which the drug might work. While it was reasonable for the company to search for such evidence, as Public Citizen pointed out at the 2005 advisory committee meeting, that was an argument for returning Iressa to the status of a pre-market experimental drug, not a case for allowing it to stay on the market (FDA 2005d, p. 96).

Iressa in Europe: If at first you don't succeed...

Meanwhile, in February 2003, AstraZeneca applied to the EMEA for marketing approval of Iressa in the EU. Unlike the FDA, the CPMP reached a negative opinion about the company's initial application because the INTACT studies did not demonstrate any survival benefit in first-line patients.[8] By December 2004, when the ISEL results became available, the EU regulatory authorities had still not approved the drug. After discussing the ISEL results with the EMEA, AstraZeneca withdrew its EU application in January 2005 after the agency indicated that Iressa would not be approved in Europe (FDA 2005c, pp. 91–2). Evidently, in the early 2000s, European regulators were willing to uphold higher regulatory approval standards with respect to Iressa than their US counterparts.

Underlining the divergent approach of the two agencies, when the EMEA and the CPMP rejected Iressa, the alternative drug, Tarceva, had not yet been granted a positive recommendation for marketing approval in the EU (Anon. 2005k). Thus, the FDA left Iressa on the US market after ISEL even though a better alternative was available, while EU regulators rejected Iressa even in the absence of that alternative.

Despite these initial setbacks, in May 2008 AstraZeneca submitted a new marketing approval application for Iressa to the EMEA. This time the firm sought approval for the drug as a treatment for locally advanced or metastatic NSC lung cancer patients who had received prior platinum-based chemotherapy. Three pivotal, phase III studies were submitted in support of the application – IPASS, INTEREST and the failed ISEL study mentioned above. As we have discussed, evidence from company-sponsored studies with Iressa suggested that if the drug *did* have efficacy, then the likelihood was that only a small subgroup of patients – probably with a particular, Iressa-sensitive genetic mutation – would benefit. It was clear during the 2005 ODAC meeting that many of the FDA's special scientific advisors believed that this was the only way to reconcile the negative results of the two INTACT trials and ISEL with the anecdotal evidence that had persuaded them to recommend approval of the drug. In addition, in 2004 non-company sponsored research had identified mutations in the EGFR gene of lung cancer patients whose tumours had shrunk when they were given Iressa (Lynch *et al.* 2004). Yet despite the fact that Dr Temple had specifically indicated to AstraZeneca at the ODAC meeting that the company would need to confirm – in a prospective, randomized trial – that patients with these specific genetic mutations derived a clinical benefit from the drug, no such trial had been undertaken by the time the new marketing application was submitted to the EMEA in 2008. Instead the company submitted two, active-comparator studies of Iressa for first- and second-line treatment of lung cancer patients and the ISEL study. AstraZeneca's unwillingness to undertake the kind of trial suggested by Temple can probably be understood as a product of the company's reluctance to limit Iressa's market potential in Europe to the very small population of patients with the identified genetic mutations.

The first of the new studies, the INTEREST trial, was a randomized, open-label, non-inferiority study comparing Iressa with docetaxel in patients previously treated with a platinum-based chemotherapy. The company claimed this trial demonstrated the non-inferiority of Iressa compared to docetaxel in prolonging survival (the primary endpoint) in the overall population of lung cancer patients. This claim (of a comparable survival advantage with docetaxel) was repeated in a publication in *The Lancet* in November 2008 (Anon. 2008h). However, the EMEA's Committee for Human Medicinal Products (CHMP) – successor to the CPMP – observed that a secondary analysis, which adjusted for patients' histology, performance status, prior therapy, smoking history, sex and racial origin, indicated that Iressa might be inferior to docetaxel at prolonging survival (EMEA 2008, pp. 42 and 83).

The second new trial – the IPASS study – was a randomized, open-label, non-inferiority study comparing Iressa with first-line paclitaxel/carboplatin doublet chemotherapy in a highly selected group of patients. The study was made up of mainly Asian patients with adenocarcinoma histology, who had either never smoked or were ex-light smokers. This was the population that had been identified in ISEL and IDEAL as deriving the greatest benefit from Iressa. However, this population was *not* representative of the majority of NSC lung cancer patients in Europe (EMEA 2009, p. 85).

The objective of the study was to compare 'progression-free survival' (the primary endpoint) for patients in the Iressa arm of the trial with patients in the paclitaxel/carboplatin (doublet chemotherapy) arm. Secondary endpoints assessed in the study were survival, quality of life and symptomatic improvement. The primary endpoint, 'progression-free survival', is a measure of the time that has elapsed from a patient starting treatment to the moment when his or her tumour is found to have 'progressed' (grown), *or* the time when the patient dies. Progression-free survival is a controversial endpoint in cancer clinical trials for two reasons. First, it does not necessarily predict whether patients will live longer or whether their symptoms will be improved. In other words, any 'benefit' from the drug is hypothetical unless clinical studies also demonstrate that the endpoint was associated with symptom improvement, improved quality of life or a survival advantage. Second, there are methodological difficulties in accurately capturing the precise moment when patients' tumours begin to progress and the validity of results can easily be undermined by missed assessments, incomplete baseline assessments, or uneven assessment in different treatment groups (Broglio and Berry, 2009; D'Agostino, 2011). Nevertheless, despite the fact that any benefit is hypothetical and there are problems with assessing the endpoint in clinical trials, the FDA and the EMEA have accepted improvement in progression-free survival as the basis for both accelerated/conditional *and* regular approval of cancer drugs (Chakravarty and Sridhara, 2008).

The results of IPASS showed that in the overall (highly selected) study population, Iressa performed better than paclitaxel/carboplatin with respect to the primary endpoint (progression-free survival) and the secondary endpoint (improved quality of life) but not with respect to symptomatic improvement. The company were also measuring patient survival as a secondary endpoint in IPASS, but at the time data were collected to submit to the EMEA, the trial was ongoing and results showing whether or not Iressa prolonged survival were not available. Consequently, it was not known at that time whether the improvement in progression-free survival seen with Iressa meant that patients would actually live longer. While the study *did* demonstrate an improvement in patients' quality of life, a result which represented a genuine clinical benefit for patients, the CHMP were uncertain whether the results of IPASS could be generalized to the EU population. This was because, as discussed above, the study population of IPASS was made up of mainly Asian patients with adenocarcinoma histology, who had either never smoked or were ex-light smokers (EMEA 2009, p. 85).

The company had also undertaken a pre-specified exploratory analysis in IPASS of patients' EGFR mutation status – that is, whether patients tested positive for EGFR mutations (EGFR M+) or whether they tested negative for these mutations (EGFR M–). This post-hoc analysis of patients' mutation status indicated that in patients who were EGFR M+, progression-free survival was prolonged in the Iressa arm of the study compared to doublet chemotherapy. That is, there was a longer period of time between the start of study and the point at which patients' tumours started to grow. Conversely, in patients with EGFR M– tumours, progression-free survival was *shorter* compared with paclitaxel/ carboplatin (EMEA 2009, pp. 32, 84). Although these results were highly suggestive, this was yet another exploratory analysis by AstraZeneca, the findings of which could only give rise to a *hypothesis* that Iressa offered significant benefits to EGFR M+ patients. This was not a randomized controlled trial *confirming* that EGFR M+ patients derived a clinical benefit from Iressa as required by current techno-regulatory standards (EMEA 2008; Brown *et al.* 2009, p. 27). Further complicating the picture was the fact that an interim analysis of *survival* according to patients' mutation status failed to demonstrate a statistically significant difference between Iressa and paclitaxel/carboplatin (EMEA 2009, p. 33). And finally, data relating to the mutation status of patients was missing for the vast majority of patients in the IPASS study (EMEA 2009, pp. 32–3).

Because of the uncertainties relating, first, to the reliability of AstraZeneca's exploratory biomarker analysis in defining the population most likely to benefit from Iressa, and second, to the relevance of trials conducted in a predominantly Asian population that had never smoked, the CHMP consulted its Scientific Advisory Group (SAG) in Oncology. The SAG expressed dissatisfaction with aspects of the design and conduct of the studies submitted – particularly the large amount of missing data with respect to the EGFR mutation status of the majority of patients – and considered the studies to be inadequate. Nevertheless the SAG judged that, with respect to the subgroup of patients who were EGFR+, Iressa had shown a consistent pattern of activity across the trials on a number of measures – namely, progression-free survival, tumour response, quality of life and symptom improvement – and that this finding was supported by both clinical and non-clinical data from a number of studies conducted independently of AstraZeneca (EMEA 2009, p. 63). On this basis, the SAG recommended that treatment with Iressa should, ideally, be limited to patients with evidence of activating EGFR mutations based on reliable diagnostic tests. Following the SAG's recommendation, the CHMP concluded:

> that EGFR mutation status *may* account for the benefit observed in patients treated with gefitinib [Iressa] ... [and] that the indication should be restricted to patients harbouring activating mutation as this subgroup of patients *appeared to derive* the most clinically meaningful benefit from gefitinib therapy. [emphasis added] (EMEA 2009, p. 63)

Consequently, on 24 June 2009, the CHMP recommended that the marketing application for Iressa should be approved for this restricted population.

In June 2011, the EMEA updated Iressa's product label to include final data on patient survival from the IPASS study. There was *no* significant difference in survival between EGFR M+ patients treated with Iressa and those treated with paclitaxel/carboplatin, even though these were the patients hypothesized to derive the most clinical benefit from the drug. On the available evidence then, the therapeutic advantage offered by Iressa – a drug that had been trumpeted as a 'wonder drug' by the company and some patients and oncologists – amounts to an improvement in quality of life that appears to be limited to a small subgroup of patients with NSC lung cancer. Mature data from the completed IPASS study indicated that the observed advantage in progression-free survival did not translate into a real survival advantage over traditional chemotherapy, even in patients with Iressa-responsive tumours.

It is not known whether the data submitted to the EMEA was also submitted to the FDA to support a new indication that would allow AstraZeneca to resume marketing of Iressa in the US beyond the restrictions imposed by the FDA in 2005. Goldberg (2009, p. 3) has suggested that IPASS would raise 'profound regulatory questions' for the FDA due to its atypical cohort and lack of demonstrated survival benefit for Iressa. In any event, the pre-2009 marketing status of Iressa in the EU and the US was completely reversed in February 2011 when AstraZeneca announced at an ODAC meeting that it would be withdrawing Iressa from the US market, making the drug available to patients responding to Iressa on a compassionate-use basis instead (Goldberg 2011a, p. 2).

In attempting to understand the regulatory history of Iressa in the EU, one possible explanation for the EMEA's early refusal to approve the drug could be that the EMEA's 'exceptional circumstances' provision was far more restrictive than either the FDA's 'accelerated approval' rule or the EMEA's later 'conditional marketing' regulation. However, this explanation seems implausible in view of the fact that between 1995 and 2005, 27 per cent of all new oncology drugs approved via the centralized procedure in the EU were approved under 'exceptional circumstances' (Boone 2011). This represents a higher proportion than the proportion of new oncology indications that received accelerated approval in the US between 1992 and 2002 (20 per cent of the total) (Johnson *et al.* 2003). Although the two cohorts are not strictly comparable, the figures indicate that the EMEA did not feel bound to restrict its regulatory actions within the letter of the law, and that it may have been using the 'exceptional circumstances' provision to accelerate approval of drugs onto the market. Nevertheless, the Iressa case-study suggests that in the early to mid-2000s, the EMEA was not characterized by an abdicatory culture to the same extent as the FDA because the European agency refused to approve Iressa when the weight of the evidence suggested that the drug had no effect. How then did the EMEA and FDA positions come to be reversed by the end of the decade? This reversal cannot be explained by the fact that the data submitted by AstraZeneca finally met the EMEA's techno-regulatory standards. Instead, it is possible that the European agency came under more intense neo-liberal pressures – particularly from the European Commission – from the early to mid-2000s. This period coincided with the European Commission's review of the EU pharmaceutical legislation, and it was clear that the Commission was primarily concerned to advance the interests and competitiveness of the European pharmaceutical industry. Since the new 'conditional marketing' regulation was first introduced in the EU in 2006, the EMEA have increased the number of drugs accelerated onto the market on the basis of incomplete and uncertain evidence. Compared to the period 1995–2005 when the agency approved 27 per cent of oncology drugs under 'exceptional circumstances', between 2006 and 2010 the EMEA approved 36 per cent of new oncology products onto the market using either the 'exceptional circumstances' or the 'conditional marketing' provisions (Boone 2011). This suggests that relative to the period prior to 2005, the EMEA has become more permissive with respect to the approval of oncology drugs on to the market in the absence of adequate data on safety and efficacy.

Conclusion

In Chapter 2, we demonstrated that the emergence of 'accelerated approval' (or 'conditional marketing') regulations in the EU and the US for drugs to treat serious or life-threatening conditions was primarily driven by neo-liberal politics and industry interests, rather than patients' interests. In this chapter we have sought to examine what consequences these developments might have had for regulatory decision-making and outcomes relating to particular drugs. Our investigation of the acceler-ated approval of the lung cancer drug Iressa in the US, suggests that, in the early 2000s it was regulated in the interests of the manufacturer, but scientific and regulatory standards to protect the interests of patients and public health were not upheld. We explained this in terms of a culture of resignation among FDA scientists, which in turn can be understood as a response to a broader environment of neo-liberal pressures and corporate bias. What the Iressa case-study demonstrates is that, in such a political-cultural environment, regulators may fail to uphold even weak regulatory standards. In this way we have shown that the neo-liberal corporate bias characterizing the establishment of accelerated approval in the US did indeed have consequences at the micro-social level of the FDA's decision-making and regulatory outcomes with respect to specific drugs. Thus, our case study of Iressa does not support the claim of neo-liberal theory that the introduction of the deregulatory reform, known as 'accelerated approval' in the US, was in patients' interests or the inter-ests of public health.

It might be argued, however, that Iressa is just one cancer drug and that the wider significance of this specific case is unknown. While it is true that case-study analysis can have limitations regarding generalization, they can be mitigated by choosing a case strategically with respect to the theoretically driven research questions, which will become apparent as we outline our conclusions. In selecting Iressa we chose what social scientists call 'the hardest case',[15] namely, cancer drugs – and indeed within that a third-line lung cancer drug. This is the hardest case because there are few areas of medicine where the need for treatment is greater, so in that context, it is *hard* to imagine that accelerated drug approval could not be in the interests of (lung) cancer patients. Conversely, one would expect the case of cancer drugs, and lung cancer drugs in particular, to provide the easiest context for the claim that accelerated drug approval has operated in patients' interests to shine through.

Nevertheless, it could still be argued that had we chosen a different cancer drug – such as Gleevec, for instance – we would have reached different conclusions with regard to the patient and public health benefits of accelerated approval. Clearly, it would be foolish to argue that accelerated approvals *never* benefit patients. If a drug turns out to have clinical efficacy then patients will have derived some benefit from earlier access to the drug. However, in evaluating the performance of the *regulators* it is also necessary to investigate drugs that 'test' the system – that is, to investigate cases where firms have failed to demonstrate a clinical benefit in post-approval trials, either because the studies were negative or because the studies had not been completed. As of July 2011, this accounted for *nearly half* (45 per cent) of all oncology indications granted accelerated approval by the FDA (Johnson *et al.* 2011). In other words, Iressa is not an exceptional case. In fact, it could be argued that Iressa provides a 'hard' case for our argument in at least one further respect and that is because the manufacturer, AstraZeneca, *did* conduct confirmatory studies in a timely manner. At the time of our fieldwork, that was true for only 25 per cent of cancer drugs granted accelerated approval by the FDA (Fleming 2005; Roberts and Chabner 2004). Moreover, as of November 2008, over 60 per cent of such post-marketing confirmatory studies completed for cancer drugs had taken over five years to complete from time of approval (US GAO 2009, pp. 20–2). One would expect accelerated approval regulations to serve patients *better* in those cases (like Iressa) where attempted confirmation of drug efficacy was forthcoming in a timely manner, rather than in the 75 per cent of cases where it had been neglected. Hence, we would suggest that the failure of accelerated approval to operate in patients' interests in the Iressa case implies a much wider failure of that neoliberal regulatory reform because Iressa was, in fact, one of its better examples. Indeed, as of September 2009, 17 years after the reform was first introduced, the FDA had never withdrawn a drug granted accelerated approval from the market despite failure by the manufacturer to demonstrate clinical benefit because phase IV confirmatory trials were either negative (as occurred with Iressa) or never conducted (US GAO 2009, pp. 32–3). Such action was taken for the first time by the FDA in late 2011 when it revoked the cancer drug Avastin's accelerated approval for metastatic breast cancer (Goldberg 2011b).

Not only does our case-study and associated evidence in this chapter imply a justifiable generalization that the regulatory outcomes of American accelerated approval prioritized industry interests over those of patients, our interviews also reveal the mediating mechanisms by which it occurred. While Chapter 2 showed how neo-liberal macro-politics generated the American accelerated approval regulations and empowered industry's demands over the FDA, our evidence in this chapter demonstrates how that filtered through the 'ranks' of the regulatory agency right down to scientific reviewers of individual drugs. This led to an abdicatory culture towards challenging pharmaceutical firms and/or senior FDA managers and advisory committees to ensure that regulatory standards to protect public health were upheld when contrary to the commercial interests of companies. Our interviews suggest that such a culture of resignation was present in the case of Iressa and within the FDA's oncology division more broadly, but they also show that it extended much more widely to other divisions and across the agency. We do not claim that it was comprehensive, but certainly significant.

It is that abdicatory culture towards strict regulation of the pharmaceutical industry that also helps to explain the findings of Congress's General Accounting Office (2009) investigation of the FDA's monitoring of drug companies' commitments to post-marketing confirmatory studies under accelerated approvals. The investigation relayed:

> FDA does not know the current status of many postmarketing [confirmatory] studies or whether they are progressing towards completion. In addition, FDA does not know whether drug sponsors [firms] are submitting complete and accurate annual status reports [about post-marketing confirmatory studies]. ... We found that sponsors [pharmaceutical firms] were late in submitting annual status reports for 12 post-marketing studies in our sample. However, FDA issued an administrative action letter to the sponsor of only one of these 12 studies. ... The agency has taken a passive approach to enforcing confirmatory study requirements. It has never exercised

its authority to withdraw a drug it approved based on surrogate endpoints under the accelerated approval process, even when such studies have been outstanding for nearly 13 years. (US GAO 2009, pp. 30, 32 and 36)

There is, then, good reason to conclude that the neo-liberal deregulatory reforms of American accelerated approval in 1988 and 1992 (Subparts E and H, respectively) were not only driven by industry interests and their allies in government, but that that corporate bias influenced the internal regulatory culture of the FDA such that industry interests gained greater priority than before over protection of patients' interests within decision-making about individual drugs. The case study implies a three stage process in which neo-liberal corporate bias facilitated a significant amount of capture at higher echelons of FDA management, which, in turn, generated an abdicatory culture among a considerable number of FDA rank-and-file scientists. We suggest, therefore, that analysis of both the creation and implementation of the American accelerated approval regulations provides some support for both corporate bias theory and capture theory.

But what of disease-politics theory with its claim that the FDA has accelerated marketing approval of new drugs because of patients' demands, rather than (or more than) industry interests? We acknowledge that the Iressa case provides some evidence that patients and patient groups may have influenced the FDA's approval decision via the agency's advisory committee. There is both some truth and some superficiality in this interpretation of events. Our research demonstrates that that influence was itself significantly shaped by the manufacturer's promissory science about the drug's effectiveness to the oncology profession and the extent to which the company determined which patient voices were heard by the FDA's advisory committee. The creation of such expectations was a mechanism by which the firm's interests could be furthered in relation to stakeholders external to, but with some influence over, the FDA. This complemented and reinforced the more direct industry influence on FDA management which mediated the changes to the internal FDA regulatory culture discussed above. We conclude that the existence of the 'patient–industry complex' means that, when explaining the FDA's approval decisions, a significant amount of 'patient demand' cannot necessarily be regarded as an explanatory factor *independent* of industry interests.

Once again, our findings with respect to disease-politics theory are likely to be generalizable beyond Iressa to other accelerated drug approvals for the following reasons. First, patients' demands were clearly present and influential in accelerating Iressa on to the market, rather than absent from the regulatory process as might well be the situation with other drug cases. Leading up to Iressa's approval in 2003, and at the first ODAC meeting, the FDA was subject to particularly intense pressure to approve the drug from *The Wall Street Journal* and from some cancer patients. Indeed, Carpenter (2004) refers to Iressa as a clear example of the increasing influence US patients and their advocates were exercising over US drug regulation. Second, in focusing on the FDA scientists working within the oncology division we have chosen the hardest case to make from an *organizational* point of view. This is because the oncology division (now the Office of Oncology Drug Products) has been subject to intense and sustained media scrutiny and organized pressure from some sections of the patient advocacy community to a far greater extent than any other group within the FDA. Consequently, if disease politics is *the* key factor shaping the way in which FDA scientists implement the 'accelerated approval' regulations, then this will be most evident in relation to the oncology division's regulation of new cancer drugs.

For these reasons, one would expect disease-politics theory to shine more in the Iressa case than in many other accelerated drug approvals, if the theory is sound. Despite this, the 'hard' version of the theory is not supported by the Iressa case because its marketing approval was not in patients' interests. The 'weak' version – that patients' demands led to its approval – is supported in only a partial and superficial way, with a deeper analysis probably favouring the manufacturer's interests as being the greatest determining factor in the drug's approval. Furthermore, industry shaping of patient testimonials has extended to other cancer drugs as well as Iressa (Goldberg 2005, p. 12). It may also be concluded from the Iressa case-study that disease-politics theory offers little to explain FDA's *maintenance* of accelerated drugs *on the market* after post-marketing studies have shown them to provide no clinical benefit and patients' demands have become muted (limited, in effect, to calls for compassionate-use availability). Here the influence of industry interests would seem to be the paramount explanation.

Our case-study of Iressa in the US in the early and mid-2000s, demonstrates that 'accelerated approval' of even a third-line lung cancer drug is not necessarily synonymous with patients' interests, and that a more cautious FDA would not have been contrary to patients' interests, even when some patients supported such approval. It also reveals that FDA decision-makers were inclined to consistently give the company the benefit of the doubt, albeit an inclination strengthened by support for the drug from some patients and expert advisers. From the perspective of promoting and protecting patients' health, the key lesson from the Iressa case would seem to be that accelerated approval based on non-established surrogate measures of drug efficacy has been an undesirable development leading to low regulatory standards that are further undermined in some cases by poor decision-making. This is because the low regulatory standards of accelerated approval may lead the company to take longer to discover which patients can genuinely benefit from the drug (as occurred with Iressa) than would be the case if clear evidence of clinical efficacy were required from the outset. Similarly, whether or not post-marketing confirmatory studies are delayed, once another 'accelerated approval' drug to treat the same condition becomes available the hypothetical benefit of 'accelerated approval' disappears since patients do not know which drug to 'choose' without sufficient data. And for patients with end-stage disease this can mean that a small window of opportunity, in which to access the most effective therapy, is missed. Thus, counter-intuitively, it may be that 'accelerated approval' regulations have actually delayed patients' access to genuinely therapeutic drugs.

In this way it can be seen that, even under conditions of optimal regulatory decision-making, any potential benefits offered by 'accelerated approval' are time-critical. As our case study of Iressa shows, these *inherent* problems are exacerbated by weak regulation of minimal standards resulting in regulatory outcomes that are unlikely to be consistent with the best interests of patients or public health. The corollary is that *more* patients might benefit if regulatory agencies, whose express purpose is to promote and protect patients' health, moved away from such 'accelerated approval' regulations and towards more extensive requirements for evidence of direct clinical efficacy and benefit before granting marketing approval. It is often argued, however, that that would delay the marketing approval of new drugs for patients with life-threatening conditions. As we have shown above, this delay may be entirely illusory. In the US, there is strong evidence that 'accelerated approval' has led to quicker access to new drugs but *slower* completion of phase III studies and, therefore, delayed access to critical information about drug efficacy, effectiveness and safety (Schilsky 2003; Susman 2004; Ellenberg 2011). This means that, in practice, patient access to the most effective drugs may be delayed. In the EU, completion of post-marketing studies has been similarly slow. Between 1995 and 2005, for the eight products granted exceptional approval, it has taken companies, on average, six years to complete their post-marketing confirmatory studies

(Boone 2011). Moreover, several EU member states have established health technology assessment (HTA) bodies, such as the UK's National Institute for Health and Clinical Excellence (NICE), to inform reimbursement decisions. Innovative therapies receiving accelerated marketing approval may not be reimbursed if the evidence base for those drugs is judged inadequate for HTA purposes (Anon., 2009d; Eichler *et al.* 2008; McCabe *et al.* 2008; Sculpher and Claxton 2005). Consequently, accelerated approval of new drugs that fail to properly define a drug's real health benefits may actually result in delayed patient access to therapies in the EU. Moreover, higher regulatory standards demanding substantial evidence of clinical benefit could incentivize pharmaceutical firms to increase efficiency and focus their R&D efforts on bringing to market a smaller number of genuinely effective therapies, rather than large numbers of cancer drugs with uncertain clinical benefit (Sculpher and Claxton 2005).

The FDA oncology division has held several meetings with its advisory committee and oncology professional groups to discuss the desirability of surrogate measures in cancer drug approvals. Some contributors to those discussions, most notably Tom Fleming, raised many concerns about accelerated approval rules, such as the low predictive value of surrogate measures in cancer drug regulation, and the lack of pressure on pharmaceutical firms to conduct post-marketing confirmatory studies (FDA 2003c; 2003d; 2003e). However, no decisions were reached or policy changes agreed at those meetings. In 2007, the FDA Amendments Act provided the agency with the authority to assess monetary penalties against firms, which had not conducted required post-marketing studies under the 'accelerated approval' process. However, as of July 2009, the FDA has never used that new authority (US GAO 2009, p. 14). Yet those modest efforts to address some of the worst aspects of American 'accelerated approval' regulations did not tackle the fundamental problem of requiring adequate clinical evidence before approval. Furthermore, FDA managers have called for an *increase* in the acceptance of surrogate measures of drug efficacy to *further accelerate* innovative pharmaceutical development and approval (FDA 2004). At least for some in the FDA, no lessons at all have been learnt from the Iressa case.

As for the EU, while the EMEA initially regulated Iressa quite strictly, the Commission failed completely to learn the lessons of US 'accelerated approval'. On the contrary, as we outlined in the first section of this chapter, in 2006, the EMEA introduced 'conditional marketing' regulations that are almost identical to American 'accelerated approval' rules. Such policy developments have no doubt been reinforced by dubious reports that advocate accelerated regulatory review of new cancer drugs on the grounds that cancer survival rates in different European countries can be positively linked to speed of access to cancer drugs (Anon. 2007g). Those developments imply that the issues raised by Iressa are even more relevant at the time of writing (2011) in Europe than they were in the early and mid-2000s. Finally, it is important to note that European 'conditional marketing' and American 'accelerated approval' policies have implications not only for current cancer patients, but also wider public health because of the longer-term effects of lowered efficacy standards on incentives of firms to develop future cancer drugs that might truly provide therapeutic benefit.

5
The Making of a Harmful 'Therapeutic Breakthrough': Crossing Regulatory Boundaries for Drug Approval

In Chapter 2 we explained that during the neo-liberal era, the FDA began to categorize some new drugs as offering significant therapeutic advance so that they could be given a faster 'priority' regulatory review within the agency. Priority review could be awarded to any new drug submitted to the agency for marketing approval, irrespective of the seriousness of the condition it was intended to treat, provided that it was deemed to promise significant therapeutic advance. The diabetes drugs, the glitazones, were discussed as case studies of that scenario in Chapter 3. In Chapter 4, we discussed the neo-liberal regulatory reforms of 'accelerated approval' and 'conditional marketing' in the US and Europe, respectively, regarding new drugs to treat serious and life-threatening conditions such as cancer. In this chapter, we consider an innovative pharmaceutical, alosetron, that began life within the US regulatory system under 'priority review', and the associated accolade of promising significant therapeutic advance in the treatment of irritable bowel syndrome (IBS), but which later came to be marketed under accelerated approval regulations as a drug to treat a serious or life-threatening condition.

The case study demonstrates in concrete terms the significance of the pharmaceutical industry's efforts in the 1990s to widen the category of drugs qualifying for accelerated approval, so that it included not only drugs for the treatment of life-threatening conditions, but also drugs to treat serious and/or debilitating diseases. It also provides additional insight into what has been permitted to count as 'therapeutic advance'

and 'therapeutic breakthrough' within neo-liberal pharmaceutical innovation and regulation, and how that affects overall decision-making
about drug risks and benefits. Our case study in this chapter adds
further weight to the concern that the official accounting categories
of 'significant therapeutic advance' and 'unmet need' to treat serious
or life-threatening conditions can mask what is, in reality, very modest
drug efficacy leading, in turn, to unjustified acceptance of risks to public
health in regulatory decisions. The alosetron case-study shows how that
acceptance takes hold and is rationalized institutionally and micro-
politically. The techno-scientific claims of therapeutic advance and
therapeutic breakthrough are interwoven with complex social interests,
involving the drug availability options determined by the commercial
interests of manufacturers; the FDA management's assimilation of, and
acquiescence to, the neo-liberal ideology of rapid drug approval, especially in the context of patient demands for access to new drugs; and
the role of patient groups, sometimes in collaboration with pharmaceutical manufacturers, in challenging regulatory science with experiential
testimony.

Irritable bowel syndrome as a disease condition

IBS is the term used to describe the symptoms of a diverse group of
patients who suffer with abdominal pain, bloating and disturbed defecation, with no identifiable structural cause. The 'syndrome' is a disorder
of the intestine, which is not associated with shortened survival, but has
real effects on those who suffer from it. However, it shows no sign of
disease that can be seen or measured. For example, its cause cannot be
detected by blood tests or X-ray. It is thought that people with IBS may
have particularly sensitive colons that react to stimuli, such as stress or
eating, that do not trouble most people. It is estimated that for about
65 per cent of people with IBS the condition is not severe enough to
prompt them to see the doctor (Cash and Chey 2003).

As there are no identified physiological abnormalities associated
with IBS, diagnosis relies on symptom-based criteria. In the late 1990s,
when alosetron was being developed, those criteria were known as
'Rome II'. The primary symptoms of IBS are recurrent abdominal
pain and discomfort. Patients also experience abnormal bowel function, which presents primarily as diarrhoea (diarrhoea-predominant
subtype), constipation (constipation-predominant subtype) or which
alternates between the two (alternating subtype). In addition, increased
sense of urgency, bloating and incomplete evacuation may be present.

Specifically, the Rome II criteria include at least 12 weeks (not necessarily consecutive) in the preceding 12 months, of abdominal discomfort or pain that is accompanied by relief with defecation, a change in the frequency of defecation, and/or a change in the form of the stool (Camilleri *et al.* 2001).

As alosetron emerged as a potential treatment for IBS, it was estimated that 8–19 per cent of the adult western population had symptoms associated with IBS. There was, and remains, a strong female predominance of around 70–75 per cent of IBS patients (Bardhan *et al.* 2000). Estimates of the syndrome's prevalence in North America vary widely from 3 to 20 per cent of the adult population with a two-to-one female predominance. Some scientists have suggested that, in the US alone, IBS affects as many as 15 million people, is responsible for 2.4–3.5 million physician-visits per year (a similar figure to asthma), and incurs costs of US$1.7 billion annually (Ladabaum 2003). It is thought that the prevalence of IBS in North America is equally divided between IBS with constipation, IBS with diarrhoea, and IBS alternating between diarrhoea and constipation. However, rigid use of those sub-types probably oversimplifies the condition because IBS with constipation may change to constipation alone or IBS alternating between constipation and diarrhoea and so on (American College of Gastroenterology Functional Gastrointestinal Disorders Task Force 2002).

The development of alosetron (Lotronex)

When the pharmaceutical company, GlaxoWellcome, began to develop alosetron in the 1990s, it was already supposed that serotonin was an important neurotransmitter in the brain-gut axis, and that it was involved in several functions of the gastrointestinal (GI) tract. Scientists believed that serotonin played a pivotal role in the modulation of multiple gut functions, such as motility, sensation and secretion. Serotonin released into the wall of the gut, it seemed, initiated both secretory and peristaltic reflexes. Serotonin's biological actions in this respect were thought to be mediated via a variety of serotonin receptor chemicals, known as 5-hydroxytryptamines (5-HTs). Laboratory studies in animals showed that the serotonin pathways in the body appeared to play a role in the gut-brain axis in health and disease. It was proposed, therefore, that serotonin pathways might be implicated in the underlying pathophysiology of functional bowel diseases via 5-HT receptors associated with motor and sensory processes in the GI tract (Anderson and Hollerbach 2004).

A chemical, known as ondansetron, which suppressed the action of the 5-HT3 receptor, was shown to delay colonic transit in healthy volunteers. Hence, scientists postulated that 5-HT3 receptors might be implicated in the underlying pathophysiology of IBS. Disturbances in colonic motor activity might account for some of the symptoms of IBS and modulation of that activity might have a beneficial effect (Gunput 1999). Scientists at GlaxoWellcome developed alosetron, a drug that selectively inhibited the serotonin 5-HT3 receptors located in the GI tract. Laboratory studies in guinea pigs, rats, cats and dogs indicated that alosetron might be valuable in affecting sensitivities of the bowel, colon and/or rectum (Barman-Balfour *et al.* 2000). According to GlaxoWellcome, in guinea pigs, alosetron did not affect normal peristaltic reflexes, but inhibited artificially accelerated reflexes of the kind believed to be present in human IBS. This led the company to believe that it was safe to give alosetron to patients without fear of paralyzing the bowel. The firm's scientists also hypothesized that excessive serotonin release to receptors linked to the brain could cause painful evacuative contractions of the colon, of the kind associated with IBS. By inhibiting that overflow of serotonin, the company believed that alosetron could prevent such painful symptoms (FDA 1999b, pp. 26–31).

Financial analysts predicted that the American IBS market would be worth US$3 billion by 2005 of which alosetron was expected to constitute US$1.1 billion because it was one of the first supposedly effective pharmaceuticals to treat the condition (Anon. 1999c). Thus, the drug was expected by the company and others to be a 'blockbuster'. Although analysts noted that, at that time, the IBS market was largely untested and undeveloped commercially. To compensate for that, GlaxoWellcome was prepared to hire several hundred additional sales representatives in the US alone to promote alosetron to both physicians and patients if granted marketing approval by regulators (Anon. 2000f).

A reality check on 'significant therapeutic advance': FDA's review and approval

On 29 June 1999, GlaxoWellcome submitted to the FDA their application to market alosetron under the tradename, 'Lotronex', in the US. Although both men and women can suffer from IBS, alosetron was found to be ineffective for men with the condition, but showed some therapeutic efficacy among women. Consequently, the firm sought marketing approval for treatment of women with diarrhoea-predominant and/or alternating IBS. Of the women with IBS in the US, about one-third have the constipation-dominant type, one-third have the diarrhoea-predominant form and one-third experience the alternating type.

Alosetron was a NME given priority review classification by the FDA on the grounds that, compared with existing therapies, it represented 'a significant therapeutic advance as a first line monotherapy for the significant population of female patients with non-constipating IBS' (Anon. 1999c, p. 18; Gallo-Torres 2002, p. 7). As we explained in Chapters 1 and 2, for regular marketing approval applications, the FDA required, and generally continues to require, at least two relatively large (phase III) placebo-controlled trials to demonstrate clinical efficacy of a new drug, designated by regulators as 'pivotal' to the case for approval (Temple and Ellenberg 2000). Accordingly, GlaxoWellcome submitted two pivotal trials, known as A3001 and A3002. In A3001, 309 women were randomized to alosetron and 317 to placebo, while in A3002, the figures were 324 and 323, respectively. These trials had identical protocols. Both involved 12 weeks of treatment (1.0 mg alosetron twice a day), and excluded women with constipation-predominant IBS.

The protocols of the trials also defined primary and secondary 'efficacy measures', which were the dimensions of effects that the clinical investigators explored with the patients during the trials in order to determine the drug's efficacy. The primary efficacy measure was relief from abdominal pain/discomfort. Thus, clinicians running the trials asked the patients about their abdominal pain/discomfort and recorded the results for analysis, often using quantifiable rating scales. For example, patients were asked: 'In the past seven days, have you had adequate relief of your IBS pain and discomfort?' A responder to the treatment was a patient who indicated 'adequate relief' for at least two weeks out of the month. Hence a patient could be a responder for any of months one, two and/or three.

In both A3001 and A3002, a statistically significantly higher proportion (41 per cent) of women treated with alosetron had adequate relief of abdominal pain/discomfort than those on placebo (26 and 29 per cent), but only for the women with IBS treated consecutively for all three months. This efficacy was mainly demonstrated among the women with diarrhoea-predominant IBS (46 per cent), but for women with alternating IBS, placebo outperformed the drug (Prizont 1999, p. 34). Both trials indicated no difference in primary efficacy between alosetron and placebo for patients who responded to treatment for only one or two months (Prizont 1999). On primary efficacy, the FDA concluded:

> Of women with diarrhoea-predominant IBS, who take Lotronex and improve, between 68 and 80 per cent improve spontaneously or due to factors not attributable to Lotronex. Many of these patients may continue to take Lotronex because of a false belief that improvement is due to a drug when, in fact, improvement is probably due to other factors. These patients are exposed, possibly chronically, because they believe that they are experiencing a drug benefit, to risks of the drug without benefit from the drug. (FDA 2000b, p. 114)

The key secondary efficacy measure was the nature and extent of abdominal pain/discomfort, with a numerical rating scale, including 1.00 = 'mild', and 2.00 = 'moderate'. In both A3001 and A3002, the proportion of women experiencing pain-free days in all three months was small for both placebo and drug, but by the third month of A3001, on average, the decrease in pain intensity was statistically significantly greater among the women taking alosetron (ranging from 1.93 to 1.05) compared with placebo (ranging from 1.97 to 1.25) (Prizont 1999, p. 22). However, the significance of this decrease was derived from the reduction in pain intensity among the patients with *alternating* IBS; there was no difference in the monthly decrease of pain intensity for the women with diarrhoea-predominant IBS. In A3002, on average, decrease in pain intensity was also statistically significantly greater among the patients taking the drug (ranging from 1.95 to 1.08) than those on placebo (ranging from 1.90 to 1.17), but in this case it was driven by the subgroup with *diarrhoea-predominant* IBS (Prizont 1999).

Thus, on the primary efficacy measure of adequate relief of abdominal pain/discomfort, alosetron showed no significant improvement over placebo among the women with alternating IBS. This is why, when the FDA approved the drug, it did so solely for the treatment of diarrhoea-predominant IBS in which alosetron showed efficacy if taken for three months. However, the significant decrease of pain intensity among women with diarrhoea-predominant IBS was demonstrated in only one pivotal trial, A3002 – it was not replicated in A3001. Moreover, the statistically significant decreases in pain intensity, over and above placebo, on average amounted to 14–16 per cent (0.14–0.16) of the difference between mild and moderate pain.

Regarding the other secondary measures of efficacy, alosteron was statistically significantly more effective than placebo at firming 'stool consistency', decreasing daily 'stool frequency', and reducing the patients' 'sense of urgency', but the drug made no difference to feelings of 'bloating'. Indeed, in A3001, the patients felt less bloated on placebo (Prizont 1999, pp. 22–3). While alosetron surpassed placebo in firming stools, this efficacy measure included making normal stools harder, which was detrimental, not beneficial, to patients. Consequently, FDA medical reviewers questioned the validity of this measure in GlaxoWellcome's protocols (FDA 1999b, p. 157). Regarding stool frequency, the superiority of the drug over placebo amounted, on average, to a reduction of 0.52 times per day, so the average patient needed to pass stools about once less every two days if on alosetron, instead of placebo. Similarly, alosetron's advantage over placebo on sense of urgency translated into an average of just two days reduction in urgency per month (from 21 to 15 days on the drug compared with 21 to 17 days on placebo).

Although the official accounting of alosetron was that it offered 'significant therapeutic advance', individual expert FDA scientists used rather different language to describe the realities of its effects on patients in clinical trials. One told us:

> I was struck by the fact that it was approved only for women. That is so rare that a drug will work for one gender and not the other in a disease that occurs in both with a sex ratio that isn't too striking [about 2 or 3 to 1]. If you actually look at the data, they [patients on clinical trials] got a mild improvement in abdominal pain, and about a third [of the patients who developed constipation] dropped out [of the trials] because of constipation, and it was only effective for the diarrhoea-predominant form. So all these caveats: not for men, not for all IBS, and a third lost because of constipation. So I concluded that the drug was only mildly effective.[1]

This scientist, who worked for the FDA, also suggested that one could regard the strategy of seeking marketing approval only for women as artificially inflating the efficacy of the drug because most men were non-responders. Eliminating men from the picture increased the possibility of the drug reaching an efficacy standard. Essentially, in his view, there were two hypotheses, either GlaxoWellcome were right because the drug worked only for women with IBS, for some unknown reason, or the company was wrong because, taking men and women with IBS as a whole, the drug simply was not effective.

Another FDA scientist involved with assessing Lotronex reflected:

> If you looked at the Lotronex NDA [new drug application], well you [the average patient] go from 2.5 bowel movements a day to 2.1. The average patient in the clinical trials didn't meet the Rome criteria for IBS, but that aside, how do you conceptualize that? Maybe there's a handful of people who really substantially benefit. Did most people [on the trials] benefit? Basically they don't. If you consider normal is once a day, then 2.1 is still abnormal. Practically speaking, it [the drug] doesn't have a big impact.[2]

The alosetron case shows that efficacy demonstrated in clinical trials need only be minimal to reach the threshold of 'significant therapeutic advance' defined by allocation to priority review within the FDA. This provides further insight into what is meant by the agency's official statistic that approximately 40 per cent of NMEs approved by the FDA offer 'significant therapeutic advance'.

Alosetron's minimal efficacy should also be seen in the context of pre-approval evidence about its potential to harm patients. The main database submitted to the FDA by GlaxoWellcome concerning the drug's pre-market safety comprised four 12-week placebo-controlled clinical studies – the two pivotal phase III trials discussed above and two dose-ranging studies in which daily doses of alosetron varied from 0.1 to 8.0 mg. Taking those four trials together, data pertinent to the safety of alosetron had been collected on 1263 patients (184 men and 1079 women) taking the drug and 834 (54 men and 780 women) who received placebo (Senior 1999).

Constipation was one adverse effect strongly associated with alosetron in those trials. In one of the dose-ranging studies, involving 345 patients on alosetron and 117 on placebo, withdrawal from the trial by patients due to constipation was dose-related and significantly more than with placebo. Only 1.7 per cent of patients withdrew due to constipation in the placebo group but, among those taking the drug, this rose steadily to 2.6 per cent of patients taking 0.1 mg/day, 6.9 per cent among those on 0.5 mg/day and 7.9 per cent of those taking 2.0 mg/day. According to FDA scientists, in the other dose-ranging study, there was a highly statistically significant dose-related increase in patients on alosetron withdrawn due to constipation – 2.5 per cent of patients on placebo compared with 10 per cent on

1.0 mg/day of alosetron, 16.4 per cent on 2.0 mg/day, 10.7 per cent on 4.0 mg and 19.1 per cent on 8.0 mg/day (Senior 1999). The increase in patients on alosetron developing constipation (but not withdrawn from study) was also found by FDA reviewers to be dose-related and highly statistically significant (placebo 6.3 per cent, 1.0 mg/day alosetron 20 per cent, 2.0 mg/day alosetron 19.4 per cent, 4.0 mg/day alosetron 20 per cent, 8.0 mg/day alosetron 29.4 per cent).

Regarding the two pivotal trials, in A3001, only 6.6 per cent of the 317 women randomized to placebo developed constipation during the study, but 25.9 per cent of the 309 women randomized to alosetron did so. While only 1.6 per cent of patients on placebo had to be withdrawn permanently from the trial due to constipation, 10.4 per cent of patients on alosetron had to be so withdrawn. Once again these differences were found to be highly statistically significant. In the other pivotal trial, A3002, 3.1 per cent of the 323 women randomized to placebo developed constipation compared with 29.8 per cent of the 324 women randomized to alosetron. Just 0.3 per cent of patients taking placebo had to be withdrawn from the trial prematurely due to constipation, while 10.2 per cent on alosetron were so withdrawn (Senior 1999).

Another much more serious adverse reaction to alosetron identified in the pre-market trial was ischaemic colitis (inflamed colon due to loss of blood supply to the bowel). It has been aptly described as like a 'heart attack' happening in the bowel, so that blood simply stops flowing to it (Moynihan and Cassels 2005, p. 158). Three cases occurred, one in one of the dose-ranging studies, and one in each of the two pivotal trials. In all three cases, the patient had no history of circulatory problems. There were no deaths due to ischaemic colitis or any other adverse reaction during the clinical trials, but there were other serious bowel problems (Senior 1999). A fourth case of ischemic colitis associated with alosetron in A3001 was reported by GlaxoWellcome to the FDA on 12 November 1999. Following their analysis of the clinical trial database on patients' liver enzyme measures, concern was also raised by FDA scientists about the possibility of rare, but serious alosetron-induced hepatitis.

Regarding the drug's safety profile, FDA scientists criticized GlaxoWellcome's presentation of the evidence in their application, commenting:

> It is very disturbing that the applicant [GlaxoWellcome] has chosen to downplay so strongly the important issue of constipation induced commonly and unpredictably by alosetron, and has totally ignored

the potentially very serious, although uncommon, problems of ischaemic colitis.... The incidence of the serious lesion of ischaemic colitis is 'buried in the fine print' and minimized by being termed rare. By their [GlaxoWellcome's] own definition it was not rare, but probably infrequent..... The serious clinical adverse event of ischemic colitis cannot be ignored. It must be dealt with constructively and thoroughly. Although only 3 cases out of 921 patients exposed were diagnosed, preliminary inspection of the adverse events reported indicates that there were several cases of unexplained and un-investigated rectal bleeding. Ischaemic colitis caused by drugs may be mild and transient if no occlusion [blockage] of major mesenteric vessels [blood vessels in the membrane attaching organs to the abdominal wall] occurs, but can be catastrophic if it does, resulting in bowel infarction [tissue death due to lack of blood supply], segmental gangrene, perforation, peritonitis [inflammation of the abdominal lining], and death if the dead bowel is not re-sected in time. Such problems may be anticipated to occur rarely. On the other hand, there may be milder cases of slight ischaemic colitis that are not recognized or diagnosed, not investigated, not treated. The index of suspicion among physicians and patients needs to be raised to deal with this uncommon but potentially very serious adverse effect of alosetron. The calculated 95 per cent confidence interval for the true incidence of ischemic colitis had an upper bound between 1 and 2 per cent of women with IBS taking alosetron for 12 weeks, based on the three cases discovered. (Senior 1999, pp. 59 and 62)

One FDA scientist characterized the risk of ischaemic colitis as 'very high' for just 12 weeks of therapy given that IBS is a chronic condition and alosetron could be taken for many months, even years.[1] However, most FDA scientists and managers, together with the agency's Gastrointestinal Drugs Advisory Committee (GDAC) took the view that these safety issues did not warrant denial of marketing approval, and that they could be adequately handled by mentioning them on the label for prescribing doctors combined with some post-marketing investigation (FDA 1999b). The FDA approved alosetron in February 2000. Marketing approval was solely for the treatment of women with diarrhoea-predominant IBS (Anon. 1999e, p. 18). At the request of the FDA, GlaxoWellcome agreed to undertake a number of phase IV post-marketing studies, including a 'large, long-term (one-year) population risk trial to assess the incidence of colitis in patients receiving alosetron'

(Raczkowski 2000, p. 2). The drug product was launched in the US one month later with the tradename, Lotronex, still expected to reach the giddy heights of billion-dollar sales (Anon. 2000a, p. 19).

Meanwhile, in Europe, an application to market alosetron in the EU was submitted to the EMEA via its centralized procedure for innovative pharmaceuticals in December 1999, according to a financial report by GlaxoWellcome issued on 27 July 2000 (GlaxoWellcome plc 2000a) and a company press release on 10 February 2000 (GlaxoWellcome plc 2000b). However, the drug was never approved by the EMEA. Due to secrecy within the EU regulatory system at that time, regulatory assessment summaries, the European Public Assessment Reports (EPARs) mentioned in Chapter 2, were publicly available only for drugs *approved* under the centralized procedure. Consequently, there is no detailed information about why EU regulators never approved the drug. However, our interviews suggested that, very broadly, European regulators were less impressed with alosetron's efficacy than their American counterparts, and concerned about safety reports emerging from the US once marketed there.

Commercial imposition of a regulatory problem

Between market launch of alosetron and June 2000, there were seven serious cases of constipation and eight cases of ischemic colitis reported post-marketing in the US. Six of the seven constipation-associated colonopathies required hospitalization and three required surgery (FDA 2000b, pp. 122–3). As a result of these cases, the FDA reconvened the GDAC on 27 June 2000 to discuss these risks. Before the meeting, the FDA had proposed that the company should place a black box warning about the risks on the Lotronex product label, but the company opposed that, so the GDAC meeting was called in an attempt to resolve the dispute (FDA 2000b, p. 17). The firm proposed a medication guide for doctors and patients to accompany each prescription. The guide, it was argued, would educate patients and physicians about the drug's risks. The GDAC trusted GlaxoWellcome's approach and did not support the introduction of a black box warning on the label.

By early November 2000, the FDA had received many more post-marketing adverse drug reaction (ADR) reports, including: 49 cases of ischemic colitis (of which 30 required hospitalization and five required surgery), 21 severe cases of constipation (of which 14 required hospitalization and five required surgery) and five deaths (Uhl *et al.* 2000, pp. 3–4).

GlaxoWellcome challenged the causal link between alosetron and all the cases involving deaths or surgery, though the FDA regarded the drug to be the 'probable' cause of three of the five fatalities (Uhl *et al.* 2000, pp. 3–4). At a meeting with FDA officials on 13 November 2000, including senior managers, such as Janet Woodcock, the Director of the FDA's Center for Drug Evaluation and Research (CDER), the company suggested that the risks of ischaemic colitis could be addressed by adding some precautions on the labelling about use in women over 65 on the grounds that ischaemic colitis was related to old age, and that it could be eliminated with guidance on proper identification and management of constipation. Three days later, Kathleen Uhl, Acting Director of the Division of Drug Risk Evaluation, along with other FDA scientists/consultants, Zili Li, Ann Corken Mackey and Paul Stolley, produced a memo, which responded to GlaxoWellcome's suggestion and assessed the risks of ischaemic colitis and severe constipation associated with alosetron.

According to Uhl *et al.* (2000), of the 49 cases of ischemic colitis, 73 per cent were *under* 65 years of age, while 57 per cent of the 21 cases of severe constipation were under 65. Two of the five cases of ischaemic colitis requiring surgery were under 65 as were two of the five cases of severe constipation requiring surgery. Thus, argued these FDA scientists, the risk strategy of limiting use of alosetron in women over 65 would fail to prevent further occurrences of ischaemic colitis or complications of severe constipation. Regarding GlaxoWellcome's argument that controlling constipation would manage the drug's risks, Uhl *et al.* (2000) pointed out that only one of the three fatal cases had constipation, while only two of the ten cases requiring surgery presented with symptoms of constipation. Indeed, of the five surgery cases that were classified as severe constipation, only one had constipation as a presenting symptom. Constipation in the remaining cases was supported by radiologic, surgical, or pathologic evidence of constipation. As the FDA scientists put it: 'Obviously some patients that had severe complications of constipation were not able to recognize the signs or symptoms of constipation' (Uhl *et al.* 2000). Hence, neither constipation nor prospective complaints of constipation accurately predicted risk in those patients who died or required surgery. More generally, of the 49 cases of ischaemic colitis, only nine (18 per cent) complained of constipation. These FDA scientists concluded:

> The warning signs and symptoms of ischaemic colitis or colonic ischemia are not always clear, not always typical, and do not always

occur. From our analysis there are no known risk factors to predict either ischaemic colitis or severe constipation, so any risk management strategy that focuses on the patient's age or the management of constipation will fail to manage the risk in the majority of patients exposed to Lotronex. (Uhl *et al.* 2000)

Consequently, the FDA met with the company on 28 November 2000 to determine a way of addressing the risks of the drug. Despite the analysis by Uhl *et al.* (2000), which implied that GlaxoWellcome's proposed methods of eliminating the drug's risks were unsound and that there was no way of marketing alosetron without putting women at serious risk, senior management officials at the FDA proposed a 'restricted distribution' scheme which would allow severely debilitated IBS patients already receiving alosetron to continue to access the drug under close monitoring, while the firm conducted further clinical research into its benefits and safe use (Moynihan and Cassels 2005, pp. 156–74). Present at the meeting was Dr Paul Stolley, who had been appointed as senior drug safety consultant at the FDA in the summer of 2000. He went public to journalists about some of his experiences of being involved in the FDA's regulation of alosetron. According to Stolley, during the meeting, GlaxoWellcome officials aggressively attacked the 16 November memo by Uhl *et al.* (2000), while FDA managers sat by and failed to defend their staff (Moynihan and Cassels 2005, p. 161).

An FDA scientist, who had been present at the meeting told us:

They [GlaxoWellcome] called it [the FDA memo] 'crappy'. They wouldn't accept that some of the cases were ischaemic colitis. They brought in their experts who threw some of the cases out. I didn't say a word at the meeting. They [FDA managers] made it clear that they didn't want anybody else to speak except the top people.... FDA managers didn't have a plan to take the drug off the market. They wanted to save the drug, but put a lot of restrictions on it.[1]

Stolley formed the opinion that FDA managers were reluctant to act against alosetron because they did not wish to offend the pharmaceutical industry. He remarked that this meeting sent the message that the FDA 'don't argue with drug companies; we listen to their distortions and omissions of evidence and we do nothing about it' (cited in Moynihan and Cassels 2005, p. 161). However, Janet Woodcock, one of the FDA managers present at the meeting, rejected Stolley's interpretation, claiming that the FDA wanted to determine a way forward, rather than argue about the details (Moynihan and Cassels 2005, p. 161).

In any case, rather than accept the FDA managers' proposal, GlaxoWellcome withdrew alosetron from the market immediately, arguing that the FDA's proposed restrictions were so onerous that they amounted to withdrawal, and required the company to undertake potentially expensive clinical trials (Anon. 2000g; Moynihan and Cassels 2005, pp. 156–74). Evidently, providing the drug to the comparatively small number of patients most likely to benefit in order to reduce the risks to other patients less likely to benefit, together with research into how those risks and benefits could be better understood, was, according to GlaxoWellcome, not commercially viable within the constraints of the US capitalist market system. Thus, the commercial interests of the manufacturer generated the framework within which risk-benefit uncertainties about the drug had to be assessed and regulatory decisions forged in the controversies that would ensue. GlaxoWellcome issued a press release stating that the FDA's regulatory approach to the drug had, in effect, forced the company to withdraw the drug from the market, while FDA officials told the media that the agency had offered the company a reasonable compromise that the firm rejected. In any case, after about 450,000 prescriptions in the US, alosetron was withdrawn from the market because of safety concerns in November 2000 (Anon. 2000g; 2000h).

It is worth pausing and reflecting on the fact that, up until this point in the FDA's regulation of alosetron, patient activism and public demands for the drug had been virtually non-existent, so cannot explain FDA management's behaviour in attempting to maintain the drug on the market, despite the evidence against that course of action marshalled by their own scientists. Disease-politics theory certainly cannot explain the FDA's handling of alosetron up to this point. Moreover, there is little or no evidence of any exceptional creation of expectations about the efficacy of the drug among the medical profession, patients, or scientists, at this stage, though the manufacturer had promoted the drug, as with all commercial pharmaceutical products. This points, at least in part, to the need to explain FDA management's approach in terms of a willingness to give the company and the drug the benefit of the doubt reflecting either regulatory capture or some type of corporate bias. Indeed, one of the FDA scientists who worked on alosetron told us that the office director responsible for handling the drug's regulation at the FDA told him that industry was the agency's client.[2]

Patient activism and the patient group-industry-FDA management nexus

After alosetron was withdrawn from the market, a number of IBS patient organizations formed the Lotronex Action Group (LAG) to campaign for the drug's return. The International Foundation for Functional Gastro-intestinal Disorders (IFFGD) joined the campaign and lobbied GlaxoWellcome and the FDA to re-market alosetron. GlaxoWellcome was a member of the IFFGD Industry Council, which supported the Foundation's activities (Anon. 2002k, p. 17). The IFFGD received substantial funding from GlaxoWellcome, though the Foundation's President never declared those financial links, despite testifying in support of alosetron's therapeutic value at three FDA advisory committee meetings (Moynihan and Cassels 2005, p. 169). The LAG received no direct funding from GlaxoWellcome, though FDA scientists, who had worked on the drug were suspicious of indirect company involvement. One commented:

> The actual Action Group was not funded by GlaxoWellcome, but the Support Group was. So they got a big grant for the Support Group. Then they spun off an Action Group and they could say: "we're independent".[1]

Another simply remarked: 'I would assume that somehow or other there was a connection between the company and the patient support group'.[2] It is clear that there was no direct financial connection between GlaxoWellcome and the LAG, but the precise nature of indirect financial links between the two, if any, perhaps via the Lotronex Support Group remains unclear. On 9 April 2001, the LAG submitted a citizen petition to the FDA requesting that alosetron should be returned to the market (Levine 2002, p. 3).

The campaign produced a 'dramatic barrage of emails'[2] to Janet Woodcock at the FDA, who responded by re-opening negotiations with GlaxoWellcome and setting up a public FDA expert advisory committee in April 2002 on how to allow the drug back on the market. In the interim, FDA management sought to neutralize the FDA scientists who had been most critical of alosetron as those critical scientists came to be seen as a threat to the management's new strategy to resurrect marketing approval of the drug. Stolley felt frozen out of discussions about alosetron and has claimed that Woodcock told him to stop relentlessly criticizing the drug's risks, while another of the agency's most senior drug safety experts, who was alarmed by the drug's risks and had been analyzing them, was told explicitly not to work on alosetron by his superiors at the FDA (Moynihan and Cassels 2005, pp. 161–2). Woodcock has denied that she told Stolley to stop criticizing alosetron in front of colleagues (Anon. 2002k, p. 16). Whatever the precise truth of those exchanges, the process of shaping the behaviour of the FDA scientists who understood the drug's risks was generally more subtle. As occurred with the Iressa case, discussed in Chapter 4, it relied on the development of an abdicatory culture in which some FDA scientists sought to persuade their colleagues to desist from further efforts to achieve regulatory decisions based on the agency's own scientific evidence and analyses. One of the FDA scientists working closely on alosetron recounted:

> Then the word came down [from FDA management] that the drug was going to come back on the market. They told Kathleen Uhl [principal author of the 16 November memo]. She asked for a transfer and left soon after. Zili Li [co-author of the 16 November memo] stayed on a while and tried to make the case that the drug was too dangerous to bring back, and he was called in by [two senior colleagues in the division who subsequently went to work for a major transnational drug firm], who said: 'Look they're [FDA management] bringing the drug back. We can't change their minds, they've made the decision. Be a team player, you're not helping our Division by fighting with them'. And Zili said: 'You know the drug's dangerous'. And they said: 'That may be but they've made a decision and we have to go ahead and work with them so stop being obstructive'.[1]

The persistence of Zili Li to which our interviewee alluded probably included a memo he sent to Florence Houn, Director of the FDA's Office of Drug Evaluation III on 2 April 2001, by which time GlaxoWellcome had merged with SmithKlineBeecham to become GlaxoSmithKline (GSK). Regarding the incidence of alosetron-associated ischaemic colitis among women, Li stated in that memo:

> Our review found that the methods used by GSK were inconsistent with the principles and commonly accepted practices in the incident rate calculation. As a result, the incidence rate of 1 in 700 persons in the product labelling is inaccurate and misleading. .... The longer a woman is on the treatment, the more likely it is that she will develop an episode of ischaemic colitis.

A three-month treatment like the regimen used in GSK clinical trials increases the likelihood of ischaemic colitis to 1 in 218. The estimate could go as low as 1 in 75 persons for a three-month treatment because this is the lower boundary of the 95 per cent confidence interval. (Li 2001, p. 1)

Nevertheless, Woodcock pursued communications and a meeting with GSK to develop a strategy for re-marketing the drug. In an email dated 26 April 2001, to three top aides (Sandy Kweder, Florence Houn, Director of the FDA's Office of Drug Evaluation III, and Victor Raczkowski, Acting Director of FDA Division of Gastro-intestinal and Coagulation Drug Products), Woodcock described reassurances that she had given to a senior GSK executive regarding some private agreement she had negotiated with senior GSK employees, as follows:

I just spoke to Tachi Yamada [a senior GSK employee]. He wanted to follow up on our conversation of the other day. They have talked about the Advisory Committee meeting and have some reservations: 1. that it [the advisory committee] would be a media circus, and 2. that the advisors may disagree with what we [Woodcock and GSK] have negotiated and put us back at square one, and 3. that it would slow things down. I told him that we are used to 1. and that it is ok, we can manage it, and that it might be better to do it this way than just make an announcement. I said I agree that 2. is a real liability, and we have to consider the vulnerability vs the benefits. For 3., I said we could do it in a hurry. He asked us to consider their concerns and if we still want a meeting, to call him back. He seemed ok with a meeting, just worried. JW. (quoted in Horton 2001, pp. 417–18)

Even though Woodcock has written publicly about her handling of the alosetron case at the FDA, she has never denied the existence or accuracy of the above email placed in the public domain by Richard Horton, the editor of the *Lancet*. Indeed, according to the pharmaceutical industry trade magazine, *Scrip*, she has confirmed its accuracy (Anon. 2002k, pp. 16–17). According to Willman (2001c) on 3 May 2001, Dr David Wheadon, another GSK employee and subordinate to Yamada, phoned Florence Houn, the Director of Office of Drug Evaluation III at the FDA, to follow-up on the conversation between their line managers, Yamada and Woodcock, respectively. Wheadon asked about which advisors the FDA might assign to the proposed forthcoming FDA Advisory Committee meeting on alosetron. Willman (2001c) has reported Houn's email summarizing their conversation as follows:

> [Wheadon] stated they [GSK] were reluctant to go [before an advisory committee] because the statements that come from an AC [advisory committee] meeting can be used to increase their product liability and are used inappropriately in other ways that are detrimental to the company.... I [Houn] stated that FDA does not want to have unintended consequences.... I told him that we [FDA] would work with them [GSK] on developing the agenda and questions. (quoted in Willman 2001c)

Woodcock maintained that her actions were guided by a response to patient demand. We quote her at length to ensure that her perspective is fully integrated into our analysis. Writing in the *British Medical Journal* in 2002, Woodcock accounted for her actions as follows:

> Thousands of individuals from all walks of life – businesspeople, military personnel, government employees, teachers, health-care workers – wrote or emailed the agency, demanding access to [alosetron which] they characterized as 'giving them their lives back'. Meanwhile, the drug's manufacturer, GlaxoWellcome, was shutting down production lines and ongoing trials, having rejected the FDA's proposal to create a limited access programme for severely affected people. Subsequently, under ongoing pressure from patients, the manufacturer opened discussions with the FDA on potential drug availability programmes. Thus began the arduous process of crafting a proposal for the re-introduction of Lotronex.... Clearly for the majority of people with IBS, the risks of Lotronex outweighed the benefits. However, for people with disabling symptoms that precluded a normal life, a greater level of risk might be acceptable. The challenge was creating a programme that provided access for these people and prevented use by individuals with less severe IBS or with non-IBS gastro-intestinal disorders.... Subgroup analyses [of trial data] revealed significant treatment responses in people reporting severe and frequent urgency. Correspondingly, many testimonials of major benefit sent to the FDA came from people suffering socially disabling urgency and faecal incontinence. (Woodcock 2002)

Our interviews provided some limited support for Woodcock's account. For example, one consumer activist involved in meetings about alosetron stated:

> It was clear to me in a discussion with Woodcock that they [FDA management] felt the pressure and they were very worried ... not just a sense of pressure, they felt concern. They said, 'look, there are people out there who say this drug has made all the difference in the world – whatever the clinical trials say. What do we do with it?' And there were people saying 'I couldn't go out, I couldn't live a normal life until this drug came along. It's changed my life.' So then the strategy became: how do we sort of walk between all these competing interests in a way that does as much good and as little harm as possible for everybody concerned.[3]

If Woodcock's account could be taken at face value, then the re-approval of alosetron would be evidence in favour of disease-politics theory because the FDA's regulation of the drug should be explained by the agency responding to demands from patients to serve their health interests. However, there are problems with Woodcock's explanation of events. In the first instance, the evidence set out before FDA managers by their own scientists (from post-marketing experience as well as clinical trials) was that there were no risk factors, which could be sufficiently identified prospectively, to prevent serious injury from ischaemic colitis or severe constipation. Or as Horton (2001, p. 417) put it: 'Is it right to expose patients to the risk of fatal complications from a drug when not one shred of proof exists to show that either a medication guide or a risk management programme is effective in diminishing the risk?'

Secondly, if Woodcock and her colleagues in FDA management were involved solely in an effort to maximize patients' health, then why were they working with the manufacturer to determine the membership and agenda of the agency advisory committee that would consider whether alosetron should be re-marketed? Why would the scenario of an independent FDA advisory committee rejecting a private plan between FDA managers and GSK be regarded as a liability? And why was there such a strongly held perception within the agency of an effort by FDA managers to neutralize criticism of the drug by agency scientists? According to one senior FDA scientist, a colleague at the agency wrote to Woodcock about the management's plans to re-market alosetron, saying: 'well, if we're going to approve the drug based on patient testimonials why even bother with clinical trials?'[2] Our interviewee's perception was that the colleague was disciplined. Although it was a reasonable question to ask, Woodcock had argued that the testimonials that came in from patients corresponded to subgroup analyses of the clinical trial data – a point to which we will return later in the chapter.

A final problem with Woodcock's (2002) account is that it assumes that patient demands are independent of pharmaceutical firms. However, once again, as we found in the Iressa case, the proposition that patient group activism can be entirely separated from industry interests is seen to be highly dubious in the alosetron case because of the emergent patient–industry complex. According to Stolley, 'we definitely know GSK wanted to get it back on the market since they were funding patient groups to get that message across' (Anon. 2002k, p. 17). The projected US market of about 185,000 women was still worth around US$100 million (Anon. 2000h; 2000i).

Some other remarks by Woodcock suggest that her account to the *British Medical Journal* in 2002 was a somewhat sanitized version of events. In May 2001, she stated on an FDA internet outlet:

> I have to work in the real, political world. There are limits to government power, especially right now. It's unlikely that this Administration, like the Clinton Administration before it, is going to support a wide expansion of FDA control over medical practice and other matters. (cited in Willman 2001c)

This suggests that her actions within the FDA were influenced by macro-political changes in the neo-liberal era that weakened the agency's capacity to sustain regulatory intervention in the interests of public health when that intervention meant keeping drugs off the market to the distaste of industry and their deregulatory allies in the Administration and Congress. On this latter account by Woodcock, she is pointing to evidence of neo-liberal corporate bias within the US drug regulatory system, albeit perhaps unwittingly. This latter account received support from FDA scientists within the agency, one of whom made the following comment regarding the construction of the FDA's expert advisory committees:

> The agency is going, ostensibly, to an external panel of experts to get their advice, but often what they're doing is they're looking for a cover so that they can pursue a particular course of action, and say, 'well we took it to the advisory committee, and this is what they

said' I guess on rare occasions the agency will do something different from what the committee recommends but often I think it's done for political reasons.[2]

In other words, FDA management may use its expert advisory committees to help it pursue political priorities beyond the boundaries of science, but in the name of expert scientific advice – a phenomenon that the American scholar, Sheila Jasanoff, has famously referred to as science-policy 'boundary-work' (Jasanoff 1990).

Quite apart from cases of ischaemic colitis, by 22 August 2001, the FDA had received reports of 77 cases of alosetron-associated serious complications of constipation, ranging in severity from faecal impaction, to colon obstruction, toxic mega-colon, and perforation of the bowel necessitating surgery. The majority, 86 per cent (66/77) required hospitalization, while 30 per cent (23/77) required major abdominal surgery. There were two deaths reported. Constipation was a presenting complaint in 63 of the cases, but was not in 14 of the patients (18 per cent). Those patients who experienced serious complications of severe 'unreported/non-presenting' constipation required hospitalization in 100 per cent of cases and surgical procedures in 57 per cent. As the FDA scientist compiling these figures put it:

Identification of a set of patients who apparently did not report constipation even though they were already impacted, represents a further challenge to the management of alosetron-induced complications of constipation. In these individuals the slight benefit (end of diarrhoea) may be indistinguishable from the risk (development of impaction). (Kress 2001, p. 2)

Undeterred by growing numbers of serious adverse reactions associated with alosetron, and reassured by FDA management about the drug's successful rehabilitation, on 7 December 2001, GSK submitted to the FDA a 'supplemental application' to allow the re-introduction of alosetron on to the market for women with diarrhoea-predominant IBS, who had failed to respond to conventional therapy. The supplemental application was submitted under the 'restricted approval' section (section 314.520) of the 1992 Accelerated Approval regulations and was accorded expedited review. The restricted approval provision allows the FDA to approve drugs to treat serious, debilitating or life-threatening conditions when the agency judges that those therapies can only be safely prescribed under restricted conditions of use or distribution.

Thus, our case study of alosetron makes plain, in concrete terms, the significance of industry's campaign in the early 1990s to expand the accelerated approval regulation from solely life-threatening conditions like HIV/AIDS and cancer to serious and debilitating diseases. As a consequence of that expansion it became possible for GSK and the FDA management to seek restricted marketing of a drug to treat IBS under Subpart H – a regulation ostensibly reserved for drugs promising a breakthrough in treatment of serious or life-threatening conditions.

Insofar as such matters cross the minds of the public or the medical profession, Subpart H was associated in the public mind with accelerating HIV/AIDS drugs on to the market in a more risky way because the need for some treatment of the epidemic was so desperate. What our case study of alosetron reveals is that this regulatory provision can be applied to risky drugs to treat a condition which is generally not serious, never life-threatening and far removed from the image of desperate need to treat AIDS in the late 1980s. However, the implication of the Subpart H route of approval was that re-marketing of alosetron must be restricted to women with severe and debilitating IBS. The Subpart H regulation also required the firm to show that the drug possessed a favourable benefit–risk ratio and addressed an unmet patient need.

The expert advisory committee hearing: patients talk and they're persuasive

In the April 2002 FDA expert advisory committee meeting, the experts heard from GSK, the FDA, and various patients and patient groups. The company claimed that a significant number of patients with IBS greatly benefited from alosetron. Seven women suffering from IBS testified that alosetron had provided enormous benefits to them. These were *unmediated* post-marketing reports, as no interpretation of the patients' accounts was made by medical professionals before presentation to regulators. Some patient organizations might have regarded this as an advantage because they considered doctors to be 'stuck inside medico-centric forms of knowledge' with an impersonal 'inability to grasp the dimension of illness that relates to patients' personal experiences' (Barbot 2006, p. 541). On the other hand, such unmediated reports lacked any control for placebo effect – something that scientifically designed studies are able to identify.

The FDA suggested to the experts that only a tiny fraction of those who met the classification for having IBS suffered a severe form, and that the vast majority had mild symptoms (FDA 2002b; Moynihan and Cassels 2005, p. 169). Even GSK acknowledged that only five per cent of people with IBS in the US suffered severe symptoms, and only 25 per cent of patients with IBS had the diarrhoea-predominant form. Hence, on GSK's figures, only about 1.25 per cent of people with IBS suffered from severe diarrhoea-predominant IBS. By contrast, the President of IFFGD who did not declare the patient group's financial links with GSK, claimed that 'between 1979 and 1999, 1,031 deaths were attributed to IBS' in the US alone, and that '39 per cent of IBS patients rated the pain of their IBS symptoms as extreme or very severe' (FDA 2002b, pp. 171–5). Lisa Kennedy, a member of the LAG testified:

Lotronex literally saved my life and livelihood, but without Lotronex I can no longer sustain a demanding work schedule, and I couldn't face life without it. (FDA 2002b, pp. 199–200)

Her sentiments were supported by the President of the Irritable Bowel Syndrome Self-Help Group (IBSSHG), who claimed to represent this 11,000-member organization. He contended:

While taking Lotronex, IBS sufferers reported a complete cessation of their symptoms. It dramatically changed their life for the better. ... IBS sufferers are prepared to accept the risks associated with its use and to work with the FDA to reduce those risks. (FDA 2002b, pp. 178–81)

The co-founder of the LAG declared that their goal was 'to regain access to Lotronex...which we feel is a miracle medicine that substantially improved the quality of our lives' (FDA 2002b, p. 182). Those experiences supported the idea that alosetron was a therapeutic breakthrough. While FDA scientists acknowledged the realities of the patients' experience, some remained sceptical. One recounted:

These people were clearly disabled by IBS and this product [Lotronex], they attest, worked for them. If there are people like that and it works for them, and they understand the risks – that's OK. ... The problem is that the people [patients] who come [to FDA public advisory committee hearings] usually they've been instigated by the company. There isn't a counterbalance. You don't have people showing up who lost their colons from ischaemic colitis or who died. Those people don't show up because there aren't organized groups that are funded well by industry to do that.[2]

As Barbot (2006, pp. 541–43) found with some French AIDS patient associations, these IBS patient groups eschewed statistical aggregation of individual situations and 'pressed for patients to have more leeway in choosing to use drugs whose effects have not been fully assessed'. However, broader-based consumer organizations at the hearing took a different view, looking to the 'state as guarantor of true science' and 'a bulwark preventing pharmaceutical firms from marketing ineffective or dangerous products' (Barbot 2006, p. 543; FDA 2002b, pp. 165–8). For example, a spokesperson for the Centre for Medical Consumers told us:

> The IBS patients would not get involved with the policy. They're not interested in science, they're interested in their own experience: 'it helped me, I'm telling you my life is better'. So the issues of what's the evidence, the strength and power of the studies, the fact that when you really saw the benefit of Lotronex, it was hardly a benefit at all, despite the emerging risk, was not paramount for them. But when a patient stands up and says, 'my life has changed', how do you process that? They're not making up a story, and those emotional appeals have an effect on FDA advisory committees because you're hearing from real people saying: 'What do you mean you're going to take it [the drug] away from me?'.[3]

Nevertheless, prompted by the testimonies and campaigns of the IBS patient groups, the FDA advisory committee was persuaded by the unmediated anecdotal evidence that the benefits of alosetron outweighed its risk for some patients. The comments of one expert on the committee, Dr Brian Strom, Professor of Biostatistics and Epidemiology at the University of Pennsylvania, summed up the way in which the majority of the committee swayed towards a re-introduction of alosetron:

> We are seeing a pretty consistent pattern of [average] response to the drug, but very modest in magnitude [from trials], and yet we are hearing [from patients testifying at the advisory committee meeting] very dramatic responses from individual patients that are clearly very convincing. Could we be having here a problem of law of averages..... Was this small, average response that we are seeing [from trials] everybody responded a little bit, or a few people responded a lot, and most people didn't respond at all? You might be able to pull out a small subgroup of people who should use the drug and, in fact, will benefit dramatically from it. (FDA 2002b, pp. 256 and 260)

The committee recommended (by a majority of 14 to 4) that alosetron should be returned to the market, even though only four of the 18 voting committee members believed that it was possible to identify a patient population for whom the benefits of the drug outweighed its risks – a condition of accelerated approval under Subpart H (Anon. 2002l). The expert advisory committee with all its scientific credentials had, in effect, turned some patients' anecdotal experiential accounts into promissory science.

It would be a mistake, however, to infer that that recommendation was shaped entirely by patient testimonials, and not by the commercial interests of the firm. While everyone agreed that some small proportion of patients with IBS seemed to benefit from alosetron, the possibility of providing that relatively small number of patients with compassionate use of the drug on an experimental basis was never even considered, even though that would provide the best clinical supervision of those patients, vis-à-vis the drug's risks (and indeed its efficacy).[4] Such a course of action would ensure that the drug was not 'taken away' from patients who felt that they had benefited from it – perhaps only several thousand (the size of a large trial). That option was not on the advisory committee's agenda because it had been ruled out by GSK, apparently due to lack of commercial viability. One FDA scientist explained this as follows:

> The fact is nobody ever turned round and said, 'well, how many of these people are there out there?' And then also because the agency knew that the company wouldn't go along with an IND [investigational/experimental new drug status], they weren't going to consider it [in the advisory committee meeting]. So industry lays the groundwork, they establish the boundaries, and then we'll [FDA] work within them, rather than sort of saying, 'well, our primary interest is in the public health, the public well-being, and we want to be fair about this, but all options are on the table'. And so, if you read the transcript [of the 2002 advisory committee meeting on alosetron], we [FDA] didn't talk about a compassionate IND, we just didn't talk about it.[2]

Regulatory science responds to anecdotal evidence and the patient–industry complex

The logical implication of the plan to approve alosetron via the Subpart H provisions was that re-marketing of the drug should be restricted to women with severe and debilitating IBS because those accelerated approval provisions are ostensibly reserved for drugs promising a breakthrough in the treatment of serious or life-threatening conditions. The advisory committee heard from patients who retrospectively self-identified great benefits from the drug, but the fundamental regulatory problem for FDA scientists, if they were to advise doctors on how to appropriately prescribe alosetron, was to identify *prospectively* a subpopulation of patients for whom the benefits outweighed the risks. Nevertheless, the FDA scientists accepted the anecdotal evidence from the patient groups, *and associated ideas from GSK*, as the framework within which to interrogate the clinical trials and post-market surveillance database. The regulatory science agenda was now largely being set by unmediated anecdotal evidence and the patient–industry complex. In the absence of prospective data from clinical trials demonstrating a positive risk-benefit profile for some women with severe IBS, FDA scientists were instructed to focus their efforts on trying to identify, *retrospectively*, subgroups of patients within the databases who had particularly benefited from alosetron and should be prescribed it. In addition, the agency sought to identify risk factors for those patients who had suffered serious or fatal adverse reactions, so that women taking the drug, and doctors contemplating its prescription, could be appropriately warned.

Yet the scientific evidence simply did not fit the agendas of GSK and the FDA management. First, GSK and the FDA turned their attention to whether there was a patient population who uniquely benefited from alosetron, and for whom the benefit might outweigh the serious risks. GSK argued that the benefit–risk balance was positive for women with diarrhoea-predominant IBS, who had failed conventional therapy (FDA 2002b, p. 66). The FDA, on the other hand, hypothesized that 'patients with the most disabling symptoms stand to benefit the most from Lotronex, and the benefit–risk balance is likely to be most favourable in those patients' (FDA 2002b, p. 154). In *theory*, that hypothesis was more consistent with use of Subpart H because the focus was on women with the most serious IBS and an apparent unmet medical need.

Yet neither of those rationalizations by GSK nor the FDA for re-marketing alosetron were consistent with the techno-scientific evidence. As Raczowski, deputy director of Office of Drug Evaluation III acknowledged, the drug had not been studied to assess if it was efficacious in patients who had failed conventional therapies, so such patients could be just as non-responsive to alosetron as other patients, or even more so (FDA 2002b, p. 148). Indeed, that reality should have raised major questions about whether it was appropriate for the FDA to regulate alosetron under Subpart H of the 1992 Accelerated Approval rules. Furthermore, when the FDA disaggregated the clinical trial results in ways largely consistent with the preferences of patient groups and some advisory committee experts for 'situational' data, instead of overall average scores, the re-analysis of baseline severity for 'abdominal pain', and 'frequency' or 'urgency' of bowel movement, revealed that the amount of benefit provided by the drug to each subgroup was very similar, irrespective of severity of condition. Hence, contrary to what has been suggested in Woodcock's account of events, there were no data enabling the FDA to identify subgroups of IBS patients who would greatly benefit from alosetron, and who might therefore be able to take the drug under conditions of a positive benefit–risk balance (FDA 2002b, pp. 249–50). Woodcock's claim (mentioned above) that the anecdotal testimonies of patients with IBS corresponded to subgroup analyses of the clinical trials was not correct – a fact that has an important bearing on sociological analysis of this case also. For it then follows that FDA management did not have the inclination or conviction to support their own scientists' analysis which implied that re-marketing alosetron was not in the interests of patients and public health, preferring instead to support the non-evidence–based demands of patients and the drug manufacturer. That, in turn, has implications for understanding the politics of FDA management.

Nevertheless, given the unmediated anecdotal evidence from patients with severe diarrhoea-predominant IBS, FDA scientists acknowledged that perhaps about ten per cent of those patients would derive major benefits from the drug – that is, a tiny proportion of all women with IBS (FDA 2002b, p. 261). However, neither GSK nor the FDA could predict *which* women severely affected by diarrhoea-predominant IBS might greatly benefit (FDA 2002b, pp. 262–63). Consequently, it was impossible to know how to limit the use of alosetron therapy to that sub-population so that the other (90 per cent) women with severe diarrhoea-predominant IBS were not exposed to serious and life-threatening risks, while deriving no benefit. At US population level, these estimates implied that between 73,000 and 145,000 patients could receive dramatic benefits from alos-etron, but at the expense of 655,000 to 1.3 million women with IBS being exposed to the drug's risks, including a projected 1,300 to 6,500 cases of ischemic colitis after just three months, without deriving adequate benefits.[5]

In theory, another way to enhance the benefit–risk profile of alos-tron was to avoid or mitigate the severe and life-threatening risks of the drug. If that could be done, then perhaps the ten per cent of women with severe diarrhoea-predominant IBS could receive substantial therapeutic benefits without putting the majority of IBS women at grave risk. To avert the risks of the drug, FDA and GSK would need to identify a set of risk factors for developing ischaemic colitis and serious complications of constipation, and then warn physicians to avoid using alosetron in women with those risk factors. However, neither GSK nor the FDA were able to identify risk factors for ischemic colitis among women with IBS taking alosetron. As one FDA scientist put it: 'everyone who takes Lotronex is at risk' (FDA 2002b, p. 154). With respect to identifying risk factors for constipation, a medical officer at the FDA pointed out that prior to alosetron's withdrawal in November 2000: 'labelling was depended on to *exclude patients with "presumed" risk factors to reduce the incidence of serious complications of severe constipation* [emphasis in original]' (Kress 2001, p. 36). Yet this precaution did not prevent post-marketing cases of severe constipation requiring hospitalization and surgery.

While GSK had failed to identify a subset of women who would respond to alosetron therapy safely, the company nevertheless argued that complications of constipation and other severe outcomes, such as ischaemic colitis, could be mitigated by early recognition of signs and symptoms together with timely interventions. In fact, the company argued that *all* of the serious adverse events associated with alosetron were due to constipation and that a risk management strategy which targeted constipation would, therefore, also reduce the risk of ischaemic colitis. This argument was flatly rejected by FDA medical officers who pointed to cases of ischaemic colitis in young women with *diarrhoea* (Uhl *et al.* 2000, p. 3). Regarding the other possible warning signs of ischaemic colitis (abdominal pain and bloody stools), as we have noted, FDA medical reviewers argued on the basis of the cases they had seen that warning symptoms were not always present, and when they were, were not always distinguishable from the symptoms of IBS. For example, while abdominal pain was reported in association with 74 per cent of cases of alosetron-associated ischaemic colitis, abdominal pain is also a symptom of diarrhoea-predominant IBS. With respect to bloody stools as a warning sign of ischaemic colitis, by the time most patients reported blood in their stools the condition was already serious, making this signal ineffective as a means of *early* detection (Mackey and Li 2002, p. 2).

As for early recognition of signs and symptoms, and timely intervention to prevent, serious outcomes from constipation, as we have seen, nearly one-fifth of the patients developing severe complications of constipation associated with alosetron had not even realized that they were constipated before the condition became extremely serious (Kress, 2001, p.4). Evidently, some patients were not able to recognize the signs and symptoms of constipation – a powerful reminder that patients' knowledge even of their own bodies can be dangerously limited without the detective capacities of medical science (Uhl *et al.* 2000, pp. 1–2).

The firm also proposed that the risks of alosetron might be mitigated if patients started by taking half the normal dose, increasing to full-dose only if the half-dose provided insufficient relief and/or if patients took 'drug holidays' – coming on and off the drug as needed, so that it was only used for brief intervals (FDA 2002b, pp. 68–82). However, the FDA's review of dose-ranging trials found that the half-dose (1.0 mg/day) was no better than placebo. Moreover, the drug was significantly better than placebo on the primary efficacy measure in pivotal clinical studies only if patients took it for all three months. Thus, alosetron might not deliver much therapeutic benefit for *any* women with IBS, including those with severe symptoms, if the dose were halved and/or patients were advised to have 'drug holidays' (FDA 2002b, p. 165).

Moreover, the notion that decreasing the dose might reduce or eliminate the risks of alosetron was highly questionable, to put it mildly. Regarding dosage, in 12 per cent of the first 70 cases of ischemic colitis associated with alosetron, the patients were taking the lower (half) dose of 1.0 mg/day (FDA 2002b, p. 167). The prevalence of ischaemic colitis among patients on the higher (full) dose of 2.0 mg/day might have suggested that lowering the dose could reduce the size of the population of patients at risk, but it could just as easily be due to more patients taking the higher dose. Evidently a significant number of women with IBS could remain exposed to the risk of ischaemic colitis at the lower dose.

In connection with the proposed re-marketing, GSK submitted ten new epidemiological studies of alosetron, but FDA reviewers found that those studies included 'no information on the risk of ischaemic colitis in association with Lotronex' (Brinker 2002, p. 1). Consequently, when a decision to re-market alosetron was made, it was not known 'which patients will develop a more severe form of ischaemic colitis and possibly require surgery' (Mackey and Li 2002, p. 2). Also no studies had been undertaken to confirm whether serious complications of constipation could be averted, though such research had been suggested to GSK by the FDA in November 2000 (Kress 2001, p. 38). For these reasons, a team of FDA Medical Officers who specialized in gastrointestinal disorders 'strongly support[ed] the need for additional efficacy and safety data ideally *before* considering re-introduction of alosetron in the marketplace [emphasis added]' (Kress 2001, p. 37). The implication of the medical officers' recommendation was that, if alosetron was to remain in use by some patients, then it needed to be approved as a drug under investigation (an IND) and provided on a compassionate-use basis only to those women with severe, diarrhoea-predominant IBS who had already benefited from it.

FDA approval to re-market alosetron

Due to the absence of a defined sub-population of patients for whom the benefits of alosetron outweighed its risks, when the advisory committee recommended the re-marketing of the drug they proposed careful monitoring of its use. In particular, the committee advised that only doctors trained in alosetron use should be allowed to prescribe it. During the 2002 committee hearing, GSK suggested that such monitoring could be achieved by doctors merely vouching for their own capabilities in prescribing the product – so-called self-attestation. The experts on the advisory committee reached an explicit consensus that GSK's plan should not be accepted, and that doctors should be required to undergo training and certification, convincing a third party that they were competent to prescribe the drug (FDA 2002b).

On 7 June 2002, the FDA approved the re-marketing of the drug for the treatment of women with severe diarrhoea-predominant IBS, who had failed to respond to conventional therapy, as the majority of the advisory committee had recommended (Anon. 2002k, p. 16). Yet, on approving the drug back on the market, the agency accepted GSK's risk monitoring plan permitting physicians, who wished to prescribe the drug, to merely self-attest to their qualifications to do so and agree to educate their patients about the drug's risks and benefits. The more stringent monitoring recommended by the advisory committee was not adopted by the FDA (Anon. 2002m). The agency's acceptance of GSK's weaker risk monitoring programme was in the commercial interests of the company because it promised a bigger market. The weaker risk monitoring was not demanded by patients and the fact that it was opposed by the advisory committee of specialist medical practitioners in gastroenterology is hardly evidence that it was demanded by the medical professional either. Several of the advisory committee members who recommended approval for re-marketing of alosetron were reported to be 'furious' at the FDA's decision to allow self-attestation because they did not think it provided adequate public health protection (Anon. 2002k). According to Moynihan and Cassels (2005, p. 163), one member stated: 'The risk-benefit ratio is not worth it, unless the use can be restricted to those who really need it, and who are likely to benefit from it – which is a very, very, small group'. The manner in which FDA management ignored the public-health protective measures recommended by its expert advisory committee is further evidence of the agency's responsiveness to the industry's interests.

Conclusion

Like the glitazones in Chapter 3, alosetron was approved on to the US market with priority review signifying that the drug promised significant therapeutic advance. As with the glitazones, our case study of alosetron demonstrates that drug innovations deemed to offer significant advance imbedded into official regulatory statistics may, in fact, offer very limited benefits. As with Iressa in Chapter 4, alosetron was also (re)approved on to the market under accelerated approval regulations for drugs to treat serious, debilitating or life-threatening conditions. By permitting the IBS drug to fall into that category, it was possible for it to be approved in the context of greater risks to patients, even though the scientific evidence available indicated that the drug was minimally efficacious, and that no subgroup of patients with severe IBS could be identified for whom the benefits of the drug outweighed the risks. This provides further evidence that the idea that accelerated approval regulations with lower standards of proof of safety (and efficacy) were introduced to rush life-saving new drugs on to the market is a myth. Moreover, FDA management approved the re-marketing of alosetron according to the risk monitoring plans of the manufacturer, rather than the more stringent recommendations of the agency's expert advisory committee.

All of those decisions by FDA management were in the commercial interests of the manufacturer, but contrary to the best interests of patients and public health. The influence of the manufacturer on the FDA was certainly evident from the fact that the firm withdrew alosetron from the market and refused, on commercial grounds, to undertake additional clinical trials to resolve outstanding concerns about safety and benefit that would have been in the interests of public health. Thus, the FDA permitted the manufacturer largely to set the parameters for how to make the drug available to patients. Before GSK's withdrawal of the drug from the market, FDA management's preference to give the drug and the firm the benefit of the doubt cannot be explained in terms of patient activism and public demand for the drug. Rather, the explanation must lie in some aspect of capture or corporate bias. Some comments by FDA management made in relation to the alosetron debacle imply that the agency had become highly responsive to the deregulatory agenda of the George W. Bush Administration, and to the fact that since 1992 in the neo-liberal era industry fees funded a large proportion of the agency's drug regulatory review activities and staff.

In our case study of Iressa we did not encounter evidence of FDA scientists complaining that they had been told not to work on that drug or been 'frozen out' of discussions about it. Regarding alosetron, there is evidence of a more authoritarian approach by FDA management towards agency scientists critical of the drug. While that difference should be acknowledged, it is probably less significant than the similarities uncovered in our investigations of those two drugs. In both cases, most FDA scientists, including some who accepted that alosetron was 'dangerous' and should not be re-marketed, abdicated responsibility when they became aware that FDA management was working with the company to bring the drug back on to the market. They became resigned to the fact that the management was going to give in to pressure from the patient–industry complex, the manufacturer, and patient testimonials, even before the FDA advisory committee had sat. Unlike the Iressa case, a minority of FDA scientists remained outspoken about their objections to the agency management's handling of alosetron, which may account for both the perception and/or reality of more authoritarian behaviour on the part of management towards that minority of scientists.

Nonetheless, the evidence is that, as with Iressa and the glitazones, the neo-liberal corporate bias established at the macro-political level of changes in legislation, agency funding, and regulatory policy did indeed filter down to the micro-sociological level of decision-making about alosetron, and even the type of analysis conducted on the data concerning use of the drug. Firstly, the fact that alosetron could be considered for restricted marketing under the accelerated approval regulations at all resulted from the neo-liberal deregulatory reforms pushed by the pharmaceutical industry in the 1990s. Secondly, there is evidence that management's response to FDA scientists who were too critical of the drug and claims made about it by the manufacturer stemmed, in part, from a view by some managers that pharmaceutical firms had become the agency's clients due to increasing dependence on industry fees – another hallmark of the neo-liberal reforms. Thirdly, the private collaborative communications/meetings between FDA management and the manufacturer to the extent of working together on shaping the membership and agenda of the agency's advisory committee was consistent with the neo-liberal reforms encouraging collaborative consultation between the FDA and the pharmaceutical industry. It should also be noted that this kind of agency–industry interaction is also an indicator of capture. Thus, this is further evidence in support of the idea that neo-liberal corporate bias served to facilitate capture. And fourthly, the neo-liberal pro-business conviction of the Administration and Congress that approving drug innovations quickly on to the market should be the priority of the FDA weakened the agency's position to defend its own regulatory science against the patient–industry complex and other (non-scientific) demands by patients. A hostile Congress critical of the agency for not approving drugs quickly enough would not look favourably upon an FDA management which was unresponsive to high-profile demands from some patients to (re-) market a drug. Thus, the wider political context made FDA management vulnerable to patient group campaigns whether orchestrated by the manufacturer or not. That, in turn, made it more likely that in the face of patients' demands, the agency would utilize its resources to trawl through the data attempting to identify, post-hoc, subgroups of patients for whom the drug's benefits might outweigh its risks, rather than simply concluding that the drug should not be returned to the market on the risk-benefit evidence available.

Unlike the glitazones and Iressa, there is much less evidence of anticipatory excitement and expectations about the promissory science of alosetron among the medical profession or regulators. However, as we saw in the Iressa case-study, some *patients* developed huge expectations about alosetron, willingly reconstructed as promissory science by the FDA's expert advisory committee. As with Iressa, patient demands were evident in the alosetron case, so it is another investigation in which disease-politics theory seems relevant. Certainly patient demands had an influence on FDA management and the alosetron FDA advisory committee (as also occurred in the Iressa case), but once again that influence cannot be entirely, or probably even mostly, attributed to an independent patient activism. This is because of the financial links between the manufacturer and some patient organizations involved and because of the collaboration of the manufacturer and FDA management in shaping the advisory committee process behind the scenes, perhaps including industry influence on the selection of the committee's membership.

To some extent, therefore, the influence of patient activism on FDA management and the advisory committee was really influence of the industry by proxy, so should be understood, at least partly, in terms of either regulatory capture or corporate bias theory, rather than disease-politics theory. However, insofar as patient activism may be regarded as having influenced the FDA's regulation of alosetron independently of industry, then the case offers support for the soft version of disease-politics theory. Our investigation of the alosetron case provides no evidence, however, to support the hard version of disease-politics theory, namely, that the direction of FDA regulation of the drug was influenced by independent patients' interests because the techno-scientific evidence at that time was overwhelming that the *re-marketing* of alosetron to *new* patients, who had not already benefited, was not in patients' interests. It is possible that compassionate use under a treatment IND for women with severe IBS, who had already experienced benefit from alosetron, would have been in those patients' interests, but that was not the course of regulatory action taken by the FDA.

6

The Regulatory Science and Politics of Risk Management: Who Is Being Protected?

The practice and implementation of post-market risk management strategies for new drugs, especially innovative pharmaceuticals was introduced in the neo-liberal period and may be regarded as part of the neo-liberal reform agenda, especially when it emerged in the US in the mid- to late 1990s (FDA 1999a). As we explained in Chapter 2, risk management policy developed at the FDA in response to an increase in the number of drug product innovations having to be withdrawn from the market on safety grounds, which was, in turn, related to the reductions in new drug review times during the 1990s. The pharmaceutical industry and the FDA both wanted the number of drug safety withdrawals reduced. The industry because such withdrawals were against companies' direct commercial interests while indirectly eroding public confidence in the industry and its products; the FDA because of pressure from industry, the Congress, and elements of the medical profession attached to prescribing a drug when withdrawn, combined with the fact that when a drug was withdrawn from the market it tended to undermine the agency's reputation.

Although the relationship between those macro-political factors and the emergence of risk management in the US is clear from our analysis in Chapter 2, it does not show whether or not the practice of risk management operated in the interests of patients and public health. Even if risk management was motivated by the commercial interests of industry and reputational concerns of the FDA, it could nevertheless also serve the interests of patients if all those interests coincided or converged. Ostensibly, the position of regulatory officials in the EU and the US, who supported and administered risk management strategies, is that its purpose is primarily to protect patients and public health, and that any other considerations, such as industry interests and regulators' reputations are merely secondary.

In this chapter, we investigate whether the development and implementation of risk management in the neo-liberal era was primarily in the interest of patients and public health. Fundamental to that investigation is the more specific question of whether or not it was in patients' interests to maintain particular drug product innovations on the market using risk management, rather than to withdraw them from the market. To help us answer that question, we consider all the NMEs between 1995 and 2003 that were approved on to the market by regulators in both the EU and the US regions, but were subsequently withdrawn from the market due to health risks in only one of the regions, while regulators in the other region decided to manage those same health risks by leaving the drug on the market. In fact, between 1995 and 2003, only three NMEs satisfied those criteria, namely, tolcapone (Tasmar), trovafloxacin (Trovan) and levacetylmethadol (Orlaam). Furthermore, we discovered that, in all three cases, the NMEs were withdrawn from the market in the EU, but maintained and 'risk managed' on the US market by the FDA – a significant finding in itself regarding EU/US comparisons.

Tolcapone (Tasmar) – predicted unpredictables

During the 1990s, Hoffman La Roche (hereafter 'Roche') developed a drug, known as tolcapone (Tasmar), to treat Parkinson's disease. Often a devastating neuro-degenerative disorder, Parkinson's disease afflicted, and continues to afflict, well over a million people in Europe and the US (Tanner and Goldman 1996). Although Parkinson's disease does not occur only in the elderly, it is much more prevalent among that age group, so its incidence is likely to increase with ageing populations in Europe and North America. Mortality is two to five times higher in patients with Parkinson's disease than in age-matched controls (Bennett *et al.* 1996; Morens *et al.* 1996).

In the 1990s, and for years before, the most important pharmacotherapy for Parkinson's disease was levodopa (a precursor of dopamine), which treated the debilitating motor dysfunction associated with the disease. However, it has been estimated that nearly half of the patients with Parkinson's disease, who have been maintained on levodopa for two to five years experience gradual deterioration after initial improvement because the effectiveness of the therapy 'wears off' (LeWitt 1992).

From the late 1980s, the reasons for the deterioration in the efficacy of levodopa therapy came to be better understood. For instance, scientists came to believe that one of the enzymes responsible for levodopa metabolism, known as catechol-O-methyl transferase (COMT) degraded levodopa over time, thus reducing the levels of levodopa reaching the brain. On that basis, it was hypothesized that inhibition of COMT could produce more stable brain concentrations of levodopa and mitigate the 'wearing-off' phenomenon (Micek and Ernst 1999).

In developing tolcapone, a new molecular entity (NME), Roche created a potentially innovative drug product that inhibited COMT in laboratory tests. On 31 May 1996, the company submitted an application to the EMEA to obtain marketing approval for tolcapone across the EU via the supranational centralized procedure. On 19 March 1997, the expert EU regulators on the Committee for Proprietary Medicinal Products (CPMP) issued a positive recommendation for marketing approval of tolcapone at daily doses of 100 mg and 200 mg. The European Commission confirmed that recommendation with official approval on 27 August 1997. The firm submitted a similar application to the FDA on 3 June 1996, which the agency assigned to 'standard review' status implying that the drug was not deemed to offer any significant therapeutic advance. Specifically, the Deputy Director of the FDA's Division of Neuropharmacological Drug Products, who reviewed the drug for the agency, concluded from the tolcapone trials that: 'the treatment appears clinically useful, although of moderate degree, on average' (Katz 1997, p. 21). A former member of the CPMP also took the view that tolcapone's new mechanism of dopamine delivery by COMT inhibition was not necessarily very clinically significant because 'the degeneration of the neurones takes away the dopamine receptors, so you [the patient] can have a lot of dopamine, but if the receptors are not there you [the patient] cannot derive the benefit'.[1] The FDA approved it on to the US market on 29 January 1998. In both Europe and the US it was marketed under the tradename, 'Tasmar'.

Before approval by regulators, risks of liver toxicity from taking tolcapone were identified by both the FDA and the EMEA from analyses of the pre-market clinical trials. Increases of more than three times the upper-limit-for-normal of liver enzymes occurred in 1–3 per cent of patients and three patients had elevated liver function tests of ten and 25 times the upper-limit-for-normal that continued to rise up to 42 days after discontinuing the treatment.[2] Also, liver injury (jaundice) was implicated in the death of a woman receiving tolcapone, who had no history of liver disease or known exposure to hepatitis. The FDA scientist reviewing the pre-market trials described the death as 'reasonably related to treatment [tolcapone]' (Katz 1997, p. 15). EU regulators did not mention this death in their approval material or the label for European doctors. They may have missed it because typically European regulators do not review primary clinical data to the same extent as FDA scientists.[1,3,4] There were also three cases of neuroleptic malignant syndrome (fever, muscular rigidity and altered consciousness) in the pre-marketing database on tolcapone patients (Katz 1997). EU regulators attributed one death during clinical trials to that syndrome (EMEA 1997, p. 15). In addition, FDA scientists identified liver toxicity in non-clinical, *six-month* animal toxicology studies, involving dogs, though the agency's pharmacologist, who reported it, concluded: 'The liver toxicity observed in dogs was successfully monitored and appeared reversible upon discontinuation of the drug' (Ellis 1997, p. 45).

Hence, both regulatory agencies approved tolcapone on to their markets despite knowledge of these risks, even though the drug was not found to offer significant therapeutic advance (Katz 1997, p. 21; La Revue Prescrire 1999). The regulators gave the drug the benefit of the doubt – trying it out on the market, rather than delaying approval and requiring Roche to provide an additional safety study, but advising doctors via labelling that patients' liver-enzyme levels should be monitored monthly when starting treatment for the first three or six months. As a member of the CPMP put it:

> Tasmar had a known problematic safety profile. However, you [the regulator] tried, you say this has good benefit, and Tasmar was one in particular where it was a new form of drug for Parkinson's disease. ....
> So it was put out to the market with a fairly restricted indication.[5]

Regarding the risks of liver toxicity, both regulatory agencies were reassuring. For example, the initial product label for European doctors stated:

> [Liver-enzyme] increases usually appeared within 6 to 12 weeks of starting treatment, and were not associated with any clinical signs or symptoms. In about half the cases, transaminases [liver enzymes] levels returned spontaneously to baseline values whilst patients continued Tasmar treatment. For the remainder, when treatment was discontinued, transaminase levels returned to pre-treatment levels. (EMEA 1997)

After tolcapone had been on the market for about a year, nine new cases of serious abnormal liver reactions, including two deaths and six cases of probable hepatitis had come to the attention of EU regulators. Two of the six hepatitis cases caused secondary disturbance to other organs – known as fulminant hepatitis. In response, on 15 October 1998, the EMEA issued a Product Safety Announcement, warning of the risk of potentially life-threatening liver reactions with the drug (EMEA 1998a). This was quite distinct from labelling changes because it was made publicly and directly to doctors, patients and the media. Roche were also required to update the label and prescribing information for doctors to reinforce the necessity, and increased frequency, of liver function monitoring. However, barely another two weeks had passed when the EMEA received reports of adverse neurological reactions to tolcapone including new cases of rhabdomylosis (destruction of muscle tissue). Furthermore, the EMEA was informed of another case of severe hepatitis. Crucially, in that case of hepatitis, there was clear evidence that liver monitoring *had been carried out on a monthly basis*, but the patient nevertheless rapidly developed liver failure and died (EMEA 1998a).

Consequently, in November 1998, an extraordinary meeting of the CPMP was convened. The CPMP concluded that Tasmar could no longer be safely maintained in normal clinical usage because 'serious hepatic reactions occur unpredictably and liver monitoring does not seem able to predict the development of severe, sometimes fatal, hepatic disease', and 'taking into consideration this hepatotoxicity, as well as possible occurrence of rhabdomylosis and neuroleptic malignant-like syndrome, the overall benefit to risk balance of tolcapone was considered unfavourable' (EMEA 1998a). In other words, the EU regulators concluded that there was no risk management strategy that could protect patients from the potential severe and/or fatal adverse effects of the drug, which offered no significant therapeutic advance. Thus, they suspended and withdrew the drug from the European market on 12 November 1998, which accounted for about 40 per cent of the drug's sales. By that time, about 60,000 patients had taken the drug worldwide amounting to global sales of approximately US$71 million (Anon. 1998b; FDA 1998c).

By contrast, in the US, senior FDA scientists, aware of exactly the same risks, decided that tolcapone should remain on the American market, and that those risks could be 'managed' by risk management strategies, especially labelling changes.[6] There is evidence from our interviews, however, that some more junior FDA scientists, who reviewed the drug, wanted it to be removed from the American market, rather than risk managed, but they were over-ruled. One very senior FDA scientist/manager, who was proud of the fact that the FDA never withdrew tolcapone, told us in 2003:

> Within the division [of FDA scientists who reviewed the drug for approval] I think they probably would have wanted to take it off [the market], but I said I didn't think we were ready yet. And I don't think we've had another fatality. There have been some other injuries, but that looks manageable.[6]

On 16 November 1998, the FDA published a 'Talk Paper' on its risk management strategy for tolcapone, and Roche issued a 'Dear Healthcare Professional' letter to American physicians, pharmacists and other healthcare professionals about the drug. The letter referred to 'reports of severe, potentially life-threatening cases of severe hepato-cellular injury, including three deaths' (Roche Laboratories Inc. 1998). Changes to the product label for physicians included recommenda-tions to increase liver monitoring to every two weeks for the first year of therapy; to restrict use to patients who were not responding to, or intolerant of, alternative therapies; to withdraw tolcapone treatment if a patient failed to show substantial clinical benefit in the first three weeks and to use an informed consent form to maximize patients' understanding of tolcapone's risks and benefits (Roche Laboratories Inc. 1998).

Regulators we interviewed on both sides of the Atlantic believed that, as a result of the FDA's risk management strategy, prescribing of tolcapone in the US fell.[7] That is a plausible hypothesis, though there is no published study to verify it. Thus, the FDA's risk manage-ment strategy of keeping the drug on the market may have reduced the size of the patient population exposed, but *it did not reduce the risks to those who were prescribed the drug.* For the FDA and Roche accepted the fundamental point, appreciated and clearly articulated by the CPMP, that liver monitoring could not protect against serious or fatal liver toxicity, as even the 'boxed warning' on the revised US labelling partly acknowledged:

> Although a programme of frequent laboratory monitoring for evidence of hepatocellular injury is deemed essential, it is not clear that baseline and periodic monitoring of liver enzymes will prevent the occurrence of liver failure. However, it is generally believed that early detection of drug-induced hepatic injury, along with immediate

withdrawal of the suspect drug enhances the likelihood of recovery. (Roche Hexagon 1998)

Whatever might have been 'generally believed', it was a clear fact that a death had occurred in one patient despite regular liver monitoring – a fact that the FDA permitted Roche to omit from their warning on the tolcapone label in the US.

A 'risk management strategy' may sound like it is a means to protect patients from risk. In fact, compared with the strategy of withdrawing the drug from the market, the FDA's risk management of tolcapone was not in the best interests of patients, who remained at risk of severe adverse effects from a drug offering no significant therapeutic advance. What the risk management strategy seemed to achieve was a major reduction in the number of patients who were prescribed the drug. Although the FDA's risk management strategy was not in patients' interests, it was in the commercial interests of the manufacturer because the drug remained on the US market, and in the institutional interests of FDA management, who were not embarrassed in front of Congress and other agency observers about having to withdraw another drug on safety grounds.

EU regulators, on the other hand, behaved more markedly in patients' interests than the FDA by suspending a drug of little therapeutic value when it became clear that patients could not be protected from tolcapone's severe and life-threatening risks. That said, in 2004, after reviewing the FDA's risk management programme for tolcapone, EU regulators lifted the suspension of the drug in the EU, giving it a new label that mimicked the American one (Anon. 2004b). Specifically, a black-box warning was added to the EU label listing the following precautions and restrictions on use:

1. Tolcapone should only be prescribed if there has been a 'complete informative discussion of the risks with the patient'.
2. Tolcapone was indicated for use only in patients, who failed to respond to, or were intolerant of, other COMT inhibitors.
3. If substantial benefits were not seen within 3 weeks of the initiation of tolcapone treatment, then it should be discontinued.
4. The drug was restricted to prescription and supervision by physicians experienced in the management of advanced Parkinson's disease.
5. A requirement that liver monitoring should be conducted every two weeks for the first year, every four weeks for the next six months and every eight weeks thereafter.

However, as with the US label, the European black-box warning from 2004 stated:

> Periodic monitoring of liver enzymes cannot reliably predict the occurrence of fulminant hepatitis. However, it is generally believed that early detection of medication-induced hepatic injury along with immediate withdrawal of the suspect medication enhances the likelihood for recovery. Liver injury has most often occurred between 1 month and 6 months after starting treatment with Tasmar. (Electronic Medicines Compendium, undated)

In other words, after six years of the US risk management strategy, there still remained no way to protect patients from its severe and life-threatening risks. Yet the EMEA let tolcapone back on the market, albeit with added restrictions. Ostensibly, the lifting of the suspension in Europe was because, by 2004, tolcapone had been shown in one study to be more effective than an alternative Parkinson's drug, entacapone. However, regarding that matter, a former member of the CPMP told us:

> Tolcapone was considered more active than entacapone, but it was mostly on the experimental side, rather than on the clinical effect. So this was one of the reasons why they [EU regulators and Roche] said this [tolcapone] was a superior drug, because of the efficacy, but it was theoretical efficacy.[1]

More significantly, the EU regulators' adoption of the American approach to regulating tolcapone probably reflected their more general re-orientation from the early 2000s towards FDA-style risk management.

Trovafloxacin (Trovan/Turvel) – marketing versus prevention

By the mid-1990s, Pfizer had developed a new broad-spectrum antibiotic in both oral and intravenous form. The oral version was known as trovafloxacin and the intravenous as alatrofloxacin (Anon. 1999d). However, for the sake of simplicity and brevity, in this book, we will refer to both as 'trovafloxacin'. The drug was indicated to treat 14 types of bacterial infection, including bacterial exacerbations of bronchitis, community-acquired pneumonia, intra-abdominal and pelvic infections, sexually transmitted diseases (gonorrhea and chlamydia) and skin infections including diabetic foot ulcers.

According to the pharmaceutical trade press, Scrip, Pfizer claimed that trovafloxacin had a key advantage over other treatments because of its therapeutic activity against anaerobic bacteria (such as those in the gastrointestinal tract), which might allow doctors to replace a combination of antibiotics with the single treatment by trovafloxacin. Apparently, the firm also claimed that it was the only oral and intravenous once-a-day drug with 'complete coverage of respiratory pathogens with clearly superior eradication of streptococcus pneumoniae and haemophilus influenzae' (Anon. 1997g). However, when the FDA came to review Pfizer's new drug application for trovafloxacin on 30 December 1996, the agency classified the drug for standard review, implying that it offered little or no therapeutic advance over existing drug products. Nevertheless, the drug was found to be efficacious, so the FDA approved it on to the US market on 18 December 1997 (Feigal 1997). It was marketed in the US by Pfizer with the brand name, 'Trovan'.

A similar application was made by Roerig Farmaceutici (Pfizer's Italian subsidiary) to the EMEA through the EU's supranational centralized procedure for innovative pharmaceuticals on 24 January 1997. Regarding Pfizer's claims about the efficacy of the drug, the CPMP seemed to take a similar view to the FDA. One former member of the CPMP told us: 'I don't believe it because my recollection is that there were no real specific benefits [of trovafloxacin over existing similar antibiotics on the market]',[1] but it did have efficacy. The EMEA approved the drug on to the EU market in July 1988 (EMEA 1998b). Pfizer and Roerig Farmaceutici jointly sponsored trovafloxacin on the European market with brandnames 'Trovan' (Pfizer) and 'Turvel' (Roerig). As the antibiotic came on to the major markets of Europe and the US, financial analysts forecast that its peak sales could be as high as US$1.0–1.6 billion of which about US$900 million would be derived from the US market alone. As it turned out, they correctly predicted that the main market would be the US, but the figures were rather optimistic. Trovan recorded worldwide sales of US$62 million in 1998 (US$54 million from the US), while in the first quarter of 1999, sales in the US reached US$54 million and US$8 million in Europe (Anon 1999d; 1999e).

After marketing approval in the US, apparently Pfizer told Scrip that trovafloxacin's most common side effects were dizziness, nausea, headache and light-headedness, which were usually mild and transient – a message reproduced by the EMEA's summary information about the drug after marketing approval in the EU (Anon. 1997g; 1998c). Strictly speaking, across all the pre-market clinical trial data, this may have been true. However, FDA scientists also identified risks of liver toxicity upon reviewing the randomized controlled trials submitted in support of the drug's marketing approval application. In a trial to support use of the drug to treat prostatitis, 15 of 140 (10.7 per cent) patients receiving trovafloxacin had liver-function-test abnormalities compared with just one of 132 (0.8 per cent) patients taking the comparator drug, ofloxacin (Alivisatos 1997, p. 227). In only five of the cases had Pfizer's clinical trial investigator considered the liver-function-test abnormalities to be attributable to trovafloxacin. However, the FDA's reviewing medical officer, Dr Alivisatos, determined that there was a consistent pattern of abnormalities in all cases, both in terms of the timing of the events and the duration of trovafloxacin therapy (Alivisatos 1997, pp. 229–230). There was also a case in that trial involving simultaneous elevation of liver-enzyme levels and bilirubin,[8] – a sign of drug toxicity capable of severe liver injury, though not present in the other clinical trials (Alivisatos 1997, p. 226; FDA 2000c).

Despite the identification of those risks, and the acknowledgement that Trovan offered little or no therapeutic advance, both regulatory agencies decided to approve this antibiotic on to their markets. Once again, the regulators gave the drug the benefit of the doubt, as illustrated by the initial product labelling approved in the US, which stated:

> The incidence and magnitude of liver function abnormalities with Trovan were the same as comparator agents except in the only study in which Trovan was administered for 28 days (chronic bacterial prostatitis)..... Patients were asymptomatic with these abnormalities, which generally returned to normal within 1–2 months after discontinuation of therapy..... Because Trovan can cause elevations of liver function tests during or soon after prolonged therapy (i.e. equal to or greater than 21 days), periodic assessment of hepatic function is advisable.

Just two months after Pfizer's February launch of trovafloxacin (Trovan) on to the US market, in April 1998, Alivisatos noted three post-marketing reports of increased liver-enzyme levels in patients taking the drug for less than 14 days. That was significant because the initial labelling had implied that liver problems were likely to occur only if taking the drug for more than 21 days. The FDA then instituted a 'Trovan-monitoring system' to capture all adverse liver reactions, which included explicit instruction to Pfizer to report all serious and non-serious cases promptly (Alivisatos 1998, pp. 3–4).

By the end of June 1998, the FDA had received 38 such reports, nine involving liver injury, though no deaths. The average duration of therapy before symptoms or liver abnormalities in these 38 cases was 11 days – much less than what was supposed at the time of marketing approval. The most severe case involved a 38-year-old man who took Trovan for 10 days for recurrent sinusitis, resulting in such serious injury that he needed a liver transplant (Alivisatos 1998).

As with tolcapone, the FDA sought to manage the risks while maintaining the drug on the market. The agency requested that Pfizer should revise Trovan's labelling to take account of these new cases, but the firm's initial proposed revisions, which emphasized uncertainty about the causal relationship between liver injuries and the drug were deemed unacceptable by the FDA and not approved for use. Pfizer's proposed (and non-approved) labelling, which was never used, read as follows:

Post-marketing experience: adverse events reported with trovafloxacin during the post-marketing period for which a causal relationship is uncertain include: anaphylaxis, hepatitis, liver failure, Stevens-Johnson syndrome. (Alivisatos 1998, p. 12)

According to FDA scientists involved in assessing the antibiotic, Trovan had proved to be very popular in the US because it was broad spectrum and had a strong marketing company behind it, namely Pfizer. One agency scientist remarked: 'This [trovafloxacin] was a big product for them [Pfizer] – and they were very concerned about FDA wanting to try to take this product off the market'.[9]

It was not until September 1998, three to six months after the liver toxicities were identified, that the agency and Pfizer produced agreed labelling changes advising doctors more fully about Trovan's risks, stating:

During the post-marketing period, Trovan-associated liver enzyme abnormalities and/or symptomatic hepatitis have occurred during short-term or long-term therapy. Liver failure has also been reported rarely. Clinicians should monitor liver function tests and pancreatic tests in patients who develop symptoms consistent with hepatitis and/or pancreatitis as clinically indicated. (Pfizer 1998)

Meanwhile, the EMEA had just approved trovafloxacin (Trovan and Turvel) on to the EU market that July, but the EU label advising European doctors about how to prescribe the drug contained little reference to trovafloxacin's potential to cause liver injury. In particular, it made no mention of the American post-marketing experience of liver injuries *within 14 days* of use – a reality known to Pfizer and the FDA since April 1998. It appears that the EU regulators did not consider that evidence at that time. The secrecy of the EU drug regulatory system prevents us from knowing whether that was because Pfizer did not fully inform them about that important American post-marketing experience at that time or because the European regulators chose not to act on the information upon receipt from Pfizer.

On 6 May 1999, Pfizer contacted the EMEA concerning 140 post-marketing reports of serious liver injury associated with trovafloxacin, including eight cases of liver failure where patients required liver transplantation or died (EMEA 1999a; 1999b, Annex 1). More cases of severe liver injury were reported over the following weeks (EMEA 1999b). Consequently, on 25 May 1999, the EMEA urgently issued a Product Safety Announcement warning:

> The occurrence of the liver injuries varied between 1–60 days after start of treatment. These data suggest that the onset and the severity of these liver injuries are *unpredictable* (EMEA 1999a, emphasis in original).

At that point, the EU regulators had become aware that the serious and life-threatening liver injuries associated with trovafloxacin were unpredictable in both onset and severity, and that there were no associated risk factors, such as dose or duration of therapy.

On 10 June 1999, the Pharmacovigilance Working Party of the CPMP re-evaluated the safety profile of trovafloxacin, while an ad-hoc expert group re-evaluated the efficacy of the drug and considered the therapeutic place of the product compared to other antibiotics. Conclusions of both these meetings were then presented to an extraordinary meeting of the full CPMP on the same day. The committee determined that there was a clear causal relationship between the use of trovafloxacin and serious liver injuries, which were more frequent and severe than with other antibiotics. The CPMP further confirmed that the occurrence of severe hepatic injuries was unpredictable and that no preventive measures had been identified.

During the course of the CPMP extraordinary meeting, Pfizer made an oral presentation in which they proposed more restricted uses of the drug. The firm also argued that trovafloxacin had potential advantages over other antibiotics, with respect to availability of oral and intravenous form, a broad spectrum of activity, specific activity against penicillin-resistant streptococcus pneumoniae and against anaerobes, together with its therapeutic potency as a single drug in place of combination therapy. However, after reviewing the original new drug application in search of evidence for superior clinical efficacy over the comparator drugs used in clinical trials, the CPMP judged that no clinically significant superiority of trovafloxacin had been demonstrated – a finding consistent with the FDA's classification of trovafloxacin as a 'standard' drug offering no significant therapeutic advance.

Moreover, the CPMP could not identify any serious indications/needs for which safer therapeutic alternatives did not exist. Logically, the committee concluded that even for the company's proposed restricted indications, suitable therapeutic alternatives existed. In August 1999, the EMEA, therefore, suspended and then withdrew Trovan from the EU market on the grounds that its risks outweighed its benefits, and that it could no longer be safely maintained in clinical usage (EMEA 1999b). By that time, over a million patients may have been exposed to the drug, perhaps over a hundred thousand of them in Europe. It has been estimated that by May 1999, 2.5 million prescriptions had been issued worldwide for use of trovafloxacin of which 200,000 were in Europe (Anon. 1999d; Feczko 1999).

By contrast, the FDA permitted the drug to stay on the market by extending its risk management strategy via more elaborated risk communication, and by restricting its use to patients with serious or life-threatening infections where therapy was initiated in an in-patient health care facility. Having received reports of over 100 cases of clinically symptomatic liver toxicity in patients taking Trovan since July 1998, on 9 June 1999, the FDA issued a Public Health Advisory to physicians stating:

FDA is aware of 14 cases of acute liver failure strongly associated with Trovan exposure. Four of these patients required liver transplant (one of whom subsequently died). Five additional patients died of liver-related illness. Three patients recovered without transplantation, and the final outcome is still pending on two patients ... Trovan-associated liver failure appears to be unpredictable. It has been reported with both short-term (as little as two days exposure) and longer-term drug exposure, therefore the efficacy of liver function monitoring in acceptably managing this risk is uncertain. (FDA 1999c)

As with Tasmar, the FDA agreed with EU regulators' assessment that liver monitoring could neither prevent nor predict serious or life-threatening adverse reactions to the drug, but despite this was willing to recommend it as a risk management strategy to maintain the drug on the market. As one FDA scientist put it:

> But with Trovan, clearly, you don't know that if someone who takes it is going to develop liver failure. But in large enough numbers, you would see that. .... After we came up with the solution of restricting the label, putting in new warning information, putting out public health advisories, having Pfizer do what it said it would do – only make it really available under special circumstances, then we said, 'now we have to monitor this and see if this risk management programme is going to work or not'.[9]

Similarly, the FDA recommended in June 1999 that Trovan therapy should not be used for longer than 14 days because the 'risk of liver injury may increase substantially with exposure beyond 14 days' (FDA 1999c). Yet the US regulatory agency fully acknowledged that severe and/or fatal liver reactions had occurred among some patients taking the drug for just a few days, so the advice given to physicians could not prevent such injuries.

One senior FDA scientist, who had analysed the agency's database on the adverse reactions to Trovan suggested that the influence of Pfizer combined with the FDA's reluctance to dismiss all of the firm's arguments, preferring to adopt a compromise with the company instead, was the key reason why the drug remained on the US market, albeit restricted to the in-patient environment. He described the FDA's internal response to reports of liver toxicity associated with Trovan as follows:

> Well, first initially there was just this flat-out rejection that there was a serious problem. The [FDA] medical officer thought there was a serious problem, but she wasn't convinced it needed to come off the market. .... [Then] the risk of liver failure increased. .... We had a couple of [internal FDA] meetings where people were talking tough about it, [saying]: 'Yes we're going to take the drug off the market', and they'd meet with the company and they wouldn't even bring it up. .... The company argued quite strenuously for it [Trovan] to stay on the market. .... And so in the end we're [FDA] sort of saying to the company: 'Well we're not killing the drug completely, we won't take it all off the market, we'll take it off the out-patient market, but we'll leave it on the in-patient market'.[10]

According to FDA officials, prescription of Trovan in the US dropped as a result of their risk management strategy, which is not surprising since the drug could no longer be prescribed outside of a hospital setting.[6,9] Although the risk management strategy caused major declines in prescription and use of Trovan in the US, as with our tolcapone case-study, what that decline demonstrated was that the risk management served primarily to reduce the number of patients exposed to the risk, rather than the risk to those who continued to be prescribed the drug.

According to the Wall Street Journal, on 6 October 1999, speaking on behalf of FDA management, Murray Lumpkin, Deputy Director of the agency's Center for Drug Evaluation and Research (CDER) stated:

> Liver toxicity was rare enough that it wasn't picked up in tests of the drug prior to approval, tests that involved 7000 patients. The problem showed up in the FDA's post-approval reporting system, after doctors had prescribed Trovan to millions of patients. (cited in Langreth and Fawcett 1999, p. B2)

However, our research reveals that *the* public representation of the history of the Trovan case by FDA management or the Wall Street Journal (or both) was not strictly true. The science of the pre-market clinical trials in 7000 patients did detect a signal of liver toxicity. Regarding the pre-market risk assessment of Trovan, the key regulatory issue was how that signal should be interpreted. In that respect, both the FDA and EU regulators chose to give the drug the benefit of the doubt and see how it fared on the market.

Levacetylmethadol (Orlaam) – heartbeats away from adherence

In the US, the Biometric Research Institute applied to market levacetyl-methadol (Orlaam) as an alternative to methadone for the treatment of opiate addiction in 1993. The FDA accorded levacetylmethadol 'priority review' status and approved it on to the US market in July 1993 (FDA 1994). In the EU, Sipaco Internacional made a similar application, where it was received less enthusiastically and approved in early 1997, only after the CPMP had raised a series of queries about the testing of the drug (EMEA 2001c, p. 4). Unlike the FDA, the CPMP judged that levacetylmethadol offered no therapeutic advantage over methadone. Reviews of the pre-market clinical efficacy trials by both regulatory agencies support the CPMP's evaluation, despite the FDA's 'priority review' classification. The main advantage claimed was that the dosing schedule for levacetylmethadol (thrice-weekly as opposed to daily dosing with methadone) might improve patients' adherence to medication. However evidence from the clinical trials did not support this. Rates of patients dropping out of the trials were similar for the two drugs (EMEA 2001c; FDA 1993, p. 4).

As occurred in our case studies of both Tasmar and Trovan, American and European regulators identified major risks with levacetylmethadol from pre-market toxicological and clinical testing. Pre-market regulatory reviews in Europe revealed that, after six months of administration of the drug to dogs, tachycardia (inappropriate rapid heart beats) and abnormal electrocardiography (ECG) were observed in the animals (EMEA 2001c). Regarding safety testing with patients, EU regulators noted the following cardiac risks:

> Clinical trials have shown that Orlaam induces QT[11] prolongation [abnormal functioning of the heart ventricle following increases in heart-beat], thereby indicating a risk of *torsades de pointes* [severe and rapid disturbance of heart rhythm]. In a pharmacokinetic study, ECGs were monitored in 31 patients who were switched from methadone to Orlaam ... There was a significant prolongation of the QT interval. (EMEA 2001c)

Despite the identification of these cardiac risks, and the fact that levacetylmethdol offered no therapeutic advantages over methadone, both regulatory agencies decided that levacetylmethdol could be approved on to the market using labelling to communicate these risks to prescribing doctors. Physicians in Europe were advised that, 'in patients in whom the potential benefit of Orlaam is deemed to outweigh the potential risk of *torsades de pointes*, an ECG should be performed prior to initiation and after two weeks of treatment to detect and quantify the effect of Orlaam on QT interval'. Yet, as European regulators later acknowledged, previous experience with several other medicinal products had demonstrated that 'the predictive value of [ECG] monitoring is quite poor' (EMEA 2001c, p. 10). Similarly, the US label recommended monitoring in patients with a history of cardiac defects, or those taking medications affecting cardiac conduction (FDA 1993, p. 8).

Levacetylmethadol was marketed in both Europe and the US under the brand name, 'Orlaam'. While on the European market, between July 1997 and December 1999, the EMEA received reports of 10 cases of life-threatening cardiac rhythm disorders associated with Orlaam. The cases included five cardiac arrests, four episodes of *torsades de pointes*, and three patients needing pacemaker insertion. Significantly, these cases occurred in young patients (median age 39 years), a population normally at low risk of cardiac disorders (EMEA 2000b). Moreover, the fact that they occurred in an unusually young population group implied that the adverse effect was also more likely to go undetected raising the specter of a much greater danger than had been reported (Anon. 2001j).

On 15 December 1999, the EMEA issued a Product Safety Announcement detailing changes in prescribing and patient information following two cases of *torsades de pointes* and one sudden death associated with Orlaam. Just 700 patients had received the drug in the EU at that time. Physicians were advised that the drug should be discontinued in patients experiencing symptoms suggesting the occurrence of a severe arrhythmia (EMEA 1999c). A year later, after further reports of life-threatening cardiac arrhythmias, the EMEA issued a second Product Safety Announcement advising prescribers not to introduce any new patients to Orlaam therapy (Anon. 2001e; EMEA 2000).

In the subsequent few months, the CPMP performed a comprehensive risk–benefit reassessment of Orlaam (EMEA 2000). Electrophysiological data submitted in the original marketing approval application and a later study, were now judged by the CPMP to confirm that Orlaam was highly pro-arrhythmic. Furthermore, in comparing the safety profile of Orlaam to the therapeutic alternatives available in the EU (methadone and buprenorphine), the Committee determined that the pro-arrhythmic potential of methadone was much lower than Orlaam (EMEA 2001c, p. 10). Specifically, the CPMP concluded that Orlaam's benefit of thrice-weekly administration instead of the daily dose required with methadone did not outweigh the risks of severe and potentially fatal ventricular arrythmias. EU regulators considered whether to restrict Orlaam to patients who were unresponsive to, or intolerant of, methadone treatment, but decided against this because there were no clinical data to support such an approach. Consequently, they suspended and then withdrew Orlaam from the EU market on 19 April 2001.

By contrast, the FDA, aware of the same evidence about the drug's adverse cardiac effects, chose to maintain Orlaam on the US market and manage the risks by labelling revisions. By that time, about 33,000 patients had been treated with the drug in the US (Anon. 2001f). The FDA required the company marketing Orlaam in the US, now Roxane, to add a 'black-box' warning about the drug's cardiotoxicity to its label and restrict its use to opiate-addicted patients who had failed on alternative treatments (Anon. 2001f). In September 2003, Roxane voluntarily withdrew Orlaam from the US market because its use had fallen dramatically since the FDA-imposed warnings. Yet the company also now acknowledged that 'the risks of continued distribution and use in the face of less toxic alternatives no longer outweighed the overall benefits' (Anon. 2003b).

Conclusion

Consistent with our discussion of accelerated approval regulations and Iressa in Chapter 4, it is instructive to compare the situation in the EU and the US. As we noted in Chapter 2, the risk management strategy introduced in the US from the late 1990s was a neo-liberal regulatory reform to avoid withdrawing drugs from the market on safety grounds, but it was not introduced into the EU until the mid-2000s. In this chapter, we have shown in stark terms the implications for patients' health of that (de)regulatory reform by comparing regulatory outcomes for the same drugs with the same known risks in the EU, where risk management strategy was not deployed as regulatory policy, with the US where it was. Our case studies provide strong evidence that from 1995 to 2003, the FDA was much more willing than EU regulators to use risk management strategies to maintain on the market pharmaceuticals, which posed severe and life-threatening dangers to patients. FDA staff, who embraced risk management, highlighted their reluctance to deny patients access to these potentially harmful drugs. That may well have been their genuine motive, rather than merely a rationalization. Nonetheless, in the cases of Tasmar, Trovan and Orlaam, it is not a credible justification because the drugs offered no therapeutic advantage of any significance.

Given the cross-Atlantic evidence, acknowledged by the scientific assessments of the EU and US regulatory agencies themselves, that the post-marketing risk management strategies put in place for these drugs could not protect patients from life-threatening dangers, one must also conclude that such permissive strategies by the FDA were not in the health interests of patients. Our investigation found that the European regulatory decisions to withdraw those drugs from the market were more robust in protecting health than the FDA's adoption of risk management policy. While the neo-liberal reform of risk management policy may well have been in the interests of industry and the FDA, the clear implication of our findings is that it was not in the interests of patients and public health. Consequently, neo-liberal theory is not supported by our investigation of pharmaceutical risk management policy.

Having said that, in two of the case studies (Tasmar and Trovan), there is evidence that some FDA scientists thought that the drugs should have been removed from the market, but senior FDA officials preferred to embrace risk management. The use of risk management strategies, therefore, seems to have been an approach that FDA managers were keener to deploy than some of their subordinates within the agency. That may be because the managers, more than the scientific officers, had an acute concern about handling the agency's reputation in the aftermath of a drug safety withdrawal. There is also evidence that the FDA deployed risk management instead of market withdrawal as a compromise in the face of opposition from manufacturers attempting to protect their commercial interests. Moreover, given that communication of risks to doctors and patients on the product label is a central part of pharmaceutical risk management, the more risk management is utilized then the more dependent regulation becomes on labelling alterations. Yet, as the Trovan case study shows, labelling changes are negotiated with the drug's manufacturer, which may delay, or even underplay, warnings to doctors as the firm's commercial interests act as a partial filter for renewed labelling.

Insofar as the development of risk management within FDA pharmaceutical regulation was an institutional response by the agency to shield itself from perceived regulatory failures and concomitant criticisms from Congress and industry, then it may well be an example of what the risk analyst, Rothstein (2006) calls 'risk colonization'. In our case studies, risk management did not manifest itself as a tool for dealing with unforeseen post-marketing risks, as is often believed in media and official depictions of pharmaceutical regulation. Rather, it was a way of bolstering regulatory decisions about known risks, which casts doubt on the idea that risk management per se was needed to cope with a real increase in unexpected, post-market drug risk to patients. Rather, the evidence tends to supports the idea that such risk management practices were characterized by neo-liberal ideology of 'risk colonization' to protect industry and regulatory institutions from public and political criticism consequent upon market withdrawal. In particular, such 'risk colonization' simultaneously achieved two institutional objectives for the FDA. It reduced the conflict with industry often provoked by market withdrawals; and it deferred and significantly abdicated responsibility to physicians within an ideology of 'consumer choice', so avoiding populist accusations of 'big government' interference that might ensue from powerful deregulatory elements of Congress and Administration in the event of mounting numbers of drug safety withdrawals. In these respects, our findings about risk management policy sit well within corporate bias theory, but only partly support capture theory as the FDA's institutional interests to shield itself from public and Congressional criticism seem to be as important as responsiveness to industry interests.

Our conclusions are supported by the concerns and research of FDA scientists themselves. For example, one fundamentally challenged the agency's justifications for risk management:

> Is risk management managing industry's risks or patients' risks? It can be used as a device to keep products on the market in a fashion that they shouldn't really be on the market either through narrowing indications or the way the drug is distributed. I believe that risk management has been used as a tactic to allow products to continue to be marketed.[10]

Another felt it was based on false premises:

> There is an approach, which is approve everything and just label it with risks. There are a number of examples where it was just devastating [for public health] to take that approach. So are you really offering people a choice if you just give them an enormous book and say: 'Here it is'. I don't think that's a real choice. It is our [FDA's] job to make judgements in the public interest and if people call that paternalistic then that's fine.... It is overly facile to say we'll just give people information and let them make a choice... that really can go too far, and indeed it has. We've already seen very bad results with that.... People expect the FDA will maintain standards and act in the public interest.[12]

While yet another observed:

> They [FDA] are very reluctant to turn drugs down now. What they're saying is that all drugs have risks but we can manage those risks by labelling, by educating the patient and physician.... the evidence suggests that once a drug with a lot of risks gets on the market it's actually hard to do that.[13]

Some of that evidence has been produced by research involving FDA scientists. In the late 1990s, FDA scientists, together with colleagues outside the agency, undertook evaluative analyses of some of the FDA's risk management practices. Graham et al. (2001) examined the effect of FDA's risk management efforts, including repeated labelling changes and 'Dear Healthcare Professional' letters, on recommended periodic liver enzyme monitoring of patients taking the diabetes drug, troglitazone. Soon after marketing approval in March 1997, troglitazone was found to cause acute liver failure and was withdrawn from markets worldwide in March 2000. Graham et al. (2001) analysed the impact of four progressively stringent liver monitoring recommendations on over 7000 patients in the US between April 1997 and September 1999. They determined that 'less than five per cent of the patients received all recommended liver enzyme tests by the third month of continuous use', and concluded that 'FDA risk management efforts did not achieve meaningful or sustained improvement in liver enzyme testing' (Graham et al. 2001, p. 831). Similarly, Smalley et al. (2000) evaluated the impact of the FDA's risk management of cisapride, a drug to treat heartburn, found to cause life-threatening cardiac arrythmias from 1997 to 1999. Warnings on the label and 'Dear Healthcare Professional' letters were issued to advise doctors against prescribing the drug in patients with particular diseases or taking other medications that, combined with cisapride, could increase the risk of heart rhythm problems. Analysing data on over 22,000 patients, Smalley et al. (2000) found that, after the risk management interventions, the number of patients in the population advised against using the drug fell by only 2 per cent compared with before the interventions, and concluded: 'The FDA's 1998 regulatory action regarding cisapride use had no material effect on contraindicated cisapride use' (p. 3036).

Despite concerns among its own scientists, our documentary case studies, confirmed by interviews on both sides of the Atlantic, show that, from 1995 to 2003, the FDA was less precautionary than its European counterparts in its application of risk management to serious post-market drug risks. Specifically, the FDA used risk communication, 'warning campaigns', and patient monitoring to manage the maintenance on the market of pharmaceuticals with severe and life-threatening risks, such as liver and cardiac toxicity, when EU regulators removed the pharmaceuticals from the market because scientific analyses demonstrated that patients could not be protected from those dangers (Keijsers et al. 2008).[5,14] Indeed, senior FDA managers have remained committed to this neo-liberal risk management approach, citing doctor and patient autonomy to choose their own risks as justifications for keeping such drugs on the market,[6,15] despite growing evidence that it did not reduce patients' risks, especially regarding the risks of liver injury and cardiac arrythmias. Notably, concerns about professional clinical autonomy also exist in Europe, but that did not prevent EU regulators from removing toxic drugs from the market (Currie et al. 2009).

As recently as 2005, senior EU regulators were sceptical about the prospects of risk managing toxic drugs in the post-marketing phase, rather than simply removing them from the market. Even though the CPMP permitted Tasmar back on the market in 2004 that was done only after observing new, albeit questionable, evidence about the drug's therapeutic advantages. One European regulator asserted:

> Historically, I think not enough follow-up has gone on. It would be difficult to find a regulator that didn't agree with that. Making some regulatory decision is one thing but seeing that it actually operates in practice and has resulted in a reduction of risk is another. .... It can often be very difficult to change both professional behaviour and patient behaviour and sometimes the only way of changing behaviour is to do something drastic in regulatory terms which actually limits access to a medicine.[14]

Another European regulator raised the wider issue of the opportunity costs to public health associated with post-marketing risk management strategies:

> Risk management is risky in itself because you must put in place a control of the management of risk to make sure that everybody does it. You must measure the cost of that versus the real benefit for things that are more useful for public health.[1]

In other words, given the very minimal effectiveness of pharmaceutical risk management strategies in protecting patients, published by FDA scientists and colleagues, regulatory resources might be better spent simply withdrawing dangerous drugs that offer little or no therapeutic advance, and concentrating on more effective regulation of safer and more therapeutically significant innovations.

There is no doubt that, between 1995 and 2003, EU regulators were more precautionary than the FDA regarding risk management in the post-market phase. In analysing environmental risks more generally, scholars have drawn attention to national styles of regulation, often characterizing American regulation of industry as precautionary compared with a more permissive approach in Europe during the 1970s and 1980s (Abraham and Millstone 1989; Brickman et al. 1985; Jasanoff 1986; Vogel 1986). However, Vogel (2003) has revised those characterizations by proposing the 'flip-flop' hypothesis, which asserts that American environmental and consumer product risk regulation was more precautionary during the 1970s and early 1980s, but since then Europe has become more precautionary than the US. Given our case studies, it is tempting to conclude that pharmaceutical risk management also adheres to the 'flip-flop' hypothesis. However, that would be a hasty oversimplification.

One of the advantages of longitudinal research of the kind presented in this book is that spatial (inter-regional) comparisons are not frozen in time. Since 2005, the EU has developed a formalized pharmaceutical risk management policy, known as the 'European Risk Management Strategy (ERMS)', and there is evidence that – as occurred in the US – risk management has been used as a means of keeping unsafe drugs on the market (International Society of Drug Bulletins and the Medicines in Europe Forum, 2009, pp. 2–3). Those developments demonstrate a number of things. First, the sharp Euro-American contrasts in pharmaceutical risk management, which existed in the period from 1995 to 2003, are fading and there is now a convergence between the EU and the US in this aspect of drug regulation. Second, with respect to the goal of promoting and protecting patients' health interests, the lessons of the 1995–2003 period have not been learned by regulators and governments – if they had been, then the FDA should have followed the EU lead, rather than vice-versa. And third, the fact that there is this particular type of convergence shows the power of neo-liberal corporate bias (in the form of the transnational pharmaceutical industry and its pro-business allies in government) to steer state regulatory intervention across both continents away from product market withdrawals towards potentially endless marketing of risky drugs, including those offering no therapeutic advantage.

Finally in our case studies of risk management, like the glitazones, but unlike Iressa and Lotronex, we found no evidence of patient activism pressing for the FDA to keep Tasmar, Trovan or Orlaam on the US market, though some FDA scientists mentioned the agency's awareness that Trovan was popular with prescribing physicians. Hence, the FDA's decisions to maintain Tasmar, Trovan, and Orlaam on the market, which was inconsistent with patients' best interests, cannot be explained by patient pressure for access to the drugs (whether as part of the patient–industry complex or not). Consequently, these case studies do not support either the soft or hard versions of disease-politics theory. Unsurprisingly, great expectations about innovation did not surround Tasmar or Trovan as they were both deemed to offer little or no therapeutic advance, though Pfizer managed to generate a considerable amount of enthusiasm for doctors about Trovan in the US. In this context, the promissory science was not of the innovative potential of the drugs, but rather the regulatory science of post-market risk management and its (unjustified) promise to protect patients from drug injury.

7
Conclusions and Policy Implications

Since 1980, the pharmaceutical industry has gained unprecedented, privileged access to the state in the EU and the US, enabling it to work in collaboration with its allies in the executive and legislative branches of government to bring about regulatory reforms in its commercial interests. This has made it possible for the industry and government to advance a pro-business deregulatory agenda in the pharmaceutical sector, including reforms of drug regulatory agencies themselves, such as appointments of more industry-friendly heads of the regulatory agencies, increased dependence of the agencies on industry fees, extension of informal consultation between regulators and firms, and responsiveness to commercial, rather than health, priorities in terms of how quickly regulatory review of new drugs is completed.

Specific key changes to regulatory standards for innovative pharmaceuticals resulted from those neo-liberal reforms in the US and the EU. As we have shown, these included the broadening of drug innovation categories that should receive fast/priority review; the widening of the types of conditions for which drugs intended to treat them may attain 'accelerated approval' or 'conditional marketing'; the use of non-established surrogate markers of drug efficacy, together with companies' post-marketing commitments to establish efficacy; and the implementation of risk management in response to safety problems emerging with innovative drugs after marketing. While those changes were in the commercial interests of pharmaceutical firms, we have found considerable evidence that they have not operated in the best interests of patients and public health. This is, in part, due to the loosening of regulatory standards themselves, and partly because of the neo-liberal political context of the reforms, which generated a regulatory culture that gave health interests insufficient priority over industry interests.

It follows from this that the ideology of 'coincidence of interest' regarding innovative pharmaceuticals, often declared by senior officials in government and industry in policy agendas, such as the US Critical Path Initiative and the European Innovative Medicines Initiative, is frequently misleading and untrue. While the Critical Path Initiative and the Innovative Medicines Initiative have increasingly encouraged regulatory agencies to be proactive in assisting pharmaceutical product innovations, little emphasis has been placed on the public health need for those innovations to offer therapeutic advance.

Many of the neo-liberal reforms, including some changes to regulatory standards, were brought about via legislation or instigated at executive policy levels beyond the regulatory agencies. Indeed, as we have shown, a significant number of the early neo-liberal reforms were opposed by regulatory agencies in the US and some European countries. For that reason, it would be misleading to characterize the neo-liberal reforms, which have shifted drug regulation further towards the interests of industry at the expense of the interests of public health, simply as regulatory capture. The main thrust of the reforms is better understood as corporate bias. However, it is also the case that many of the reforms put in place as a result of corporate bias during the neo-liberal period have made the agencies more vulnerable to capture by the interests of industry, and even facilitated such capture (see later).

In the US, but not Europe, there is evidence of some additional, independent drivers of those reforms, namely AIDS patient activism, but such influence has been overstated in both scholarly and non-scholarly work, sometimes for ideological reasons. Although AIDS activism caused the FDA to accept more uncertainty in its risk-benefit assessments of the first AIDS treatments, the formalization of lower and looser standards for drug approval into new regulations, well beyond life-threatening diseases, let alone AIDS, resulted from interventions by the pharmaceutical industry and its allies within the Reagan, Bush, and Clinton Administrations. In particular, the President's Task Force on Regulatory Relief, followed by Quayle's White House Council on Competitiveness exerted continual pressure on the FDA throughout the 1980s and into the early 1990s to 'streamline' the drug approval process for the benefit of industry. Furthermore, it is, perhaps, also worth noting that it is possible even AIDS activists miscalculated the real interests of public health at that time. As we found in our interviews, in subsequent years and with hindsight, some of the prominent activists from the period expressed the opinion that they were wrong to demand earlier access to drugs whose benefit and safety had not been established because, in the long-run, that probably slowed access to truly effective therapy (Hilts 2003, pp. 304–5).

In Europe, patient activism demanding accelerated approval of new drugs was rare during the neo-liberal reforms of the 1980s and 1990s. Although patient activism in Europe has grown during the 2000s, it has been primarily concerned with gaining access to drugs within the healthcare systems of individual European countries after marketing approval by the EMEA or a national regulatory agency. For example, this has been evident in the UK where patient activism has pressed the National Institute of Heath and Clinical Excellence (NICE) to recommend that some innovative pharmaceuticals, which had already received marketing approval from the EMEA, should be freely available to patients on the National Health Service (NHS) in England and Wales. Hence, insofar as patient activism has been significant within Europe in the 2000s, it cannot account for the neo-liberal framework that shaped the emergence of supranational EU pharmaceutical regulation during the 1980s and 1990s. Nor can it explain the deregulatory measures adopted during the 2000s to accelerate drug development and approval at the *supranational EU* level. Moreover, one would be hard-pressed to find any patient or consumer group in Europe or the US that supported or demanded increased dependence of drug regulatory agencies on industry fees.

The significance of disease-politics theory is, therefore, quite limited both spatially (to the US) and temporally to the late 1980s and early 1990s in terms of explaining regulatory reforms. Even then, patient activism and culture was only one of a constellation of factors. Rather than hailing the arrival of a new disease politics, a more sober analysis of recent history suggests that patient activism, and subsequent patient-group campaigns about other drugs, can be understood within an interests-framework with which social scientists and others are already familiar. The demands of the American AIDS activists for weaker regulatory standards to expedite approval of AIDS drugs in the late 1980s and early 1990s provided public legitimation and FDA rationalization for that goal, as patients themselves appeared to be asserting a coincidence of interests between industry and patients' health. As we have shown, the nature of the regulatory relationship between the FDA and industry had already shifted considerably before the AIDS crisis.

The ideological representation of coincidence of interests between industry and patients was then fostered by the organizational strategies of the pharmaceutical industry, especially in forging the emergence of the patient–industry complex, and the discourses of the 'expert patient' and the 'informed patient'. All of these served as levers with which to create the impression that the commercial interests of industry were indistinguishable from advancement of patients' health, and that weakening regulatory standards was a liberatory development in the best interests of patients. Yet, the evidence suggests overwhelmingly that, irrespective of the motivations and ideological representations behind the deregulatory reforms, the number of innovative pharmaceuticals offering modest or significant therapeutic advance to patients actually went into decline afterwards.

Such findings provide no evidence to support the claims of neo-liberal theory that the deregulatory reforms were, across the pharmaceutical sector as a whole, in patients' health interests. Indeed, our macro-political findings suggest that the neo-liberal reforms have undermined the capacity of pharmaceutical regulation to promote and protect the best interests of patients and public health. Furthermore, our case studies show that even those innovative pharmaceuticals designated as offering significant therapeutic advance may be only very marginally more effective than placebo. If nothing else, that should lead us to question exactly what judgements underpin the official FDA data suggesting that 40 per cent of NMEs offer modest or significant therapeutic advance. Our in-depth, case-study research supports those surveys that put the figure much lower. Van Luijn *et al.* (2010, p. 445), for instance, have reported that of the 122 NMEs they identified between 1999 and 2005, only 13 (ten per cent) were shown to be superior to already available medicines. Evidently, not only it is an ideological misperception to conflate innovation with therapeutic advance, but it is also mistaken to conflate official FDA categorizations of 'modest and significant therapeutic advance' with actual evidence of therapeutic advance. This can lead to distortions in culture, discourse and policy about the correct priorities for pharmaceutical innovation and regulation, such as overemphasis on rapid marketing approval at the expense of thorough pre-market checks on risk or timely withdrawal of harmful drugs. As we saw in the Lotronex case, not only this is an ideological misrepresentation, but it is also dangerous because it can encourage patients, citizens and healthcare professionals to accept higher risks and incidences of harm and injury due to the misplaced conviction that it is a worthwhile sacrifice ostensibly in exchange for a valuable therapy.

Our case study of the glitazones shows why the neo-liberal deregulatory reforms were so dysfunctional for the task of increasing the number of innovative pharmaceuticals offering significant therapeutic advance. The whole thrust of those reforms was to expand the use of surrogate measures even to those that were 'non-established', to shorten approval times and, through increased industry consultation, to reduce the risk of non-approval in response to industry's demands. Consequently, the reforms completely failed to address the public health problem that, for most NMEs, pharmaceutical regulation has required only that drug innovations have greater efficacy than placebo. Moreover, the glitazones case study shows that regulatory reliance on some 'established' surrogate measures of drug efficacy (such as blood-glucose control) instead of reliance on measures of real clinical benefit (such as reduction in heart attacks and strokes) can lead to widespread use of drugs that are less safe for patients than existing medications. Patients' interests with regard to the glitazones would have been better served by more extensive pre-market testing of real health outcomes compared to available therapy. But the neo-liberal deregulatory environment in both the EU and the US militated against tackling that problem because the real solution lay in *increasing* regulatory demands on industry. Instead, a succession of deregulatory reforms encouraged more permissive regulation in which pharmaceutical firms were required to provide less conclusive evidence of drug efficacy before marketing and increasingly allowed to 'confirm' the effectiveness and/or safety of their products in post-marketing studies. Consequently, doctors and patients have had less secure knowledge about whether the drugs they are prescribing and using actually work, let alone whether they offer therapeutic advance.

Detailed analyses of the case studies also show that the neo-liberal corporate bias that characterized the establishment of the deregulatory reforms did indeed have an effect on the micro-social level of decision-making and regulatory outcomes with respect to specific drugs. The Iressa case does not support the claim of neo-liberal theory that the introduction of the deregulatory reform, known as 'accelerated approval' in the US, was in patients' interests or the interests of public health. Rather, the case implies that requiring companies to provide better and more targeted evidence about efficacy, if appropriate, is much more likely to be in patients' interests, even if that process takes longer initially. It also reveals that FDA decision-makers were inclined to consistently give the company the benefit of the doubt, albeit an inclination strengthened by support for the drug from some patients and expert advisers. Notably, in the EU, where the impetus for neo-liberal reform was less intense than it was at the time of Iressa's approval in the US, European regulators were more cautious about the drug's efficacy and did not approve it on to the market until some (albeit weak) evidence of benefit for a specific target population was established.

The American 'accelerated approval' regulation, whose applicability industry had pushed beyond drugs to treat life-threatening conditions, also permitted FDA management to (re)approve alosetron on to the US market. By permitting the irritable bowel syndrome (IBS) drug to fall within that regulatory provision, it was possible for it to be approved in the context of greater risks to patients, even though the scientific evidence available indicated that the drug was only mildly efficacious, and that no subgroup of patients with severe IBS for whom the benefits of the drug outweighed the risks could be identified. This provides further evidence that the idea that accelerated approval regulations with lower standards of safety (and efficacy) were introduced solely to rush life-saving new drugs, like AIDS drugs, on to the market is a myth.

The implementation of risk management, as practised by drug regulators in the US, also primarily served the interests of manufacturers by enabling their products to remain on the market, despite severe and/ or life-threatening risks and the availability of equally effective alternatives. The expectation and hope of some experts in pharmacovigilance that risk management might be used to proactively generate data to better characterize the safety profile of medicinal products (Waller 2006) is not supported by our investigations, which indicate that risk management was used as a means to maintain on the market drugs with known risks. In the neo-liberal era, in which regulators were encouraged to collaborate with industry about nurturing innovative pharmaceuticals, and the ideology of coincidence of interests between industry and patients was enthusiastically promoted, risk management was politically and institutionally a much more congenial option than withdrawing drugs from the market on safety grounds. In the US, drug safety withdrawals provoked conflict with industry and potentially criticism from Congress and some elements of the medical profession, not to mention some patients and the patient–industry complex. Risk management, acting as a form of 'risk colonization', evaded those uncomfortable political and institutional tensions (Rothstein 2006). Our detailed analyses of specific drug case studies, however, such as Tasmar, Trovan and Orlaam, demonstrate that the decisions by EU regulators to withdraw those drugs from the market was more in the interests of public health than risk managing them on the market, as the FDA chose to do.

In all of those case studies, the regulatory decisions taken consistently awarded the benefit of the doubt to the drug firm and facilitated the approval and continued marketing of the innovative pharmaceutical involved. The neo-liberal deregulatory reforms since 1980 are not the only explanation for that, but there is considerable evidence that they accentuated and facilitated it. Notably, the extent to which this occurred was often greater in the US than in the EU during a period when the deregulatory agenda was more intense and extensive in the US, especially regarding accelerated approval regulations and risk management. While many scholars have observed that the neo-liberal reforms since 1980 have changed the risk–benefit calculus for regulators making decisions about new drugs, they have not followed up their observation and examined whether or not such a change has been in the interests of patients and public health. What marks out our analysis is that we have done this in our case studies. We have found that the change has not been in the interests of public health by reference to the standards of the regulatory science supposed to inform the institutions' decision-making. In that respect, our case studies are consistent with aspects of corporate bias and capture theories, but not neo-liberal theory.

This leads us to the question of how regulatory agencies, such as the FDA, which are supposed to promote and protect public health, have ended up making decisions inconsistent with that goal. While Chapter 2 showed how neo-liberal macro-politics generated deregulatory reforms, our case studies provide important insights into how that filtered through the 'ranks' of the regulatory agency right down to scientific reviewers of individual drugs at the FDA. First, neo-liberal corporate bias within the state facilitated a significant amount of capture at higher echelons of FDA management. This involved political appointments of industry-friendly top managers and the promotion by agency management of an organizational culture which – due to increasing dependence on industry fees – viewed pharmaceutical firms as FDA's 'clients'. Second, the instigation and perpetuation of a pro-business ideology by the US Administration and Congress – the ideology that approving drug innovations quickly on to the market necessarily benefited public health and should therefore be the FDA's main priority – weakened the agency's ability to defend its own regulatory science against the patient–industry complex and other (non-scientific) demands by patients. Thus, the wider political context made FDA management vulnerable to campaigns by patient groups – whether orchestrated by the manufacturer or not. That in turn, made it more likely that FDA scientists would, for example, undertake regulatory science trawling post-hoc for subgroups of patients for whom the drug's benefits might outweigh its risks, rather than simply concluding that the drug should not be marketed on the risk–benefit evidence available.

Thus, the relative capture of FDA management, manifested by their commitment to a pro-drug approval ideology, was not accomplished by industry alone, but was nonetheless consistent with industry capture. Further evidence that the neo-liberal reforms served to facilitate capture is manifest by the increase in private collaborative communications/meetings between FDA management and drug manufacturers to the extent of working together to shape the agenda of the agency's advisory committees. Such capture of FDA management, in turn, generated what we call an *abdicatory culture* among a considerable number of FDA rank-and-file scientists. This abdicatory culture was characterized by an *acculturated resignation* towards the futility of challenging pharmaceutical firms, senior FDA managers and advisory committees to ensure that regulatory standards to protect public health were upheld when contrary to the commercial interests of companies. For example, as we saw in Chapter 5, most FDA scientists, including some who accepted that alosetron was 'dangerous' and should not be re-marketed, abdicated responsibility when they became aware that FDA management was working with the company to bring alosetron back on to the market. They became resigned to the fact that the management was going to give in to the pressure from the patient–industry complex, the manufacturer, and patient testimonials, even before the FDA advisory committee had sat.

Although the neo-liberal reforms facilitated capture of the FDA management, which then filtered down the agency to engender an abdicatory culture, it is important to appreciate that it was rank-and-file FDA scientists who often documented many of the problems with companies' new drug applications. At conferences and in publications, it is sometimes argued by cynical analysts of pharmaceutical regulation that regulators are so deeply in the pocket of the pharmaceutical industry that society would be better off without drug regulation and/ or that one should concentrate one's investigations on the industry because that is where the real power lies, rather than with regulators, who make no difference. Our research clearly shows that this cynical view is a flawed oversimplification. The thorough and commendable work of rank-and-file FDA scientists frequently raised concerns about whether a new drug should be granted marketing approval or remain on the market, if already approved. In fact, an overwhelming finding from the case studies is the extent of evidence against approval which was amassed by agency scientists, but then ultimately set aside by FDA management to make way for marketing approval. Hence, with a less captured management, FDA regulation could have been substantially more effective in protecting public health, even in the context of reduced review times.

It should also be noted that not all rank-and-file agency scientists abdicated responsibility to challenge pharmaceutical firms, FDA management, advisory committees, and the patient–industry complex. Some continued to defend a sceptical, arms-length relationship with pharmaceutical firms and rejected the idea that the drug industry's commercial interests in product marketing necessarily coincided with the health interests of patients. However, those scientists tended to find themselves in frequent conflict with FDA management.

Similar capture of management occurred in many of the drug regulatory agencies of individual western European countries during the late 1980s and 1990s at the behest of neo-liberal reforms. As we explained in Chapter 2, that then significantly shaped the agenda of supranational EU pharmaceutical regulation, which, until 2010, fell within the responsibility of the Commission's trade and industry division, DG Enterprise, rather than DG Health. Thus, within the EU too, neo-liberal reforms created a context conducive to industry capture of EU drug regulation. For example – as in the US with PDUFA – the 'quid pro quo' for industry funding of EU regulatory agencies during the late 1980s to early 1990s was acceleration of the drug review process. Moreover, there is evidence that industry funding of EU regulatory bodies, coupled with optionality in the regulatory system, has further facilitated industry capture by creating an environment in which agencies and individual regulatory scientists are forced to compete for fees. The preservation, at industry's behest, of a decentralized procedure for drug approvals following the creation of a new, centralized drug regulatory agency ensured first, that national agencies had to compete against each other for fees in an internal EU market (Abraham and Lewis, 2000); second, that the EMEA has had to compete with national agencies (Garattini and Bertele, 2001); and third – since companies can express a preferences as to which rapporteurs they would like to review their applications – that CPMP members are under pressure to compete for rapporteurships and a portion of the application fee (Garattini and Bertele, 2001).

Another important factor influencing the regulation of innovative pharmaceuticals in the neo-liberal era was the role of expectations and promissory science. There are at least two quite distinct levels at which promissory science have operated. First, at the meso-level, promissory science was built into some of the neo-liberal deregulatory reforms themselves. For example, the extension of marketing approval of some drugs based on non-established surrogate measures of efficacy under 'accelerated approval' and 'conditional marketing' provisions in the US and EU, respectively. The very possibility of such approvals created expectations and generated promissory science about, for instance, what followed for cancer patients' survival from the finding that a new drug could shrink tumours. While there might be much doubt about whether the promissory science could deliver any clinical benefits for patients, one thing was certain: pharmaceutical firms would get their drugs on the market earlier. Similarly, risk management promised to protect patients from drug risks and fostered expectations among doctors and patients that the dangers of risky pharmaceuticals had been tamed. There are enormous question marks about whether the promissory science of risk management has lived up to those expectations, but one thing is certain: it enabled drug companies to maintain their products on the market. Evidently in the scenarios of both accelerated approval and risk management, promissory science can be associated with a neo-liberal ideology of serving patients, but in reality it has promoted the commercial interests of the pharmaceutical industry devoting little serious attention to the ramifications for patients' and public health.

Promissory science, therefore, has performed neither a neutral nor an entirely external, role in the shaping of pharmaceutical regulation in the neo-liberal era. For example, although the surrogate efficacy paradigm has been partly rooted in pursuits endogenous to scientific and professional exploration, its resilience in regulation stemmed from its congruence with quick and cheap drug development, which is in industry's interests and is consistent with ideological expectations that mistake technological novelty for public health benefit. Consequently, crucial information about the efficacy and safety of new drugs came to be collected after, not before, doctors and patients medicated illness, as occurred with the trials of the glitazones' macrovascular effects.

The second level at which promissory science has functioned is the one more familiar to readers, namely, the micro-level related to the expectations constructed around individual drugs. This often occurs in connection with the manufacturer's promotion of apparently hopeful clinical trial results in the lead-up to a new drug application for marketing approval. As we showed in many of our case studies, both the hopeful results and the novelty of the new drug's hypothesized mechanism of therapeutic efficacy may be elevated to great clinical significance to an expectant medical profession, patient population and stock market. The creation of such expectations has enabled manufacturers' interests to be expressed throughout a regulatory agency's external environment, sometimes via an emergent patient–industry complex and expert science advisory committees. This complemented and reinforced the more direct industry influence on regulators already discussed.

Something that is clear from our case studies, but is often neglected in sociological analyses, is that the *dominance of particular expectations is structured around interests*. For example, the expectations about the glitazones that took centre-stage in the minds of the medical profession and regulators (and of course the firms promoting them) related to the drugs' innovative mechanism and its potential for therapeutic advance in diabetes treatment, even though clinical evidence to substantiate that hypothesis was absent. Yet, there was other evidence at the time of the drugs' approval – evidence relating to carcinogenic risks with pioglitazone and cardiac risks with rosiglitazone and the consequent potential for major harm to patients. However, expectations about the possible realization of those risks were entirely absent from the discourse about the drugs around the time of their approval and early marketing because no organized interest sought to foster expectations about the drugs' risks. As we saw in Chapter 3, it was, in fact, the risks that turned out to be real, while the promissory science of therapeutic advance proved to be no more than wishful thinking.

While all our case studies reveal regulatory decisions in Europe and/ or the US that awarded the claims made by drug manufacturers the benefit of the doubt by approving or maintaining innovative pharmaceuticals on the market, in most cases (Actos, Avandia, Tasmar, Trovan and Orlaam) that decision-making occurred in the absence of any significant patient activism demanding the drugs. The fast-tracking of the glitazones, for instance, went ahead together with the creation of expectations about their innovative mechanism and concomitant claims about therapeutic advance with little or no involvement of patient activism on either side of the Atlantic. There was also precious little evidence of patient activism pressing for the FDA to keep Tasmar, Trovan or Orlaam on the US market via risk management, though some FDA scientists mentioned the agency's awareness that Trovan was popular with prescribing physicians. Hence, the FDA's decisions to maintain Tasmar, Trovan and Orlaam on the market, which was inconsistent with patients' best interests, cannot be explained by patient pressure for access to the drugs. Consequently, most of our case studies do not support either the soft or hard versions of disease-politics theory. In those cases, drug regulatory decisions took the trajectory that they did because of the influence of industry interests, Congress, and neo-liberal corporate bias, rather than because of a new politics of patient lobbying and patient-activist culture.

By contrast, in the cases of Iressa and Lotronex, patient activism clearly played a significant role in the regulatory decision-making process in the US, mainly via the public fora of FDA advisory committee hearings. Yet even in those two cases, the influence of patient activism on FDA management and the advisory committee was, to some extent, really influence of the industry by proxy, so should be understood, at least partly, in terms of either regulatory capture or corporate bias theory, rather than disease-politics theory. Insofar as patient activism demanding access to Iressa and Lotronex influenced the FDA's decisions about the drugs, independently of industry, then those cases support the soft version of disease-politics theory. Nevertheless, the pharmaceutical firms in both cases still played an enormous role in shaping the regulatory options according to their commercial interests, quite separately from patients' activism and interests. Even in the cases of Iressa and Lotronex, our investigation does not support the hard version of disease-politics theory because, given the techno-scientific evidence available at that time, the marketing approval of Iressa and the re-marketing of Lotronex to new patients, who had not already benefited on trials or access programmes, was not in patients' interests. Taking the seven cases together in Europe and the US, one led to the conclusion that disease-politics theory, which grew out of analyses of American AIDS activism, actually provides a fairly limited explanation of drug regulatory trajectories even in the US, let alone the EU. Nonetheless, in view of cases like Iressa and Lotronex in the US, corporate bias theory should be modified to appreciate the full spectrum of actors shaping regulatory change evident from our empirical investigations. Specifically, corporate bias theory should progress to recognize that the regulatory state tends to embrace demands for accelerated drug approvals from patients and the medical profession because those demands help to cement a smooth partnership with industry, especially in regulatory agencies where a political culture of timidity about antagonizing manufacturers has percolated down to agency scientists.

The EU–US comparison

One lesson we have learnt from our investigation of pharmaceutical innovation and regulation over the last 30 years in Europe and the US is that general statements about the two regions that hold for the whole period should be made with caution. For instance, major neo-liberal reforms began in Europe, particularly with increases in the extent to which regulatory agencies became dependent on industry funding, changes in the senior management of the agencies to increase 'efficiency' for industry, and shortening of drug approval times. However, it would be a mistake to erect an inter-regional comparative thesis that neo-liberal deregulation has been more extensive in Europe because, by the late 1990s, neo-liberal deregulatory reforms in the US had, in many respects, 'overtaken' the 'light-touch' regulation which had come to characterize much of Europe. Yet there is evidence that even the more recent comparative trajectory may be temporary since by the mid-2000s, many of the flagship neo-liberal deregulatory reforms in the US, such as accelerated approval provisions for drugs to treat serious or life-threatening conditions and risk management policies, had been emulated in Europe. The main macropolitical conclusion to be drawn about Europe and the US since 1980 is that there has been a convergence of trajectories around neo-liberal deregulatory reforms, but which region has led the trend has varied over time, and according to the specific regulatory reform in question.

Nonetheless, our case studies do reveal some differences, which suggest that the EMEA initially developed a more precautionary patient-protective approach to regulation after 1995 than existed in the US over the same period. In our case studies, the EU regulators took much more seriously the implications of the presence or absence of substantiated evidence of clinical outcomes in relation to companies' promissory science claims about drugs' efficacy. This is illustrated by the stricter way in which the EU regulated Iressa and the glitazones. Moreover, until the mid-2000s, EU regulators were much less willing to embark on risk management strategies that kept on the market pharmaceuticals posing severe and life-threatening dangers to patients when equally effective alternative therapies existed.

It is well-known that the FDA was a tougher drug regulator than most European countries in the 1970s and 1980s. Thus, the inter-regional comparative dimension of our research on the regulation of pharmaceutical product innovation implies some support for Vogel's (2003) 'flip-flop' hypothesis regarding consumer product regulation, namely that US regulation was more precautionary during the 1970s and 1980s, but since then roles have reversed with Europe becoming more precautionary than the US. Insofar as that is the case, it can be explained primarily by two factors, both arising out of the particular structures of EU governance. First, EU regulators may have been able to resist more effectively than FDA reviewers the influences of neo-liberal corporate bias and inducements to industry capture, which were clearly present in both regions. Significantly, this explanation draws support in our interviews not only from European regulators, but also many FDA scientists who were not managers, and relates to the extent of multi-layered autonomy of regulatory science from political ideology and institutional interests. FDA scientists are directly accountable institutionally and career-wise to FDA managers, who, in turn, work under the FDA Commissioner, who is a political appointment made by the US Administration of the day. Furthermore, the US Congress controls the FDA's budget and has the power to regularly review and, to some extent, define the agency's performance targets. Hence, the FDA's regulatory science and decision-making, whether precautionary or permissive, is vulnerable to direct influence by the political ambitions of the US Administration and Congress. In the neo-liberal era of industry-friendly government, the manifestation of that influence has been to push the FDA towards negotiated consensus with drug companies and increasingly permissive regulatory science.

By contrast, the organizational structure of the EMEA is relatively horizontal and diffuse. As far as scientific assessment of new drug applications is concerned, the CHMP members are the decision-makers, but we encountered no examples of those members complaining about how they were managed by the EMEA. CPMP members are not full employees of either the EMEA or the European Commission, and their careers are not primarily determined by either institution, so they retain considerable autonomy while also wielding greater power than FDA advisory committee members. Consequently, it may be much more difficult for managers at the EMEA to over-rule or neutralize scientific assessors than it is for FDA management to control its scientists and medical officers. In fact, we did not find evidence of such a process, though we acknowledge that it would be much more difficult to uncover within the relatively opaque EU drug regulatory system.

The second explanation relates to the political environment within which the EMEA has operated during the period studied. Although the EU Commission – particularly DG Enterprise – has consistently pushed an industry-friendly, neo-liberal agenda for pharmaceuticals regulation, the orientation and influence of the other key political institutions within Europe have been more diverse. While the EMEA is accountable to the EU Commission, the Commission, in turn, is accountable to a European Council of (initially 15, now 27) Heads of EU Member States with a less unified political ideology than the US Administration, and to a European Parliament that has been less active than the US Congress in pressing for neo-liberal deregulatory reforms. Indeed, the EU Parliament acted as a moderating influence over the Commission's neo-liberal agenda during the 2001 legislative review, and played a key role in shaping the outcome of that process towards one that was less oriented to the needs of industry and more oriented towards the health needs of EU citizens. Consequently, the intrusion of neo-liberal political demands into the content of pharmaceutical regulation was less intense in the EU than the US during the 1990s and early 2000s.

However, we would not want to push that comparative contrast very far. It is important to understand spatial (inter-regional) comparisons in the context of temporal trends. As we noted in Chapter 2, in the mid- to late 2000s, the EMEA elected to follow the FDA's approach to both accelerated approval and risk management, and these approaches have continued to dominate debates about the future of pharmaceutical regulation in both regions (Breckenridge et al., 2012). Moreover, as in the US, the drivers for those developments were rooted in a mix of political concerns, ambitions and ideology, namely, concerns to boost the economic competitiveness of European companies and fulfill the ambitions of the EU supranational state to compete with the US in the context of a globalized, knowledge-based economy (Lofgren and Benner, 2011), coupled with a neo-liberal ideological perspective which assumes that the route to achieving this aim is to reduce alleged regulatory barriers to technological innovation. Subsequent adoption by EU regulators of the regulatory mechanisms, concepts and language of the post-PDUFA and post-FDAMA FDA with respect to innovation, accelerated drug development and post-marketing risk management suggest that the sharp Euro-American contrasts in regulation of innovative pharmaceuticals, which existed in the period from 1995 to 2003, have *diminished* and there is now mainly convergence between the EU and the US in those, and many other, aspects of drug regulation.

Lessons and changes needed

A presumption in much pharmaceutical policy analyses has been that the therapeutic benefits of pharmaceutical innovations are large because they are characterized as offering significant therapeutic advances or breakthroughs. It was accentuated culturally and politically by the experience of American AIDS activism and anti-retroviral drug innovation in the late 1980s and early 1990s. Upon that presumption, a high tolerance of drug risks and harm was accepted, even promoted in the academic-industry-policy nexus. The first lesson from our investigation of pharmaceutical regulation in the neo-liberal era is that drug product innovation should not be confused, or conflated, with medicines that offer significant therapeutic advance. Moreover, in many cases, those new drugs given priority review by regulators on the grounds that they promise significant therapeutic advance should not be so characterized given the scientific clinical evidence in support of their efficacy. This suggests that regulators need to introduce a much more sophisticated system of differentiating between innovative pharmaceuticals, so that prioritization of regulatory review serves the interests of public health more effectively. In fact, before the neo-liberal reforms, the FDA had a more sophisticated system in the 1970s, as we mentioned in Chapter 2. The introduction of that kind of system in the EU and US would be a step forward and a good starting point from which to address the problem that there is no explicit prioritization of regulatory review in the EU, while in the US, on some estimates, three-quarters of NDAs given priority review do not offer significant therapeutic advance.

Ostensibly, the publicly declared aim of the neo-liberal deregulatory reforms since the 1980s was to reduce regulatory barriers and stimulate pharmaceutical innovations needed by patients. What we have learnt is that the number of pharmaceutical product innovations, including those offering significant therapeutic advance, has declined in the aftermath of deregulation. That reality undermines the claim that over-regulation was responsible for difficulties in producing drugs really needed by patients. Rather, it suggests that the challenges of developing new drugs offering significant therapeutic advance lay elsewhere -in the sheer biological intractability of some diseases, for example, or in the commercial priorities of the industry. Moreover, the fact that the number of pharmaceutical innovations offering significant therapeutic advance declined after the neo-liberal deregulatory reforms, while pharmaceutical markets and sales expanded, is strongly suggestive that the primary consequence of deregulation was to permit the industry to widen the gap between meeting its commercial objectives and meeting the health interests of patients in accessing genuinely valuable new drug therapies.

Broadly speaking, the lesson from the neo-liberal era is that regulatory standards need to be raised and extended in the interests of public health, rather than lowered and loosened. In particular, the narrowly construed definition of regulatory efficiency as speed of regulatory review and marketing approval during the neo-liberal era has been misguided from the perspective of the interests of patients and public health, though it has served the commercial interests of industry. Our case studies have demonstrated why that has been the case across a number of key neo-liberal deregulatory reforms and disease conditions. In particular, our research has shown why the terms of debate about early fast approval should not be construed solely as early potential benefit versus heightened risk of harm, but also as early potential benefit versus increased risk of harm and marginal, or even absent, benefit.

The overwhelming lesson from our case studies is that the goal of public health benefit would have been better achieved by gathering and analyzing more and better data on drugs' clinical efficacy and outcomes before marketing approval. Speed of approval is not the most important criterion of regulatory performance so far as health is concerned. Hence, in some respects the reforms needed to improve pharmaceutical regulation for health since 1980 should have gone in the opposite direction and encouraged different priorities to the priorities pursued under neo-liberal deregulation. For example, much progress could have been made towards strengthening the evidence base for new medicines had regulators required new drugs to be tested against existing gold-standard therapies (where they exist), rather than merely placebo. Often gold-standard therapies would be medicines, but not necessarily so. For example, in the case of IBS, new drugs might also be systematically compared with dietary interventions. The problem, of course, during the neo-liberal era is that far too frequently, before marketing approval, the efficacy of NMEs has not been systematically tested against other effective drugs, let alone gold-standard non-drug therapies. Where use of a placebo is ethical, three-arm trials (involving the experimental new drug, gold-standard therapy and placebo) provide the best evidence for patients, doctors and healthcare systems and encouraging this kind of clinical trial design should have been a priority of the regulatory reform agenda. It is a more challenging standard for industry, and more expensive since it would require larger clinical trials. Its neglect, relative to the enormous efforts marshalled to speed up marketing approval during the neo-liberal era, further highlights the extent to which the regulatory state in the EU and the US prioritized the commercial interests of industry over health interests.

Equally important for the construction of a regulatory system that aims to achieve optimal health benefits for patients is the crucial question of how companies are required to demonstrate and measure drug efficacy in clinical trials. For some disease conditions, such as diabetes, where established surrogate measures of drug efficacy have been used, the shorter drug development times made possible by regulatory acceptance of surrogate endpoints have not been in the medium- or long-term interests of public health. As our case study of the glitazones makes clear, regulatory approval based on surrogate measures of drug efficacy has enabled the marketing of drugs that may harm more patients than they save. This dilemma was well understood by regulators by the early 1990s yet the surrogate efficacy paradigm remained largely unchallenged during the neo-liberal era until the late 2000s because requiring pharmaceutical firms to provide at least some evidence of a beneficial impact on long-term clinical outcomes ran counter to the deregulatory focus on cutting regulatory review times and rushing innovative pharmaceuticals on to the market. The neo-liberal 'solution' to the hazard of approving drugs with a negative benefit-harm profile has been to require companies to conduct additional studies during the post-marketing period. This approach has proved erroneous for public health. Pharmaceutical firms lack the incentive to conduct meaningful or timely studies since their drugs are already making profits on the market, yet doctors and patients have little knowledge about whether the drugs in question will actually deliver clinically significant benefits in some of the most important health dimensions of the disease, or whether those benefits will outweigh longer term harms.

Given the findings from our research on diabetes drug regulation, we recommend that, for such conditions, regulatory standards should be raised to require pharmaceutical companies to design trials with long-term clinical outcomes as the study endpoint. At a minimum such trials should be underway (perhaps for a few years depending on the details of the condition) before marketing approval so that regulators, physicians, and patients can have more confidence that the benefits of the new drugs will outweigh their risks. This requirement would also generate a strong incentive for drug firms to conduct such studies to a high and appropriate standard in a timely manner. Interim data from these long-term clinical-outcome studies should be submitted at the time of a marketing approval application with complete results reported to regulators in the post-marketing phase in the event of marketing approval. Ideally, however, outcome studies would be completed prior to marketing. This should be mandatory either where there is no reason to believe that the new drug offers an advance over existing therapies for any specific patient population, or where there are grounds for believing that the new drug may pose the risk of serious harm – defined as any demonstrated or suspected harm, the extent or nature of which would outweigh the hypothesized benefit in the intended patient population.

As we have seen with the introduction of the 'accelerated approval' regulations in the US from 1992, and from our case study of Iressa, far from limiting regulatory reliance on surrogate measures of drug efficacy, the neo-liberal reforms expanded such reliance to non-established surrogate markers, such as tumour shrinkage in the case of cancer. Nor was the neo-liberal reduction in pre-market evidence about efficacy confined to drugs intended to treat life-threatening conditions, as we saw in our case study of the IBS drug, Lotronex. A key lesson from the Iressa case and from general experience of American accelerated approval involving non-established surrogate measures of efficacy, is that this type of accelerated drug development and review may be a false economy for most patients and public health for several reasons. First, the majority of patients who took Iressa did not benefit from the drug but were exposed to the drug's toxicities. Had AstraZeneca been required to identify the sub-population of patients that could benefit from the drug prior to marketing authorization this would have dramatically decreased the risk of patients being exposed to a toxic, but ineffective drug. Second, as the number of drugs marketed on the basis of non-established surrogate measures in a therapeutic field grows, the more difficult it becomes for physicians and patients to choose which drug is best for the individual patient because many, or all, of them are on the market with unsound evidence of efficacy. Third, in Europe, marketing approval based on non-established surrogate measures of efficacy increases the likelihood that the drug will not be regarded as cost-effective by national healthcare systems, so the drug will be available only to private patients on a full-cost basis. Consequently, for most patients, access to therapy may actually be delayed by such 'accelerated approval'. Fourth, the longer companies take to complete post-marketing studies, the greater the number of patients likely to suffer these 'disbenefits' – that is, of being treated with ineffective, suboptimal or toxic treatments, or of not being able to access treatments because they have been refused reimbursement. And, as we have seen, the time taken by companies to complete post-marketing studies can far outstrip the 'delays' involved in developing a sound evidence base for new drugs prior to marketing.

Unfortunately, the EU drug regulatory state learnt nothing from the failures of 'accelerated approval' in the US because, by 2006, it had introduced an almost identical version of this regulation in the form of 'conditional marketing approval'. This occurred in the aftermath of the 2001–2004 EU legislative review of pharmaceutical regulation, during which the pharmaceutical industry lobbied for a loosening of regulatory standards and the European Commission was only too keen to accommodate this on the grounds that the US industry was significantly outperforming the EU in the production of NMEs and pharmaceutical innovations. Of course, as is clear from the foregoing analysis what counts for public health is whether those innovations contribute any evidence-based therapeutic advance. Instead, the European Commission was focused on concerns about the EU's international performance compared with the US and its capacity to attract pharmaceutical industry investment in member states. Likewise in the US, it appears that no lessons at all have been learnt from the Iressa case, since some senior managers in the FDA have called for an increase in the acceptance of surrogate measures of drug efficacy to further accelerate innovative pharmaceutical development and approval, apparently irrespective of sound evidence that the drugs will work (FDA 2004).

In fact, what needed to happen in the interests of patients' and public health was that the misguided American deregulatory reforms of the neo-liberal era needed to be largely reversed, rather than copied. In particular, a requirement to demonstrate efficacy based on clinical outcomes or symptom improvement needed to be re-established as a condition of marketing approval. However, that need not imply that some patients who have found benefit during surrogate efficacy trials would be denied future use of the drug once trials were completed. Regulations should guarantee continued access for them on a compassionate basis, if appropriate.

To meet the needs of patients with life-threatening conditions who are unable to access clinical trials and have run out of treatment options, pharmaceutical companies should be expected to offer their experimental drugs to patients on a compassionate-use basis under pre-market expanded access programmes after a positive risk–benefit profile has been demonstrated in phase II studies. This is justified by the fact that companies benefit from patients' high expectations, expectations that are in turn created by companies' early release of study results and various promissory claims about their drugs. Our proposal would serve patients and public health better than the status quo, given the vast array of problems that have arisen with accelerated/conditional marketing approval based on non-established surrogate measures of efficacy. Furthermore, higher regulatory standards demanding substantial evidence of clinical benefit before marketing approval might not delay the availability of valuable drugs at all because pharmaceutical firms would be incentivized to focus their R&D efforts on bringing to market a smaller number of genuinely effective drug therapies, rather than a large number of new drugs offering doubtful or marginal efficacy with insufficient safety data.

Nor, it appears, has the EU regulatory state learnt the lessons of the FDA's misadventures with risk management as a means of regulating innovative pharmaceuticals with severe and life-threatening adverse effects instead of simply withdrawing them from the market. We suggest that there is highly persuasive evidence that pharmaceutical risk management policy in the US was a form of risk colonization, and that its implementation regarding specific drugs served merely to reduce the patient population exposed to the risks, rather than preventing or significantly mitigating the risks to those who continued to be exposed. Risk management policy, therefore, masked the need to withdraw some products from the market. In the US, the key policy lesson is that the conceptualization of risk management as an alternative to market withdrawal was misguided, assuming the FDA's mission to protect public health is to be taken seriously. It may be that risk management plans have the potential to enhance drug safety, but they should be constructed primarily as tools for proactively monitoring drug safety, or restricting drug access to a defined patient population where the risk–benefit balance of a drug is demonstrably positive and there are no alternative treatments, rather than mapped out as alternatives to market withdrawal. Thus, European regulators should have persisted with their sceptical, evidence-based approach to risk management after the completion of the legislative review of EU pharmaceutical regulation in 2004. Instead, since the mid- to late-2000s, the EU regulatory state appears to be following the US approach to pharmaceutical risk management as a means of maintaining unsafe drugs on the market (International Society of Drug Bulletins and the Medicines in Europe Forum 2009, pp. 2–3).

Yet, the changes to pharmaceutical regulation in the neo-liberal era were not confined to reforms of regulations themselves, whether in terms of priority reviews, accelerated approvals, or risk management. As we have explained, those changes were also about changing the political and institutional culture of regulation. They entailed the evolution of a regulatory philosophy that viewed pharmaceutical firms as partners, collaborators, and even clients in the development and successful marketing approval of pharmaceutical innovations. The interests of the regulatory agencies became increasingly intertwined with those of the industry, as agencies' dependence on industry fees and agency-industry consultation led to growing opportunities for regulatory capture. This is partly why we have seen such a consistent trend in regulators awarding the benefit of the doubt to manufacturers regarding uncertainties about the efficacy and/or safety of their innovative drugs, rather than to patients and public health (albeit more so in the US than the EU in our case studies).

To counteract that tendency towards capture and corporate bias in decision-making about regulating innovative pharmaceuticals, the government agencies need to be fully independent of the industry. If government regulatory agencies in the EU and US are to subject companies' claims about their products to serious scrutiny in order to promote and protect public health (as their mission statements still proclaim even during the neo-liberal era), then they must not regard pharmaceutical firms as their clients. The implication of this is that drug regulatory agencies should be funded directly from general taxation, so that regulators have stability in their employment and sufficient autonomy from industry's commercial interests. The government could recover the revenue from the pharmaceutical industry by means of a separate tax, so that the industry and not the taxpayer continues to bear much of the costs of drug regulatory review. New appointees as heads of the regulatory agencies should have a clear record of independence from the industry throughout their career. The liberal democracies of Europe and the US take great pride in their political systems' 'separation of powers' between the judiciary, the legislature, and the executive arms of government. However, they have been catastrophically remiss in asserting a 'separation of responsibilities and duties' between the government bureaucracy and capitalist industry.

To further ensure the independence of the science of pharmaceutical evaluation in the interests of public health from commercial influence, ideally, much of the pre-market innovative drug testing would be conducted by scientists and clinicians appointed or employed by regulatory agencies, rather than the drug manufacturer. That would be a radical departure from the status quo and monumentally difficult to realize at present. However, as a practical first step, there is a realistic case for at least one pivotal clinical trial per new innovative pharmaceutical to be carried out by scientists wholly independent of the manufacturer. The single, independent pivotal trial could be conducted by scientific investigators selected and contracted by the regulators, who would in turn charge the manufacturer the cost of the trial so that there would be no added cost to the government purse.

This arrangement would have several advantages over the current regulatory system. First, it would help to establish a body of drug-testing clinical experts who are independent of the pharmaceutical industry – something desperately needed in Europe and the US. Second, it would mean that the regulators could draw on the advice of a group of scientists as close to the clinical-testing data as the manufacturer, but independent of the firm sponsoring the drug. That would be valuable not only during final regulatory assessment, but would also enable regulators to put into better perspective the expectations and promissory science often generated by companies about their new drugs during clinical development, as we saw with Iressa and the glitazones. Third, it would significantly counteract the problems of clinical trial design and execution associated with the influence of manufacturers' commercial self-interest. And fourth, it would create a database controlled and owned by the regulatory state, which could be made available to other regulatory agencies (to minimize duplication), the medical and scientific professions, and the public in order to maximize their knowledge and vigilance regarding future pharmaceutical innovation. In this context, the patent system might facilitate earlier release of information about new drugs without undermining research incentives. As we alluded to in Chapter 1, the main drawback of the patent system has been monopoly prices, but it may be that this is best addressed by pricing regulation and exceptional legal provisions to protect public health, such as compulsory licensing to generic manufacturers, rather than the abolition of all intellectual property protections. We do not advocate such abolition until it is clear that an alternative system could be put in place that would not damage research incentives to discover new medicines by the commercial industry, state sector research organizations or universities.

The mission statement of the EMEA and the FDA also needs to be altered to make clear that, while the interests of patients' health and the commercial interests of the pharmaceutical industry may sometimes coincide, that is not always the case, and when it is not the case the regulatory agencies' uncompromising duty is to prioritize the interests of patients and public health over and above the commercial interests of the industry. A revised mission statement should also emphasize that, while some pharmaceutical product innovations have provided significant therapeutic advances for patients and doctors, in most of the last 20 years that has been true for only a small minority (perhaps ten per cent) of such innovations. Not only might this improve the regulatory culture within government agencies, but it might also discourage the perpetuation of misleading expectations about pharmaceutical innovation and concomitant ideologies of promissory science.

Familiar by-words of the neo-liberal era have been 'freedom of choice'. While most heavily emphasized in the libertarian political discourse of the New Right led by Reagan, the Bush Administrations, Gingrich, and Thatcher, 'consumer choice' has also been championed by Blair, Clinton, and the European Commission. As we saw in Chapter 2, it was such rhetoric that lay behind the Commission's unsuccessful proposals for legalization of direct-to-consumer advertising of prescription drugs in the EU. Alongside this ideology of 'consumer choice' sit the discursive constructs, 'the informed patient' and 'freedom of information'. Although neo-liberalism ostensibly championed 'freedom of choice' for consumers, its record in promoting freedom of information – the basis of informed consumer choice – is hardly commendable. Apart from the advent of the internet, during the neo-liberal era, in the US, no significant advances were made in the extent to which the public could access FDA files about pharmaceutical regulation. On the contrary, citizens have often had to wait longer to have their requests processed. The FDA continues to refuse to make its regulatory reviews of drugs denied marketing approval in the US routinely available for public scrutiny, even though that might help to protect citizens elsewhere and contribute to the collective medical project of pharmaceutical science.

Meanwhile, in the EU, there has been some progress in extending freedom of information about pharmaceutical regulation to citizens, but starting from a very low base. For the first half of the neo-liberal era since 1980, drug regulation in Europe was conducted behind a cloak of intense secrecy, with the exception of Sweden. While European governments, such as the Thatcher Administration espoused a mythology of 'consumer choice', that same government actively defended laws that prohibited citizens having any access to the techno-scientific basis upon which prescription drugs were approved on to the UK market by regulators, thereby denying citizens the ability to make informed choices. Although the supranational EU regulation of innovative pharmaceuticals via the centralized procedure has managed since 1995 to greatly improve on those dark years of secrecy, often as a result of pressure from consumer organizations, a culture persists within EU regulatory institutions of resisting disclosure of information to citizens in order to protect the commercial interests of pharmaceutical firms. Yet, real and informed patient choice depends on being able to access crucial information about new drugs that is held by regulatory agencies.

The degree of secrecy surrounding pharmaceutical regulation in the EU and the US is detrimental to both science and democratic accountability. Citizens and professionals in the medical, clinical and pharmaceutical sciences cannot fully participate in debates about how to maximize the role of pharmaceuticals in healthcare if regulatory information is either concealed or unavailable in a timely manner. All regulatory reviews for new drug marketing applications should become publicly available on request after the end of phase II (by which time patents would be secured), phase III, and marketing approval/non-approval. This would empower patients, professionals and others to subject companies' promissory science claims about their new drugs, while still under pre-market clinical development, to more informed scrutiny and public evaluation. Given that, during the neo-liberal era, many pharmaceutical firms have, in effect, begun to 'promote' (or 'raise awareness' about) their new drugs to doctors and patients before marketing approval, regulatory evaluations also need to be available at that earlier stage in the innovation process. This is particularly important since medical journals and the published literature often fall short in providing such scrutiny because they, unlike regulators, cannot legally require drug companies to submit all data.

While democratic accountability and constructive citizen-participation in regulatory decision-making depends on information being made fully available to all stakeholders, we would also argue for a more critical understanding of such participation. As we have shown, the invocation of 'the patient' and a concept of 'patients' rights' has been used by the state, industry and regulatory agencies to justify earlier marketing of medicinal products on the basis of an incomplete and inadequate evidence base. Sometimes this was done in the absence of real demand by patients. At other times, some sections of the patient community did support such aims. More generally, the goal of increased public participation has been promoted by most stakeholders as well as academics. Our historical and case study analyses signal the pressing need for a more sophisticated, critical and differentiated understanding within pharmaceutical and health policy of 'patients' needs', 'patients' rights', 'the public' and the relationship between these and public health.

First, we would argue that regulatory agencies and policy-makers must be more explicit about the fact that just because an individual patient or patient group demands something, this does not mean it is necessarily in the best interests of patients or public health. Individual patients and patient advocates may demand access to drugs because they are vulnerable and have been unduly influenced by the promissory science of industry, regulators and academic scientists. Or they may be unaware of a drug's uncertain risk–benefit profile. But patients' demands are not always or necessarily consistent with their objective health interests – that lesson goes all the way back to AIDS patient activism of the late 1980s and early 1990s. Second, patients – even patients with the same disease – do not necessarily share a unified set of interests. For example, while a very small number of patients may benefit from earlier marketing of a new drug, this may have been achieved at the expense either of the greater number of patients who do *not* benefit yet run the risk of severe drug toxicity, or of future populations of patients who, as a consequence of early marketing and late (or non-) completion of confirmatory trials, lack crucial information about the safety and efficacy of those drugs relative to other therapies. Patients – both present and future – need a strong and independent regulator willing to make those distinctions and weigh up potentially competing needs in the objective interests of public health.

Furthermore, in the current context of restricted healthcare budgets and an emergent patient–industry complex, it is not in the general interests of patients, or equity, when particular sections of the patient community, with well-resourced representative organizations, are able to mobilize political support and force reimbursement bodies to fund hugely expensive drugs offering minimal benefits. These potential conflicts have been obscured by neo-liberal discourse and ideology, with those patients whose demands are more closely aligned with industry interests most likely to win out in the 'participation' arena. In other words, broad discussions by social scientists and policy-makers about 'public participation' in regulatory decision-making may be misleading. What matters in understanding regulatory processes and outcomes is which types of 'publics' are involved and, crucially, how they relate to other relevant organized interests. Particular instances of 'public partici-pation' and the invocation of 'patients' rights' need to be subjected to critical scrutiny in the interests of public health and social justice, just as much as the claims of other protagonists. First, transparency with respect to potential conflicts of interests and the funding of patient and other civil society groups needs to be improved. Our case studies provide further support for research showing that current mechanisms for ensuring transparency and guarding against conflicts of interest are inadequate (Perehudoff and Alves 2010). Second, the impact of public participation in relation to specific regulatory outcomes needs to be scrutinized, rather than simply assumed. In making this point, we are *not* arguing that pharmaceutical regulation should be left solely to scientists. Consumer organizations, public health advocacy groups and patient groups can, and should, be involved in regulatory decision-making. However, our case studies of Lotronex and Iressa point to the health risks involved when regulators base decisions to allow the marketing of drugs on unmediated anecdotal reports by patients. In these cases, selective public accounts of the miracle benefits offered by the drugs led to outcomes that were in the commercial interests of companies but not in the interests of public health.

Notes

1 Putting Pharmaceutical Regulation to the Test: From Historical Description to a Social Science for Public Health

1. During the late 2000s, there has also been publication of some bioethical studies of the organization of US drug trials in the neo-liberal era, most notably Fisher (2009). This work touches on regulation, but the primary concern is with the treatment of trial subjects and trial conduct, rather than *drug* regulation.

2. Carpenter (2010a) also has a brief chapter on the international context of the FDA, including Europe. While commendable, this chapter is somewhat of an afterthought and cannot be regarded as a substantial investigation of EU drug regulation. Indeed, his discussion of Europe draws mostly on publications by Abraham, Daemmrich, and Dukes.

3. By 2010, Carpenter seems to distance himself somewhat from this earlier conviction, stating merely that 'there is considerable truth in this narrative' (2010a: 394). Our comment is equivocal because his revised, and more sophisticated, views of 2010 are themselves ambivalent. On the one hand, he writes: 'The changes [to US pharmaceutical regulation] struck during the apex of American AIDS politics in the 1980s and early 1990s were lasting. They transformed the agency's image and they sculpted anew its direction, gate-keeping, and conceptual powers. These changes would not have happened or come as they did had AIDS never surfaced' (Carpenter 2010a, p. 394). This is a more moderate formulation than his declarations of 2004. However, it remains overstated and misleading by exaggerating the transformational role and newness of regulatory developments occurring in the US during the 'AIDS crisis' of the 1980s and early 1990s. Indeed, we believe that the characterization of such regulatory changes around a pivot of 'AIDS politics' by Carpenter and some other scholars has itself been misleading. On the other hand, later in his book, he acknowledges: 'Yet there is much that is misleading about the received legend of the FDA's response to AIDS, and much that it misses. The agency was changing and furiously adapting its regulatory processes several years before [AIDS] activists made their most public and memorable protests' (Carpenter 2010a, 428–9). We strongly endorse that statement, though would have to add that Carpenter's (2004) pronouncements contributed to the legend, which he now partly disowns. The misleading nature of previous accounts of regulatory reform in the US during the 1980s explains why we have felt it so important to revisit the topic in this book.

4. Some writers who subscribe to a philosophy known as relativist constructivism, sometimes regarded as part of the field of 'science and technology studies', argue that analysts attempting to explain our world should not distinguish between techno-scientific factors and social/political factors

(Latour 1987; see Irwin 2001, pp. 161–87). For example, on this view, the realm of explanation for, say, a conflict-of-interest of an expert adviser to a regulatory agency is held to be no different from the realm of explanation for, say, the relationship between heart attack and mortality. We disagree and suggest that such relativist constructivism is flawed. It is beyond the scope of this book to explore these philosophical issues, which need not detain us here, but readers interested in a fuller explanation of why the relativist constructivist outlook should be rejected may wish to consult Abraham (1995a, pp. 1–35; 2002; 2007), Bunge (1999); Lukes (1982), Potter (1999) and van Zwanenberg and Millstone (2000).

5. Many scholars following a relativist or constructivist approach to social analysis advocate adopting neutrality even about judging whether knowledge-claims are 'true' or 'false'. This takes value-neutrality to extremes never dreamed of by positivists.

6. In his exposition on science, Bhaskar (1975) asked: 'what must the natural world be like in order for science to be possible?' He coined this approach, 'transcendental realism'.

7. It is important not to confuse 'non-neutrality' with 'bias' – a mistake sometimes made by positivists and constructivists alike. Non-neutrality does not necessarily tell us anything about the truth-value of a knowledge-claim because knowledge can be pursued from a range of stances (though whether any can be adequately described as 'neutral' is debatable). Bias, however, leads to deviation from, or in extreme cases, even corruption of, the truth (though not all deviations from the truth are derived from bias, some may be simple error).

8. In 2010 the EMEA changed its name to the European Medicines Agency (EMA).

9. Until 2004, the CHMP was known as the Committee for Proprietary Medicinal Products (CPMP).

10. Until the mid-1990s, PhRMA was known as the Pharmaceutical Manufacturers' Association (PMA).

11. Although we began our research on the neo-liberal era and the science of pharmaceutical innovation and regulation eight years ago, it is worth noting that a growing number of scholars have begun to research neo-liberalism and science across various aspects of society and the economy (Lave *et al.* 2010).

12. Interview with senior scientist and manager at FDA's Centre for Drug Evaluation and Research (CDER), Rockville, Maryland, 1 August 2003.

13. Interview with senior scientist and drug regulator at FDA, Rockville, Maryland, 28 August 2003.

2 The Political Economy of 'Innovative' Drug Regulation in the Neo-Liberal Era

1. Interview with former senior official at FDA CDER, Rockville, Maryland, US, 16 May 2001.

2. Interview with former FDA manager at CDER, Bethesda, Maryland, US, 24 May 2001.

3. Interview with former Professional Staff Member of US Congressional Committees, Washington DC, 4 May 2001.
4. Interview with senior FDA scientist and drug reviewer, Rockville, Maryland, 31 July 2003.
5. Interview with senior representative of a medical consumer organization, New York, 10 February 2005.
6. Interview with senior representative of public health advocacy organization, Washington DC, 5 September 2003.
7. Interview with a former senior director of the Patients and Consumer Coalition, Washington DC, 11 September 2003.
8. Interview with AIDS patient representative and senior member of Patients and Consumers Coalition, Washington DC, 16 February 2005.
9. In the UK, the neo-liberal Conservative Prime Minister, Margaret Thatcher, was famously and staunchly Eurosceptic. However, by 1992, when the UK signed up to the Maastricht Treaty, John Major had succeeded Thatcher as Conservative Prime Minister. While he shared her neo-liberal conservative politics, he was a Europhile.
10. In most countries, including the UK and the US, pharmaceutical regulation has always come under the responsibility of a Department of Health, rather than a Department of Trade and Industry. In 2008, EU pharmaceutical regulation was switched from DG Enterprise to DG Health.
11. The Patients' Coalition was later joined by a number of consumer health advocacy groups such as Public Citizen, the Center for Medical Consumers and the National Women's Health Network and became the 'Patient and Consumer Coalition'.
12. Interview with former supranational EU expert scientist and regulator, Southern Europe, 20 January 2005.
13. Interview with former supranational EU regulator, UK, 8 March 2004.
14. Interview with senior official at European Commission Pharmaceuticals Unit of DG Enterprise, London, 15 March 2003.
15. Interview with senior scientist and drug regulator at FDA, Rockville, Maryland, 28 August 2003.
16. Interview with senior regulator and scientist at FDA, Rockville, Maryland, 5 August 2003.
17. Interview with former supranational EU regulator, Scandinavia, 12 May 2004.
18. Interview with another former supranational EU regulator, North Europe, 12 May 2004.
19. Interview with former FDA biostatistician FDA, Maryland, 15 August 2003.
20. Interview with FDA expert science advisor, Maryland, 15 February 2005.
21. Interview with senior representative of medical consumers group, New York, 10 February 2005.
22. Interview with senior FDA manager, CDER, Rockville, Maryland, 26 August 2003.

3 Designs on Diabetes Drugs

1. Interview with senior FDA scientist, Office of Drug Evaluation, Rockville, August 2003.

2. Interviews with representatives of Consumer Health Advocacy Group, Washington DC, September 2003; former CPMP Member, May 2004; FDA Drug Safety/Risk Management Advisory Committee Member, February 2005; FDA Medical Officer, Oncology, Rockville, August 2003.
3. Interview with FDA Drug Safety/Risk Management Advisory Committee Member, February 2005.
4. Interview with FDA Medical Officer, Oncology, August 2003.
5. Interview with FDA Medical Officer, Endocrinologic, Rockville, August 2003.
6. Interviews with FDA Drug Safety/Risk Management Advisory Committee Member, Gaithersburg, Maryland, February 2005; former CPMP Members, Uppsala (Sweden) and Dublin, May 2004.
7. Interview with former CPMP Member, Dublin, May 2004.
8. Interviews with former CPMP Members, May 2004.
9. Interview with former CPMP Member, Uppsala, May 2004.
10. Interviews with Member of European Parliamentary Committee on Environment, Public Health and Consumer Policy, Paris, March 2005; former CPMP Member, Milan, January 2005; Medicines in Europe Forum Member, Paris, March 2005.
11. Interview with FDA scientist, Office of Drug Safety, Rockville, Maryland, July 2003.
12. Interview with former FDA Division Director, Rockville, August 2003.

4 Desperate Regulation for Desperate Cancer Patients: The Unmet Needs of Accelerated Drug Approval

1. Interview with FDA oncologist, Rockville, Maryland, 26 August 2003.
2. Interview with former FDA pharmacologist, Washington DC, 11 September 2003.
3. We interviewed AstraZeneca, but the firm insisted that Iressa could not be discussed during the interview as a condition for being interviewed.
4. The correlation between newspaper coverage and accelerated new drug approval constructed by Carpenter (2010a) is interesting, but we would urge caution about attributing causal explanation to that correlation, especially a unidirectional one positing that the media causes the FDA to act to protect its reputation. Our scepticism is because there is no underpinning investigation of the news stories and their sources. Often media stories reflect knowledge and insights from insiders who have advance 'intelligence' of how regulators are already approaching a new drug, so the direction of causality may be from insiders to the press, or the press may merely pick up on what is already a prevailing trend in the drug regulatory world with other causes, quite separate from the press.
5. Interview with senior FDA scientist, Office of Drug Evaluation, Rockville, 28 August 2003.
6. Interview with senior representative of a Medical Consumer Organization, New York, 10 February 2005.
7. Interview with senior representative of Consumer Health Advocacy Organization, Washington DC, 5 September 2003.

8. Interview with former CPMP member, Milan, 20 January 2005.
9. Interview with deputy director of a division within FDA's Center for Drug Evaluation and Research, 1 August 2003, Rockville, Maryland.
10. Interview with senior FDA scientist, Oncology Division, Rockville Maryland, 18 February 2005.
11. Interview with Medical Officer, FDA Cardio-Renal Division, Rockville, Maryland, 18 August 2003.
12. Interview with former biostatistician at FDA's Center for Drug Evaluation and Research, Germantown, Maryland, 15 August 2003.
13. Interview with FDA endocrinologist, Rockville, Maryland, 29 August 2003.
14. Interview with senior FDA scientist, Rockville, Maryland, 31 July 2003.
15. In sociology of science, scholars who wished to investigate the social factors in scientific knowledge often referred to mathematics as 'the hardest case' among the sciences because it seemed so difficult to imagine how social factors could be involved in the production of mathematical knowledge – an apparently self-contained system of technical rules and logic. Similarly, here, it is more difficult to argue that accelerated drug approval has not been in patients' interests when the patients suffer from lung cancer, than if they suffer from almost any other disease.

5   The Making of a Harmful 'Therapeutic Breakthrough': Crossing Regulatory Boundaries for Drug Approval

1. Interview, former FDA epidemiologist, Washington DC, 22 July 2003.
2. Interview, senior FDA scientist, Rockville, 31 July 2003.
3. Interview, senior representative of Medical Consumer Organization, New York, 10 February 2005.
4. Given that alosetron had already been approved for the indication, diarrhoea-predominant IBS, in order for it to become an experimental of 'investigational new drug' (IND) under FDA regulations, it would have to have IND status for a different indication, such as severe diarrhoea-predominant IBS only (see Anon. 2002k, p. 17).
5. This calculation assumes the US population is approximately 300 million, and accepts GSK's claim that diarrohea-predominant IBS affects 5–10 per cent of that population. It is assumed that five per cent of those women have severe symptoms of whom 10 per cent could receive great benefits. The projections for ischemic colitis are derived from the FDA's estimates 1/307 to 1/218 cases per three-months exposure.

6   The Regulatory Science and Politics of Risk Management: Who Is Being Protected?

1. Interview with former CPMP member, Milan, 20 January 2005.
2. The elevation of liver enzymes, such as transaminases, in liver function tests indicates liver-cell damage (FDA 2000b).
3. Interview with former biostatistician at FDA's Center for Drug Evaluation and Research, Germantown, Maryland, 15 August 2003.

4. Interview with EU–UK regulator, UK Medicines and Healthcare products Regulatory Agency (MHRA), London, 23 August 2004.
5. Interview with former CPMP Member, Dublin, 19 May 2004.
6. Senior FDA scientist, Office of Drug Evaluation, Rockville, 28 August 2003.
7. Interviews with former CPMP Members, Uppsala *(Sweden)*, 12 May 2004.
8. Bilirubin is a yellow breakdown product of red blood cells passed to the liver.
9. Interview with deputy director of a division within FDA's Center for Drug Evaluation and Research, Rockville, Maryland, 1 August 2003.
10. Interview with senior FDA scientist, Rockville, Maryland, 31 July 2003.
11. Q and T refer to points on electrocardiograph readings.
12. Interview with FDA endocrinologist, Rockville, Maryland, 29 August 2003.
13. Interview with former FDA epidemiologist, Washington DC, 22 July 2003.
14. Interview with official from European Commission, Pharmaceuticals Unit, DG Enterprise, London, 15 March 2005.
15. Interview with FDA manager in CDER, Rockville, 26 August 2003.

Bibliography

Abraham, J. (1995a) *Science, Politics and the Pharmaceutical Industry*. London: Routledge.

Abraham, J. (1995b) The production and reception of scientific papers in the medical industrial complex: the clinical evaluation of new medicines, *British Journal of Sociology*, 46: 167–90.

Abraham, J. (2008) Sociology of pharmaceuticals development and regulation: a realist empirical research programme, *Sociology of Health & Illness*, 30: 869–85.

Abraham, J. (2009) Partial progress: governing the pharmaceutical industry and the NHS, 1948–2008. *Journal of Health, Politics, Policy and Law*, 34: 931–77.

Abraham, J. (2010) Pharmaceuticalization of society in context: theoretical, empirical and health dimensions, *Sociology*, 44: 1–20.

Abraham,J. and Davis, C. (2005). A comparative analysis of drug safety withdrawals in UK and US, 1971–1992, *Social Science & Medicine*, 61: 881–92.

Abraham, J. and Davis, C. (2006) Testing times: the emergence of the Practolol disaster and its challenge to British drug regulation, *Social History of Medicine*, 19: 127–47.

Abraham, J. and Davis, C. (2007) Deficits, expectations and paradigms in British and American drug safety assessments, *Science, Technology & Human Values*, 32: 399–431.

Abraham, J. and Lewis, G. (2000) *Regulating Medicines in Europe: Competition, Expertise and Public Health*. London: Routledge.

Abraham, J. and Lewis, G. (2002) Citizenship, medical expertise and the capitalist regulatory state in Europe, *Sociology*, 36: 67–88.

Abraham, J. and Millstone, E. (1989) Food additive controls: some international comparisons. Food Policy, 14: 43–57.

Abraham, J. and Reed, T. (2001) Trading risks for markets: the international harmonization of pharmaceuticals regulation, *Health, Risk & Society*, 1: 113–28.

Abraham, J. and Reed, T. (2002) Progress, innovation and regulatory science in drug development: the politics of international standard-setting, *Social Studies of Science*, 32: 337–69.

Abraham, J. and Reed, T. (2003) Reshaping the carcinogenic risk assessment of medicines: international harmonization for drug safety, industry/regulator efficiency or both? *Social Science & Medicine*, 57: 195–204.

Abraham, J. and Sheppard, J. (1999) *The Therapeutic Nightmare: The Battle over the World's most Controversial Sleeping Pill*. London: Earthscan.

AIDS Action Council (1996) 'FDA Reform' Proposals in Congress Pose Great Danger to People Living with HIV, *AIDS Action Network Alert*, 24 May 1996. Available at: http://www.thebody.com/content/art33554.html. Accessed 15 October 2011.

Alivisatos, R. (1997) FDA Medical Officer's Review of NDA 20–759 (Trovan), 20 July.

Alivisatos, R. (1998) FDA Medical Officer's Review. Post-Marketing Adverse Events NDA 20–760 (Trovan), 31 August.

American College of Gastroenterology Functional Gastrointestinal Disorders Task Force (2002) Evidence-based position statement on the management of IBS in North America, *The American Journal of Gastroenterology*, 97 (Suppl.): S1–S5.

Anderson, V. and Hollerbach, S. (2004) Reassessing the benefits and risks of alosetron, *Drug Safety* 27: 283–92.

Andersson, F. (1992) The drug lag issue: the debate seen from an international perspective, *International Journal of Health Services*, 22: 53–72.

Angell, M. (2004) *The Truth about the Drug Companies: How They Deceive Us and What to do About It*. New York: Randon House.

Angell, M. (2010) FDA: this Agency can be dangerous – Review of 'Reputation and Power' by Daniel Carpenter. *New York Review of Books*, 30 September.

Anon. (1981a) Change of attitude at FDA called 'essential' by Reagan advisor, *Scrip*, 558 (21 January): 12.

Anon. (1981b) US 'Drug Lag' Commission starts work, *Scrip*, 611 (27 July): 10.

Anon. (1981c) Some FDA officials avoiding past adversarial practices, *Scrip*, 573 (16 March): 9.

Anon. (1982a) Orphan drugs bill criticised, *Scrip*, 675 (15 March): 7.

Anon. (1982b) US NDA rewrite published: foreign data policy described, *Scrip*, 739 (25 October): 8.

Anon. (1986a) PMA/FDA co-operation, *Scrip*, 1096 (23 April): 19.

Anon. (1986b) Dr Young outlines Action Plan progress, *Scrip*, 1074 (5 February): 16.

Anon. (1987) US PMA meeting's hidden concerns, *Scrip*, 1213 (12 June): 19.

Anon. (1988a) New UK licence fees proposed, *Scrip*, 1369 (14 December): 1.

Anon. (1988b) US FDA discussing new legislative proposals, *Scrip*, 1336 (19 August): 16.

Anon. (1988c) US FDA's new policy on importing AIDS drugs, *Scrip*, 1331 (3 August): 17.

Anon. (1988d) Bush calls for speedier US approvals, *Scrip*, 1335 (17 August): 16.

Anon. (1988e) FDA's expedited drug approval plan, *Scrip*, 1340/1 (2/7 September): 18.

Anon. (1988f) US FDA's expedited drug plan unveiled, *Scrip*, 1356 (28 October): 18.

Anon. (1988g) US PMA comments on expedited drug plan, *Scrip*, 1346 (23 September): 17.

Anon. (1988h) US FDA on expedited drug approvals, *Scrip*, 1337 (24 August): 16.

Anon. (1988i) ABPI's blueprint for Europe, *Scrip*, 1330 (29 July): 7–8.

Anon. (1989a) UK MCA sets targets, *Scrip*, 1415 (26 May): 2–3.

Anon. (1989b) UK MCA director named, *Scrip*, 1392 (8 March): 4.

Anon. (1989c) UK licence fee agreed, *Scrip*, 1400 (5 April): 4.

Anon. (1989d) US FDA/NCI reconcile differences, *Scrip*, 1450 (27 September): 20.

Anon. (1989e) US 1988 approvals summary, *Scrip*, 1380 (25 January): 18.

Anon. (1989f) US FDA battle NIAID on AIDS, *Scrip*, 1450 (27 September): 19.

Anon. (1989g) US FDA and drug approval, *Scrip*, 1378 (18 January): 17.

Anon. (1989h) NCI/FDA clash on cancer approvals, *Scrip*, 1382 (1 February): 16.

Anon. (1989i) US FDA/NCI reconcile differences, *Scrip*, 1450 (27 September): 20.

Anon. (1990a) US 1989 NME approvals summary, *Scrip*, 1484 (31 January): 30–31.

Anon. (1990b) Dr Young's achievements at US FDA, *Scrip*, 1478 (10 January): 19.

Anon. (1990c) Mixed review for AIDS/cancer panel report, *Scrip*, 1546 (5 September): 18.

Anon. (1990d) AIDS activists urge DDI/DDC approval, *Scrip*, 1557 (12 October): 30.

Anon. (1990e) Barriers to faster FDA review times, *Scrip*, 1528 (4 July): 16.

Anon. (1991a) MCA launched as 'Next Steps' agency, *Scrip*, 1635 (19 July): 2–3.

Anon. (1991b) 11 US NMEs approved at year end, *Scrip*, 1581 (11 January): 30.

Anon. (1991c) Quayle defends role in FDA reforms, *Scrip*, 1679/80 (25/25 December): 18.

Anon. (1991d) US FDA discusses conditional approvals, *Scrip*, 1598 (13 March): 17.

Anon. (1991e) US PMA on conditional approvals, *Scrip*, 1624 (12 June): 18.

Anon. (1991f) FDA reforms: applause and opposition, *Scrip*, 1672 (27 November): 16.

Anon. (1991g) FDA reforms include recognition of non-US approvals, *Scrip*, 1670 (20 November): 19.

Anon. (1991h) User fees in 1992 FDA budget, *Scrip*, 1590 (13 February): 17.

Anon. (1991i) FDA work force at 17,000 in 1997? *Scrip*, 1631 (5 July), 14.

Anon. (1991j) FDA reform – US implications, *Scrip*, 1671 (22 November): 19.

Anon. (1992a) FDA clearing NDA backlog, *Scrip*, 1686 (24 January): 26–27.

Anon. (1992b) PMA discussing user fees with FDA, *Scrip*, 1742 (7 August): 13.

Anon. (1993a) US FDA begins to collect user fees, *Scrip*, 1844 (4 August): 16.

Anon. (1993b) No clear strategy for US industry? *Scrip*, 1802 (12 March): 15.

Anon. (1994a) Debate over the aims of FDA user fees, *Scrip*, 1957 (13 September): 17.

Anon. (1994b) Kessler and the new Congress: how big a challenge, *Scrip*, 1985 (20 December): 15.

Anon. (1994c) Change afoot for the FDA, *Scrip*, 1983 (13 December): 18.

Anon. (1995a) PhRMA's principles for FDA reform, *Scrip*, 2002 (24 February): 17.

Anon. (1995b) PhRMA's FDA reform plan, *Scrip*, 2040 (7 July): 16.

Anon. (1995c) US group proposes private drug review, *Scrip*, 2038 (30 June): 14.

Anon. (1995d) US bill to speed FDA approvals, *Scrip*, 2034 (16 June): 17.

Anon. (1995e) FDA bill reflects industry views, *Scrip*, 2043 (8 July): 14.

Anon. (1995f) Outline of Senate FDA reform bill, *Scrip*, 2052 (18 August): 11.

Anon. (1995g) Kassebaum alters FDA reform bill, *Scrip*, 2077 (14 November): 17.

Anon. (1995h) FDA responds to WLF attack, *Scrip*, 2112 (31 March): 20.

Anon. (1996a) Third party drug review to replace FDA? *Scrip*, 2104 (20 February): 18.

Anon. (1996b) Kassebaum proposals unrealistic says Kessler, *Scrip*, 2107 (1 March): 13.

Anon. (1996c) US off-label use proposals debated, *Scrip*, 2108 (5 March): 14.

Anon. (1996d) Call for bipartisan FDA reform, *Scrip*, 2095 (19 January).

Anon. (1996e) Time running out for FDA reform? *Scrip*, 2122 (23 April): 18.

Anon. (1996f) FDA and industry negotiates user fee bill, *Scrip*, 2184 (26 November): 13.

Anon. (1997a) FDA and industry agree on user fees, *Scrip*, 2205 (February): 14.

Anon. (1997b) Industry unveils new US FDA reform proposals, *Scrip*, 2224 (18 April): 14.

Anon. (1997c) US Congress fails to pass FDA reform bill, *Scrip*, 2256 (8 August): 14.

Anon. (1997d) Senate approved FDA reform bill, *Scrip*, 2272 (3 October): 15.

Anon. (1997e) Congress passes FDA reform bill, *Scrip*, 2284 (14 November): 13.

Anon. (1997f) Shift towards more data disclosure in EC, *Scrip*, 2283 (11 November): 2.

Anon. *(1997g)* First approvals for Pfizer's Trovan, Scrip, 2296/97 (26 December): 17.

Anon. (1998a) EMEA's pre-submission guidance, *Scrip*, 2391 (27 November): 2.

Anon. (1998b) EC CPMP recommends Tasmar suspension, Scrip, 2389 (20 November): 17.

Anon. (1998c) Trovan launched in Germany, Scrip, 2359 (7 August): 19.

Anon. (1999a) Firms missing EC submission deadlines, *Scrip*, 2487 (5 November): 3.

Anon. *(1999b)* EU body attacks delays to single market, Scrip, 2492 (24 November): 4.

Anon. (1999c) Glaxo Wellcome's Lotronex backed by FDA panel, *Scrip,* 2491 (19 November):18.

Anon. *(1999d)* Severe liver injury with Pfizer's Trovan, Scrip, 2441 (28 May): 20.

Anon. *(1999e)* Trovan restricted in US, suspension recommended in Europe, Scrip, 2446 (16 June): 23.

Anon. (2000a) ICH progress on single dossier, *Scrip,* 2522 (15 March): 17.

Anon. (2000b) First European launches for Actos, Scrip, 2591 (10 November): 16.

Anon. (2000c). SmithKline Beecham's Avandia advertisements misleading, Scrip, 2594 (22 November): 18.

Anon. (2000d) Comparative trials no threat to public health, Scrip, 2599 (8 December): 3.

Anon. (2000e) EC public health group looks at value, Scrip, 2532 (19 April): 3.

Anon. (2000f) First launch for Glaxo Wellcome's Lotronex, *Scrip,* 2526 (29 March):19.

Anon. (2000g) Glaxo Wellcome withdraws Lotronex, *Scrip,* 2597 (1 December): 20.

Anon. (2000h) Lotronex facing more restrictions, *Scrip,* 2593 (17 November) :21.

Anon. (2000i) Lotronex re-approved with restrictions, *Scrip,* 2754 (12 June): 20.

Anon. (2001a) EC review goes to Parliament, *Scrip,* 2699 (28 November): 2.

Anon. (2001b) New global dimension for EMEA, *Scrip,* 2662 (20 July): 3.

Anon (2001c) EU centralised procedure does not mean a drug is 'innovative', *Prescrire International,* 10: 53.

Anon. (2001d) Iressa IDEAL for lung cancer? *Scrip,* 2691 (31 October): 26.

Anon. *(2001e)* More restrictions for Orlaam, Scrip, 2605/06 (3/5 January): 21.

Anon. *(2001f)* Orlaam suspended in EC, Scrip, 2637 (25 April): 20.

Anon. (2002a) EMEA failure rate linked to optionality, *Scrip,* 2765 (19 July): 3.

Anon. (2002b) Call for change in governance of EMEA, *Scrip,* 2729 (15 March): 3.

Anon. (2002c) EMEA comments on application failures, *Scrip,* 2765 (19 July): 3.

Anon. (2002d) G10 Medicines Group nears consensus, *Scrip,* 2717 (1 February): 2.

Anon. (2002e) French group wants radical change to EC review, *Scrip,* 2750 (29 May): 3.

Anon. (2002f) Sharp fall in applications to the EMEA, *Scrip,* 2788 (9 October): 2.

Anon. (2002g) IDEAL 2 data support Iressa in lung cancer, *Scrip,* 2751 (31 May): 26.

Anon.(2002h) First approval for AstraZeneca's Iressa, *Scrip,* 2762 (10 July): 18.

Anon. (2002i) Iressa fails in NSCLC combination trials, *Scrip,* 2775 (23 August): 21.

Anon. (2002j) More Iressa ADRs in Japan, *Scrip,* 2794 (30 October): 23.

Anon. (2002k) Lotronex US re-approval under attack, *Scrip,* 2782 (18 September): 16–17.

Anon. (2002l) FDA panel supports restricted return of Lotronex, *Scrip,* 2741 (26 April): 20.

Anon. (2002m). Lotronex re-approved, with restrictions, *Scrip,* 2754 (12 June): 20.

Anon. (2003a) First glitazones approved as monotherapy in the *EU,* Scrip (5 September).

Anon. (2003b) Roxane to discontinue Orlaam in the *US,* Scrip (17 September).

Anon. (2004a) Tarceva gets first approval for NSCLC, *Scrip,* 3007 (24 November): 22.

Anon. *(2004b)* Tasmar EU suspension to be lifted, but with closer monitoring, Scrip (30 April).

Anon. (2005a) EU prepares guidance on 'serious risk to public health', *Scrip,* 3032 (25 February): 4.

Anon. (2005b) EMEA to clarify which products must use centralized procedure, *Scrip*, 3066 (24 June): 2.

Anon. (2005c) EU preparing guidelines on new legislation, *Scrip*, 3054 (13 May): 4.

Anon. (2005d) European pharmaceutical forum to improve EU competitiveness, *Scrip*, 3113 (7 December): 3.

Anon. (2005e) New EU drug applications rise, *Scrip*, 3038 (18 March): 3.

Anon. (2005f) Pharma industry research has tripled in 25 years, *Scrip*, 3076 (29 July): 22.

Anon. (2005g) European Commission to take action on innovation and safety, *Scrip*, 3038 (18 March): 2–3.

Anon. (2005h) EU to adjust industry fees to pay for EMEA's new tasks, *Scrip*, 3046 (15 April): 3.

Anon. (2005i) EU to take steps to improve pharmacovigilance, *Scrip*, 3055 (18 May): 2.

Anon. (2005j) US FDA restricts Iressa, *Scrip*, 3065 (22 June): 24.

Anon. (2005k) Tarceva gets EU nod for lung cancer, *Scrip*, 3067 (29 June): 22.

Anon. (2006a) New EU accelerated assessment procedure for unmet therapeutic needs, *Scrip*, 3119/20 (4/6 January): 2.

Anon. (2006b) EMEA defines 'commercially confidential information', *Scrip*, 3180 (4 August): 2.

Anon. (2006c) EU launches new, less stringent conditional approvals procedure, *Scrip*, 3145 (5 April): 3.

Anon. (2006d) EU group wants less talk and more innovation, *Scrip*, 3126 (27 January): 2.

Anon. (2006e) Paving the way for better R & D in Europe, *Scrip*, 3193 (20 September): 2.

Anon. (2006f) EMEA expects rise in applications this year, *Scrip*, 3121 (11 January): 4.

Anon. (2006g) EC consults on post-marketing safety monitoring, *Scrip*, 3143 (29 March): 2.

Anon. (2006h) EMEA template to help companies prepare risk-management plans, *Scrip*, 3201 (18 October): 3.

Anon. *(2006i)* ADOPT positive but barriers remain to first-line Avandia. Scrip, 3216 (8 December): 18.

Anon. (2007a) EU introduces fines for firms that fail their obligations, *Scrip*, 3271 (27 June): 4.

Anon. (2007b) Europe goes ahead with plans to improve the drug development process, *Scrip*, 3261 (23 May): 3.

Anon. (2007c) Risk management takes centre stage, *Scrip*, 3269 (20 June): 2.

Anon. *(2007d)* Actos still bright for Takeda in first quarter, Scrip, 3282 (3 August): 11.

Anon. *(2007e)*. Primary endpoint in diabetes trials should be changed, urges US FDA panel chair for *Avandia*. Scrip, 3285 (15 August): 22.

Anon. (2007f) Iressa leads to inferior survival in US study, *Scrip*, 3269 (20 June): 24.

Anon. (2007g) Study draws criticism, but opens debate, *Cancer Drug News*, 283: 1–2.

Anon. (2008a) US prescription drug sales growth, *Scrip*, 3346 (26 March): 16.

Anon. (2008b) Centralized assessment soon compulsory for viral and autoimmune diseases, *Scrip*, 3328 (18 January): 2.

Anon. (2008c) Still scope for more centralized guidance in Europe, *Scrip*, 3345 (19 March): 7.

Anon. (2008d) Boosting innovation in the EU, *Scrip*, 3354 (18 April): 23–6.
Anon. (2008e) EMEA and FDA agree on renal safety biomarkers, *Scrip*, 3365/66 (28/30 May): 3.
Anon. (2008f) F-D-triple-A steps out to mixed reception, *Scrip*, 3384 (1 August): 26–9.
Anon. (2008g) Life under FDAAA: four months of post-approval studies and REMS, *Scrip*, 3385 (6 August): 22–4.
Anon. (2008h) Last chance for Iressa? *Cancer Drug News*, 342: 1–2.
Anon. (2009a) EC transfers pharma policy to health commissioner, *Scrip*, 3474 (4 December): 21.
Anon. (2009b) Iressa set to make it as second asking, *Cancer Drug News*, 361: 1–2.
Anon. (2009c) Tarceva filed for new lung cancer use, *Scrip*, 3438 (27 March): 11.
Anon. (2009d) Risk sharing: Pharma's failure to prove its value, *Scrip*, 3476 (18 December): 7.
Anon. (2010a) US report finds flaws in US FDA oversight of foreign clinical trials, *Scrip*, 3503 (2 July): 23.
Anon. (2010b) US Commissioner renews calls for 'strengthening regulatory science', *Scrip*, 3502 (25 June): 21.
Applbaum, K. (2007) *The Marketing Era: From Professional Practice to Global Provisioning*. London: Taylor & Francis.
Argiris, A. and Schiller, J.H. (2004) Can current treatments for advanced non-small-cell lung cancer be improved? *Journal of the American Medical Association*, 292: 499–500.
Association Internationale de la Mutualite (AIM), European Social Insurance Platform (ESIP), Health Action International (HAI), International Society of Drug Bulletins (ISDB) and Medicines in Europe Forum (MiEF) (2009) Legal Proposals on 'Information' to Patients by Pharmaceutical Companies: a Threat to Public Health, *Joint Briefing Paper*, 6 March.
Ault A. (2003) McClellan's FDA: boon to industry, consumers, or both?, *The Lancet*, 36: 379–80.
Avorn, J. (2005) Torcetrapib and Atorvastatin – should marketing drive the research agenda? *New England Journal of Medicine*, 352: 2573–6.
Baldwin, J. (2002) Demand grows for early access to promising cancer drugs, *Journal of National Cancer Institute*, 94: 1668–70.
Bangemann, M. (1992) Welcome and introduction: welcome address by Mr Martin Bangemann, Vice-President of the Commission of European Communities, in P.F. D'Arcy and D.W.G. Harron (eds) *Proceedings of the First International Conference on Harmonization*. Belfast: Queens University Press.
Barbot, J. (2006) How to build an 'active' patient? The work of AIDS associations in France, *Social Science & Medicine*, 62: 538–51.
Bardelay, D. and Kopp, C. (2002) Commentary: Concern over drug industry's influence on regulatory policy in Europe, *British Medical Journal*, 325: 1167–68.
Bardhan, K.D., Bodemar, G., Geldof, H., Schutz, E., Heath, A., Mills, and Jacques, LA. (2000) A double-blind, randomized, placebo-controlled dose-ranging study to evaluate the efficacy of alosetron in the treatment of IBS', *Alimentary Pharmacological Therapy*, 14: 23–34.
Barman-Balfour, J.A., Goa, K.L. and Perry, C.M. (2000) Alosetron, *Drugs*: 511–18.
BBC News24 (2007) Alzheimer's drugs remain limited, 10 August, news.bbc.co.uk

Beck, U. (1992) *Risk Society: Towards a New Modernity*. London: Sage.

Bennett, D.A., Beckett, L.A. and Murray, A.M. 1996 Prevalence of Parkinsonian signs and associated mortality in a community population of older people, New England Journal of Medicine, 334: 71–6.

Bernstein, M. (1995) *Regulating Business by Independent Commission*. New Jersey: Princeton University Press.

Bhaskar, R. 1975. *A Realist Theory of Science*. Sussex: Harvester Press.

Boone, H. (2011) 'Conditional Marketing Authorisations in the European *Union*', Presentation to the FDA ODAC meeting, 8 February, Silver Spring. Available at: *http://www.fda.gov/downloads/AdvisoryCommittees/CommitteesMeetingMaterials /Drugs/OncologicDrugsAdvisoryCommittee/UCM243231.pdf. Accessed 9 January 2012.*

Bourcier, T. (2010) Advisory Committee Non-clinical Briefing Document NDA *21–071*, Division of Metabolism and Endocrinology Products.

Braithwaite, J. (1986) *Corporate Crime in the Pharmaceutical Industry*. London: Routledge and Kegan Paul.

Breathnach, O.S., Freidlin, B. and Conley, B. (2001) Twenty-two years of phase III trials for patients with advanced non-small-cell lung cancer: sobering results, *Journal of Clinical Oncology*, 19: 1734–42.

Breckenridge, A., Mello, M., and Psaty, B.M. (2012) 'New horizons in pharmaceutical regulation', *Nature Reviews Drug Discovery*, 11: 501–02.

Brickman, R., Jasanoff, S. and Ilgen, T. (1985) Controlling Chemicals: The Politics of Regulation in Europe and the US. Ithaca, NY: Cornell University Press.

Brinker, A. (2002) 'Summary comments on 10 epidemiological studies submitted in an efficacy supplement under Lotronex NDA', FDA Memo, 26 March.

Britten, N. (2008) *Medicines and Society: Patients, Professionals and the Dominance of Pharmaceuticals*. Basingstoke: PalgraveMacmillan.

Broglio, K.R. and Berry, D.A. (2009) Detecting an overall survival benefit that is derived from progression-free survival, *Journal of the National Cancer Institute*, 101: 1642–9.

Brown, N. and Michael, M. (2003) A sociology of expectations: retrospecting prospects and prospecting retrospects. *Technology Analysis & Strategic Management*, 15: 3–18.

Brown, N. and Webster, A. (2004) *New Medical Technologies and Society: Re-ordering Life*. Cambridge: Polity.

Brown T., Boland A., Bagust A., Oyee J., Hockenhull J., Dundar Y., Dickson R., Ramani, V.S. and Proudlove C. (2009) *Gefitinib for the First-Line Treatment of Locally Advanced or Metastatic Non-Small Cell Lung Cancer (NSCLC): A Single Technology Appraisal*. LRiG: The University of Liverpool.

Bruce, F. (2011) French Actos risk not seen with other diabetes drugs and only in bladder cancer, Scrip, 10 June.

Bunge, M. (1999) *The Sociology-Philosophy Connection*. New Brunswick: Transaction Publishers.

Callon, M. and Rabeharisoa, V. (2008) The growing engagement of emergent concerned groups in political and economic life: lessons from the French Association of Neuromuscular Disease Patients, *Science, Technology & Human Values*, 33: 230–61.

Camilleri, M., Chey, W.Y., Mayer, E.A, Northcutt, A.R., Heath, A., Dukes, G.E., McSorley, D., and Mangel, A.M. (2001) A randomized controlled clinical trial of the serotonin type 3 receptor antagonist alosetron in women with diarrhea-predominant IBS, *Archives of Internal Medicine*, 161: 1733–40.

Carpenter D.P. (2004) The political economy of FDA drug review: processing, politics and lessons for policy, *Health Affairs*, 23: 52–63.

Carpenter, D. (2010a) *Reputation and Power: Organizational Image and Pharmaceutical Regulation at the FDA*. Princeton: Princeton University Press.

Carpenter, D. (2010b) Response to Marcia Angell, *New York Review of Books*, 30 September.

Carpenter, D. and Fendrick, A.M. (2004) Accelerating approval times for new drugs in the US, *The Regulatory Affairs Journal – Pharma*, 15: 411–17.

Cartwright, A.C. and Matthews, B.R. (eds) (1991) *Pharmaceutical Product Licensing: Requirements for Europe*. Chichester: Ellis Harwood.

Cash, B.D. and Chey, W.D. (2003) Advances in the management of IBS, *Current Gastroenterology Reports*, 5: 468–75.

Cawson, A. (1986) *Corporatism and Political Theory*. Oxford: Basil Blackwell.

Centre for Medicines Research (CMR) International (2002) *CMR International News*, 20 (1).

CMR International (2005) 'Innovation on the wane?' Latest News. Available at http://www.cmr.org

Chakravarty, A. and Sridhara, R. (2008) Use of progression-free survival as a surrogate marker in oncology trials: some regulatory issues, *Statistical Methods in Medical Research*, 17: 515–18.

Charles River Associates (2004) *Innovation in the Pharmaceutical Sector: A study undertaken for the European Commission* available at http://pharmacos.eudra.org /F2/pharmacos/docs/Doc2004/nov/EU%20Pharma%20Innovation_25–11–04. pdf. Accessed 16 May 2005.

Ciardiello, F., De Vita, F., Orditura, M. and Tortora, G. (2004) The role of EFGR inhibitors in non-small-cell lung cancer, *Current Opinion in Oncology*, 16: 130–5.

CMS Cameron McKenna and Andersen Consulting (2000) *Evaluation of the operation of Community procedures for the authorisation of medicinal products*, October 2000, European Commission, Directorate-General Enterprise, Pharmaceuticals and Cosmetics. Available at: http://www.pharmacos.eudra.org/F2/pharmacos /docs/DOC2000/nov/reportmk.pdf. Accessed 13 May 2005.

Code of Federal Regulations (1987) *Treatment use of an investigational new drug*, 21 CFR 312.34.

Code of Federal Regulations (1988) *Subpart E – Drugs intended to treat life-threatening and severely debilitating illnesses*, 21 CFR 312.80–312.88.

Code of Federal Regulations (1992) *Subpart H – Accelerated approval of new drugs for serious or life-threatening illnesses*, 21 CFR 314, Subpart H.

Code of Federal Regulations (2010) *Applications for FDA approval to market a new drug: adequate and well-controlled studies*, 21CFR314.126.

Cohen, D. (2010) Rosiglitazone: what went wrong? *British Medical Journal*, 341: 530–7.

Cohen, M.H. (2002) 'Briefing document,NDA-21399' FDA Oncology Division Medical Review, September.

Cohen, S. (1962) Isolation of a mouse submaxillary gland protein accelerating incisor eruption and eyelid opening in the new born animal, *Journal of Biological Chemistry*, 237: 1555–62.

Cohen, S., Carpenter, G. and King, L. (1980) Epidermal growth factor – receptor – protein kinase interactions: co-purification of receptor and epidermal growth factor-enhanced phosphorylation activity, *Journal of Biological Chemistry*, 255: 4834–42.

Collier, J. (1989) *The Health Conspiracy: How Doctors, the Drug Industry and Government Undermine our Health*. London: Century.

Conrad, P. and Leiter, V. (2008) From Lydia Pinkham to Queen Levitra: direct-to-consumer advertising and medicalization, *Sociology of Health & Illness*, 30: 825–38.

Cortes-Funes, H. and Soto Parra, H. (2003) Extensive experience of disease control with gefitinib and the role of prognostic markers, *British Journal of Cancer*, 89 (suppl. 2): S3–8.

CPMP (1996) *Accelerated evaluation of products indicated for serious diseases (life-threatening or heavily disabling diseases)* CPMP/495/96. Available at: http://www. emea.eu.int/pdfs/human/regaffair/049596en.pdf. Accessed 13 May 2005.

Currie, G., Humphreys, M., Waring, J. and Rowley, E. (2009) Narratives of professional regulation and patient safety: medical devices in anaesthetics, Health, Risk and Society, 11: *117–35*.

Czoski-Murray, C., Warren, E., Chilcott, J., Beverley, C., Psyllaki, M.A. and Cowan, J. (2004). Cost effectiveness of pioglitazone and rosiglitazone. Health Technology Assessment 8(13).

Daemmrich, A. (2004) *Pharmacopolitics: Drug Regulation in the US and Germany*. Chapel Hill: University of North Carolina Press.

Daemmrich, A. and Krucken, G. (2000) Risk versus risk: decision-making dilemmas of drug regulation in the US and Germany, *Science as Culture*, 9: 505–34.

D'Agostino, R.B. (2001) Changing endpoints in breast-cancer drug approval – the Avastin story, *New England Journal of Medicine*, 365: e2. Available at: http://www. nejm.org.ezproxy.sussex.ac.uk/doi/full/10.1056/NEJMp1106984. Accessed 30 January, 2012.

Daniels, C.E. and Wertheimer, A.I. (1980) Therapeutic significance of the drug lag, *Medical Care*, 18: 754–65.

Dehue, T. (2010) 'Clinical trials as marketing instruments'. Selling Sickness Conference, October, University of Groningen.

Delbaldo, C., Michiels, S., Syz, N., Soria, Jean-Charles, S., Le Chevalier, T. and Pignon. J-P. (2004) Benefits of adding a drug to a single-agent or a 2-agent chemotherapy regimen in advanced non-small-cell lung cancer, *Journal of American Medical Association*, 292: 470–84.

Demortain, D. (2008) From drug crises to regulatory change. *Health, Risk & Society*, 10: 37–51.

Dennis, M. (2007) More spending has little effect on lung cancer, *Cancer Drug News*, 289 (24 October): 1–2.

Dennis, M. (2008) Cancer to lead death toll in 2010? *Cancer Drug News*, 34 (10 December): 1–2.

DHHS Office of Inspector General (2003) *FDA's Review Process for New Drug Applications: A Management Review*, March, OEI-01–00590.

Diamant, M. and Heine, R.J. (2003) Thiazolidinediones in type II diabetes mellitus. *Drugs*, 63: 1373–1405.

Doran, E., Henry, D., Faunce, T.A. *and* Searles, A. (2008) Australian pharmaceutical policy and the idea of innovation, *Journal of Australian Political Economy*, 62: 39–61.

Dormandy, J.A., Charbonnel, B., Eckland, D.J. for PROactive Study Group (2005) Secondary prevention of macrovascular events in patients with type 2 diabetes in the PROactive Study, *Lancet*, 366: 1279–89.

DREAM Investigators (2006) Effect of rosiglitazone on the frequency of diabetes in patients with impaired glucose tolerance or impaired fasting glucose: a randomized controlled trial, *Lancet*, 368: 1096–1105.

Dukes, M.N.G. (1985) *The Effects of Drug Regulation: A Survey Based on European Studies of Drug Regulation*. Lancaster: MTP Press.

EBE-EFPIA (2004) *Accelerated Approval System*, EBE Member Companies' Position Paper, February 2004. Available at: http://www.ebe-efpia.org/docs/pdf/Acceleratedapp_position.pdf. Accessed 13 May 2005.

Edgar, H. and Rothman, D.J. (1990) The challenge of AIDS to the regulatory process, *Milbank Quarterly*, 68 (Suppl.1): 111–42.

EFPIA (2000) *Evolution of Standards*, June 2000. Available at: http://www.efpia.org/4_pos/scI_regu/evolstandard_.PDF. Accessed 13 May 2005.

Eichler, H., Pignatti, F., Flamion, B., Leuftkens, H. and Breckenridge, A. (2008) Balancing early market access to new drugs with the need for benefit/ risk data: a mounting dilemma, *Nature*, 7: 818–26.

Electronic Medicines Compendium (undated) Summary of Product Characteristics, Tasmar 100 mg Tablets. Available at http://www.medicines.org.uk/emc/medicine/15900. Accessed 30 January 2012.

Ellenberg, S. (2011) Accelerated approval of oncology drugs: can we do better? *Journal of the National Cancer Institute*, 103: 616–17.

Ellis, A.L. (1997) FDA Review and Evaluation of Pharmacology and Toxicology Data, Trovan, 15 December.

EMEA (1997) European Public Assessment Report (Tasmar).

EMEA (1998a) Recommendation for the Suspension of the Marketing Authorisation for Tasmar (tolcapone) – EU/1/97/044/001–006, CPMP/2457/98, 17 November.

EMEA (1998b) European Public Assessment Report (Tovan), 3 July, CPMP/1302/98.

EMEA (1999a) Public Statement on Trovan (Trovafloxacin): Serious, Severe and Unpredictable Liver Injuries, EMEA/15770/99, 25 May.

EMEA (1999b) Public Statement on Trovan (Trovafloxacin): Recommendation to Suspend the Marketing Authorisation in the European Union, EMEA/18046/99, 15 June.

EMEA (1999c) Public Statement on Levacetylmethadol (Orlaam) – for Life-Threatening Cardiac Rhythm Disorders, EMEA/38436/99, 15 December.

EMEA (2000) Public Statement on Levacetylmethadol (Orlaam) – Life-Threatening Ventricular Rhythm Disorders, EMEA/38918/00, 19 December.

EMEA (2001a) EMEA/CPMP Position Statement on the Use of the Placebo in Clinical Trials with Regard to the Revised Declaration of Helsinki, London, 28 June 2001. EMEA/17424/01. London: EMEA.

EMEA (2001b) European Public Assessment Report (Actos/pioglitazone). London: EMEA.

EMEA (2001c) European Public Assessment Report, Revision 4 (Orlaam): CPMP Scientific Discussion, CPMP/169/97.

EMEA (2002a) Committee for Proprietary Medicinal Products 15–17 October 2002. Plenary Meeting Monthly Report, EMEA/CPMP/4979/02. London: EMEA.

EMEA (2002b) CPMP Note for Guidance on Clinical Investigation of Medicinal Products in Treatment of Diabetes, May. London: EMEA.

EMEA (2003) European Public Assessment Report (Avandia/rosiglitazone). London: EMEA.

EMEA (2004a) CPMP Draft Points to Consider on the Choice of Non-Inferiority Margin, London, 26 February 2004. CPMP/EWP/158/99draft. London: EMEA.

EMEA (2004b) *Ninth Annual Report 2003*. London: EMEA.

EMEA. (2004c) CPMP Draft Points to Consider on Choice of Non-inferiority Margin, February. London: EMEA.

EMEA (2008) Reflection Paper on Pharmacogenomics in Oncology. Doc Ref: EMEA/CHMP/PGxWP/128435/2006. London: EMEA.

EMEA (2009) Assessment Report for Iressa. Doc Ref: EMEA,/CHMP/563746/2008, London: EMEA.

EPPOSI (2004) Report of EPPOSI Workshop on Value of Innovation, June. Brussels: EPPOSI.

Epstein, S. (1996) *Impure Science: AIDS, Activism, and the Politics of Knowledge*. Berkeley: University of California Press.

Epstein, S. (1997) Activism, drug regulation, and the politics of therapeutic evaluation in the AIDS era: A case study of ddC and the 'Surrogate Markers' debate, *Social Studies of Science*, 27: 691–726.

EURODIS (2005) EURODIS comments on the draft regulation on conditional marketing authorisation for medicinal products. Available at: http://www.eurordis.org/IMG/pdf/eurordis_comments_conditional_approval_jan05.pdf. Accessed 13 May 2005.

Europa (2000) 2281st Council Meeting – Health. Press Release, 1 September. PRES/00/25. Available at: http://europa.eu/rapid/pressReleasesAction.do?reference=PRES/00/225&format=HTML&aged=0&lg=fi&guiLanguage=en. Accessed 1 February 2012.

European Cancer Patient Coalition (2008) Comments on the EU Pharmacovigilance Strategy, 1 February. Available at: http://ec.europa.eu/health/files/pharma-covigilance/docs/2007_02_26/3.pdf. Accessed 5 February 2012.

European Commission (1965) Directive 65/65/EEC.

European Commission (1995) Council Directive 540/95.

European Commission (1996) Rules Governing Medicinal Products in the European Community, Vol. IIA (Notice to Applicants). Brussels: EC.

European Commission (2001a) Procedures for Granting Marketing Authorisations for Medicinal Products in Regulation (EEC) No 2309/93, October. Brussels: EC.

European Commission. (2001b). Proposal for Directive of European Parliament and of Council amending Directive 2001/83/EC on the Community code relating to medicinal products, November. Brussels: EC.

European Commission (2002) Draft regulation. Available at: http://pharmacos.eudra.org/F2/pharmacos/docs/Doc2005/02_05/Penalties%20-%20Public%20consultation%2002%202005.pdf. Accessed 13 May 2005.

European Commission Enterprise Directorate-General (2000) *Pharmaceuticals in the European Union*. Luxembourg: Office for Official Publications of the European Communities.

European Commission Enterprise Directorate-General (2004) Available at: http://europa.eu.int/comm/dgs/enterprise/activit_goals_En.htm. Accessed 13 May 2005.

European Council (1975) Directive 75/319/EEC of 20 May 1975 on the approximation of provisions laid down by law, regulation or administrative action relating to medicinal products, *Official Journal of the European Communities* L 147, 9.6.75: 13.

European Council (1993) Regulation (EEC) 2309/93 of 22 July 1993 laying down Community procedures for the authorization and supervision of medicinal products for human and veterinary use and establishing a European Agency for the Evaluation of Medicinal Products, *Official Journal of the European Communities* L 214, 24.8.93.

European Council (2004) Directive 2001/83/EC of the European Parliament and of the Council of 6 November 2001 on the Community code relating to medicinal products for human use, *Official Journal of the European Communities* L 311, 28.11.2004.

European Parliament and Council (2004) Regulation (EC) No 726/2004 of the European Parliament and of the Council of 31 March 2004 laying down Community procedures for the authorization and supervision of medicinal products for human and veterinary use and establishing a European Medicines Agency. *Official Journal of the European Communities* L 136, 30.04.2004.

European Public Health Alliance (2003) Public interest group on medicines launched. Update 63. Available at: http://www.epha.org/a/509. Accessed 5 February 2012.

Faigen, N. (2010) EMA suspends while FDA severely restricts Avandia after data review, Scrip (24 September).

Feczko, J.M. (1999) 'Dear Healthcare Professionals' Letter from Senior Vice President, Worldwide Medical & Regulatory Operations, Pfizer Pharmaceuticals Group, 10 June.

Federal Register (1985) New drug and antibiotic regulations, 22 February, Final Rule (52 FR 7452). Available at: http://www.fda.gov/ScienceResearch/SpecialTopics/RunningClinicalTrials/ucm120020htm. Accessed 2 August 2011.

Federal Register (1987) New drug, antibiotic and biologic drug product regulations, 19 March, Final Rule (52 FR 8798). Available at: http://www.fda.gov/ScienceResearch/SpecialTopics/RunningClinicalTrials/ucm120111htm. Accessed 2 August 2011.

Federal Register (1988) Procedures for Drugs Intended to Treat Life-threatening and Severely Debilitating Illnesses, 21 October, Interim Rule (53 FR 41523).

Federal Register (1992) New Drug, Antibiotic, and Biological Drug Product Regulations, Accelerated Approval, 15 April, Proposed Rule (57 FR 13234).

Federal Register (2000) Prescription Drug User Fee Act (PDUFA); Public Meeting. Notice of Public Meeting, 4 August (65 FR 47993).

Feigal, D.W. (1997) FDA Approval Letter, NDA *20–759/760* (Trovan), 18 December.

Feigal D. W. (1999) *Standard of Evidence for Drug Approval: An FDA Perspective,* CDRH Presentation, 9 December 1999. Available at: http://www.fda.gov/cdrh /meetings/ibc.pdf. Accessed 13 May 2005.

Ferris, M.J. (1992) A review of the Japanese regulatory system, in J.P. Griffin (ed) *Medicines: Regulation, Research and Risk.* Belfast: Queen's University Press.

Fisher, J. 2009. *Medical Research for Hire: The Political Economy of Pharmaceutical Clinical Trials.* New Brunswick: Rutgers University Press.

Fishman, J.R. (2004) Manufacturing desire: the commodification of female sexual dysfunction, *Social Studies of Science,* 34: 187–218.

Fitzgerald, G.A. (2010) Confusion and conclusion: Avandia and the FDA, Scrip (30 September).

Fleming, T.R. (2005) Surrogate endpoints and FDA's accelerated approval process, *Health Affairs,* 24: 67–78.

Fleming T.R. and DeMets D.L. (1996) Surrogate end points in clinical trials: are we being misled? *Annals of Internal Medicine,* 125: 605–13.

Food and Drug Administration (FDA) (1991) 21 Code of Federal Regulations, 314.126(b)(2).

FDA (1992) 57 Federal Register, 57, *13234–35.*

FDA (1993) Approved Label, NDA 20–315 (Orlaam).

FDA (1994) FDA Consumer, November. *Available at: www.fda.gov/bbs/topics /CONSUMER/CON0291c.html.*

FDA (1997a) 'FDA Innovation Efforts Rewarded', press release November 20, 1997. Available at: http://www.fda.gov/cder/fdaonnivate.htm. Accessed 13 May 2005.

FDA (1997b) Final performance report: Prescription Drug User Fee Act of 1992, fiscal year 1997 report to Congress, 1 December, 1997. Available at: http:// www.fda.gov/ope/pdufa/report97/pdufa97.htm. Accessed 13 May 2005.

FDA (1997c) Statement by David A. Kessler, M.D., Commissioner, Food and Drug Administrtation, Department of Health and Human Services before the Committee on Labour and Human Resources, United States Senate, 21 February 1996. Available at: http://www.fda.gov/ola/nktest.html. Accessed 13 May 2005.

FDA (1998a) Guidance for Industry: Fast Track Drug Development Programs – Designation, Development, and Application Review. US Department of Health and Human Services, September.

FDA (1998b) *Guidance for Industry: Approval of New Cancer Treatment Uses.* Rockville: HHS.

FDA (1998c) FDA Talk Paper: Tasmar (tolcapone).

FDA (1999a) *Managing the Risks from Medical Product Use: Creating a Risk Management framework.* Washington DC: DHHS.

FDA. (1999b). Transcript of Gastro-intestinal Drugs Advisory Committee, Gaithersburg, Maryland, November 16.

FDA (1999c) Public Health Advisory – Trovan, 9 June. Available at: www.fda.gov /cder/news/trovan/trovan-advisory.htm. Accessed 27 January 2009.

FDA. (2000a) Prescription Drug User Fee Act: FDA and Stakeholders Public Meetings. Transcript, 15 September.

FDA. (2000b). Transcript of Gastro-intestinal Drugs Advisory Committee (GDAC), 'NDA 21–107 Lotronex (Alosetron), Glaxo Wellcome', Gaithersburg, Maryland, June 27.

FDA (2000c) CDER-PhRMA-AASLC Conference 2000, Clinical White Paper, November 2000.

FDA (2001a) Anti-retroviral drugs used in the treatment of HIV infection. Available at http://www.fda.gov/ForConsumers/byAudience/ForPatientAdvocates /HIVandAIDSActivities/ucm118915.htm. Accessed 2 August 2011.

FDA (2001b) Prescription Drug User Fee Act (PDUFA) FDA and stakeholders public meeting. Executive Summary, December 7, 2001. Available at: http://www.fda. gov/oc/pdufa/meeting2001/ExecSumm.html. Accessed 13 May 2005.

FDA (2001c) Public meeting: Prescription Drug User Fee Act (PDUFA). Transcript of meeting, 7 December.

FDA (2002a) Oncological Drugs Advisory Committee (ODAC) Meeting, 'Iressa NDA', 24 September.

FDA. (2002b). Transcript of GDAC and Drug Safety and Risk management Subcommittee of the Advisory Committee for Pharmaceutical Science, Bethesda, Maryland, April 23.

FDA (2003a) FY 2003 Performance report to the President and the Congress for the Prescription Drug User Fee Act of 1992 as reauthorized and amended by the Food and Drug Administration Act of 1997 and the Public Health and Security Bioterrorism Preparedness and Response Act of 2002. Available at http://www. fda.gov/oc/pdufa/report2003/default.htm. Accessed 13 May 2005.

FDA (2003b) Approval package, Iressa, FDA Center for Drug Evaluation and Research, 1 April.

FDA (2003c) ODAC Meeting, 'Accelerated Approval Process', 12 March.

FDA (2003d) ASCO/FDA Lung Cancer Endpoints Workshop, 15 April.

FDA (2003e) ODAC Meeting, 'Endpoints in Clinical Cancer Trials and Lung Cancer Clinical Trials', 16 December.

FDA (2004) Innovation or stagnation: challenge and opportunity on the critical path to new medical products, US Department of Health and Human Services, March 2004. Available at: http://www.fda.gov/oc/initiatives/criticalpath/white-paper.html. Accessed 13 May 2005.

FDA (2005a) Available at: http://www.fda.gov/oashi/aids/miles81.html. Accessed 13 May 2005.

FDA (2005b) Prescription Drug User Fee Amendments of 2002. Available at: http://www.fda.gov/oc/pdufa/amendments.html. Accessed 13 May 2005.

FDA (2005c) ODAC Meeting, 'Iressa', 4 March.

FDA (2007). Joint meeting of Endocrinologic and Metabolic Drug Advisory Committee and Drug Safety Management Advisory Committee, 30 July. Gaithesburg, Maryland: PaperMill Reporting.

FDA (2008a) CDER Approval Times for Priority and Standard NDAs and BLAs, 1993–2008. Available at: http://www.fda.gov/downloads/drugs/developmen-tapprovalprocess/howdrugsaredevelopedandapproved/drugandbiologicap-provalreports/ucm123957.pdf. Accessed 3 August 2011.

FDA. (2008b). Endocrinologic and Metabolic Drug Advisory Committee, 1–2 July. Silver Spring, Maryland: BetaCourt Reporting.

FDA (2009a) HIV/AIDS historical timeline 1981–90. Available at http://www.fda. gov/ForConsumers/ByAudience/ForPatientAdvocates/HIVandAIDSActivities /ucm151074.htm. Accessed 2 August 2011.

FDA (2009b) Expanded access and expedited approval of new therapies related to HIV/AIDS. Available at: http://www.fda.gov/For Consumers/ByAudience

/ForPatientAdvocates/HIVandAIDSActivities/ucm134331.htm. Accessed 2 August 2011.

FDA (2009c) PDUFA III Five Year Plan. Available at: http://www.fda.gov /ForIndustry/UserFees/PrescriptionDrugUserFee/ucm173542.htm.306 Accessed 3 August 2011.

FDA (2011a) Background – PDUFA II Goals – FY 2001 Update (April 2001). Available at: http://www.fda.gov/ForIndustry/UserFees/PrescriptionDrugUserFee /ucm174507.htm. Accessed 3 August 2011.

FDA (2011b) Transcript of the Oncologic Drugs Advisory Committee, Gaithersburg, Maryland, 8 February.

Food and Drug Administration Modernization Act (1997) Available at: http://www. fda.gov/RegulatoryInformation/Legislation/FederalFoodDrugandCosmeticAct FDCAct/SignificantAmendmentstotheFDCAct/FDAMA/FullTextofFDAMAlaw/ default.htm#SEC.%20406. Accessed 2 August 2013.

Friedman M.A., Woodcock, J., Lumpkin, M.M., Shuren, J.E., Hass, A.E. and Thompson, L.J. (1999) The safety of newly approved medicines: do recent market removals mean there is a problem? *Journal of the American Medical Association*, 281: 1728–34.

Gale, E.A.M. (2001). Lessons from the glitazones, *Lancet*, 357: 1870–5.

Gallo-Torres, H.E. (2002) Lotronex: sponsor's submission of December 2001: FDA Perspectives on Safety, FDA GDAC and Drug Safety and Risk management Subcommittee Background Package, 23 April.

Gambardella A., Orsenigo L. and Pammolli F. (2000) *Global Competitiveness in Pharmaceuticals. A European Perspective*. Report Prepared for the Directorate General Enterprise of the European Commission, November 2000.

Garattini, S. (2004) Designing the most favourable study design, *European Journal of Clinical Pharmacology*, 61: 85–6.

Garattini S. and Bertele V. (2001) Adjusting Europe's drug regulation to public health needs, *Lancet*, 358: 64–7.

Garattini S. and Bertele V. (2002) Efficacy, safety and the cost of new anticancer drugs, *British Medical Journal*, 325: 269–71.

Garattini S. and Bertele V. (2003a) Efficacy, safety and cost of new drugs acting on the central nervous system, *European Journal of Clinical Pharmacology*, 59: 79–84.

Garattini S. and Bertele V. (2003b) Efficacy, safety and cost of new cardiovascular drugs: a survey, *European Journal of Clinical Pharmacology*, 59: 701–06.

Glaxo Wellcome plc (2000a) Announcement of Half-Year Results for the six months ended 30 June 2000, 27 July. Available at: http://www.gsk.com/financial/reports/gw_2qearnings.pdf. Accessed 23 July 2003.

Glaxo Wellcome plc (2000b) Available at: http://www.gsk.com/press_archive /gw/2000/press_000210pr.htm#. Accessed 23 July 2003.

Goldberg, P. (2005) An 'insurgency' targets randomized trials, *The Cancer Letter*, 31: 10–15.

Goldberg, P. (2009) Iressa authorisation in Europe clarifies differences between US and EU standards, *The Cancer Letter*, 35: 1–4.

Goldberg, P. (2011a) Sponsors yank indications for Iressa, Celebrex in oncology, *The Cancer Letter* 37: 2.

Goldberg, P. (2011b) FDA revokes Avastin's accelerated approval in metastatic breast cancer indication, *The Cancer Letter*, 37: 1 & 7–10.

Goozner M. (2004). *The $800 Million Pill: The Truth Behind the Cost of New Drugs.* London: University of California Press.

Gotzsche, P. and Jorgensen, A. (2011) Opening up data at the European Medicine Agency, *British Medical Journal*, 342: 1184–6.

Graham, D.J., Drinkard, C.R., Shatin, D., Tsong, Y. and Burgess, M.J. (2001) Liver enzyme monitoring in patients treated with troglitazone, *Journal of the American Medical Association*, 286: 831–4.

Greene, J. (2007) *Prescribing by Numbers: Drugs and the Definition of Disease.* Baltimore: Johns Hopkins University Press.

Gunput, M.D. 1999 Clinical pharmacology of alosetron, *Alimentary Pharmacological Therapy*, 13 (Suppl. 2): 70–6.

Hammersley, M. (1995) *The Politics of Social Research.* London: Sage.

Hancher, L. (1989) *Regulating for Competition: Government, Law, and the Pharmaceutical Industry in the UK and France.* University of Amsterdam: PhD Thesis.

Hancher, L. (1990) *Regulating for Competition: Government, Law and the Pharmaceutical Industry in the UK and France.* Oxford: Clarendon.

Harris, G. (2010a) Diabetes drug maker hid test data, New York Times, 12 July.

Harris, G. (2010b) FDA panel votes to restrict Avandia, New York Times, 14 July.

Hedgecoe, A. (2004) *The Politics of Personalized Medicine: Pharmacogenetics in the Clinic.* Cambridge: Cambridge University Press.

Herbst, R.S. and Bunn, P.A. (2003) Targeting the epidermal growth factor receptor in non-small-cell lung cancer, *Clinical Cancer Research*, 9: 5813–24.

Herxheimer, A. (2003) Relationships between the pharmaceutical industry and patients' organizations, *British Medical Journal*, 326: 1208–10.

Hilts, P. (2003) *Protecting America's Health: The FDA, Business, and One Hundred Years of Regulation.* New York: Knopf.

't Hoen, Ellen (2009) *The Global Politics of Pharmaceutical Monopoly Power*, Diemen, The Netherlands: AMB Publishers.

Holston S (1997) *The Value of Patients' Perspective in FDA's Decision Processes*, 10th IMS International Symposium Brussels, November 3, 1997. Available at: http://www.fda.gov/oashi/cancer/value.html. Accessed 13 May 2005.

Home, P.D., Pocock, S.J., Beck-Nielsen, H. for RECORD Study Team (2009). Rosiglitazone evaluated for cardiovascular outcomes in oral agent combination therapy for type 2 diabetes, *Lancet*, 373: 2125–35.

Horton, R. (2001) The FDA and *The Lancet*: an exchange, *Lancet*, 358: 417–18.

International Federation of Pharmaceutical Manufacturers' Association *(2000). The* Value *and* Benefits of ICH to Industry. Geneva: IFPMA.

IMS Health (2003) Available at: http://www.ims-global.com/insight/news_story/0306/news_story_030611.htm. Accessed 13 May 2005.

Institute of Medicine (2007) Committee on the Assessment of the US Drug Safety System, *The Future of Drug Safety, Promoting and Protecting the Health of the Public.* Washington DC: National Academy Press.

International Society of Drug Bulletins and the Medicines in Europe Forum (2009) Pharmacovigilance in Europe: The European Commission's proposals endanger the population. Brussels, October 2009. Available at: http://www.isdbweb.org/documents/uploads/press/En_PharmacovigBriefingNoteOct2009.pdf. Accessed 4 February 2012.

Irwin, A. (2001) *Sociology and the Environment.* Cambridge: Polity.

Jasanoff, S. (1986) *Risk Management and Political Culture.* New York: Russell Sage Foundation.

Jasanoff, S. (1990) *The Fifth Branch: Science Advisers as Policymakers.* Cambridge, MA: Harvard University Press.

Johnson, J.R., Ning, Y.M., Farrell, A., Justice, R., Keegan, P. and Pazdur, R. (2011) Accelerated approval of oncology products: the Food and Drug Administration experience, *Journal of the National Cancer Institute,* 103: 636–44.

Johnson, J.R., Williams, G. and Pazdur, R. (2003) Endpoints and FDA approval of oncology drugs ,*Journal of Clinical Oncology,* 21: 1404–11.

Kahn, S.E., Haffner, S.M., Heise, M.A. for ADOPT Study Group (2006) Glycemic durability of rosiglitazone, metformin, or glyburide monotherapy, New England *Journal of Medicine,* 355: 2427–43.

Kaitin, K.I and Brown ,J.S. (1995) A drug lag update, *Drug Information Journal* 29: 361–73.

Kaitin, K.I. and Di Masi, J. (2000) Measuring the pace of new drug development in the user fee era, *Drug Information Journal,* 24: 673–80.

Kaitin, K.I, Mattison, N., Northington, F.K. and Lasagna, L. (1989) The drug lag: an update of new drug introductions in the United States and in the United Kingdom 1977 though 1987, *Clinical Pharmacology and Therapeutics,* 46: 121–38.

Karnofsky, D.A., Abelmann, W.C., Craver, L.F. and Burchenal, J.H. (1948) The use of nitrogen mustard in the palliative treatment of carcinoma with a particular reference to bronchogenic carcinoma, *Cancer,* 1: 534–56.

Katz, R. (1997) FDA Supervisory Review of NDA 20–697, Tasmar, 6 April.

Keat, R. (1981) *The Politics of Social Theory.* Oxford: Basil Blackwell.

Keijsers, L., Bossong, M.G. and Waarlo, A.J. (2008) Participatory evaluation of a Dutch warning campaign for substance-users, *Health, Risk and Society,* 10: 283–95.

Kessler D.A. (1989) The regulation of investigational drugs, *The New England Journal of Medicine,* 320: 285–6.

Kessler D.A., Rose J.L., Temple R.J., Schapiro R. and Griffin J.P. (1994) Therapeutic-class wars: drug promotion in a competitive marketplace, *New England Journal of Medicine,* 331: 1350–53.

KPMG Consulting (2002) *Reanalysis of 1993 Standard Costs for the Process for the Review of Human Drug Applications As Required Under the Prescription Drug User Fee Act,* McLean, Virginia: KPMG.

Krentz, A.J., Bailey, C.J. and Melander, A. (2000) Thiazolidinediones for type 2 diabetes, *British Medical Journal,* 321: 252–53.

Kress, S. (2001) Review of Lotronex (alosetron)-associated serious complications of severe constipation reported post-marketing, FDA Medical Officer, Division of Gastro-Intestinal and Coagulation Drug Products, 9 October.

Kris, M.G., Natale, R.B. and Herbst, R.S. (2003) Efficacy of gefitinib, an inhibitor of the epidermal growth factor receptor tyrosine kinase, in symptomatic patients with non-small-cell lung cancer, *Journal of the American Medical Association,* 290: 2149–58.

Ladabaum, U. (2003) Safety, efficacy and costs of pharmacotherapy for functional gastrointestinal disorders, *Alimentary Pharmacological Therapy,* 17: 1021–30.

Lakoff, A. (2005) *Pharmaceutical Reason: Knowledge and Value in Global Psychiatry.* Cambridge: Cambridge University Press.

Langreth, R. and Fawcett, A. (1999) Pfizer and FDA urge limits on use of Trovan due to liver risk, *Wall Street Journal*, 6 October.

La Revue Prescrire (1999) A review of new drugs, *Prescrire International*, 8: 5–20.

La Revue Prescrire (2002) Reorienting European Medicines Policy, June 2002. Available at: http://www.prescrire.org/docus/euMedPolEn1.pdf. Accessed 17 September 2013.

La Revue Prescrire (2004) Medicines in Europe: the most important changes in the new legislation, *Prescrire International*, 13: 158-1-158–8.

La Revue Prescrire (2009) Legal obligations for transparency at the European Medicines Agency: Prescrire's assessment over four years, *Prescrire International*, 18: 228–33.

Lasser K.E., Allen P.D., Woolhandler S.J., Himmelstein D.U., Wolfe S.M. and Bor D.H. (2002) Timing of new black box warnings and withdrawals for prescription medicines, *Journal of the American Medical Association*, 287: 2215–20.

Latour, B. (1987) *Science in Action*. Cambridge, MA: Harvard University Press.

Lave, R., Mirowski, P. and Randalls, S. (2010) STS and neo-liberal science, *Social Studies of Science*, 40: 659–76.

Levine, P. (2002) 'Highlights of regulatory history' FDA Alosetron Background Package for Gastro-Intestinal Drugs Advidosy Committee and Drug Safety and Risk Management Committee, 23 April.

Lexchin, J. (2006) The pharmaceutical industry and pursuit of *profit. In* J.C. Cohen, P. Illingworth, & U. Schuklenk (eds.), *The Power of Pills*, 11–24. London: Pluto.

LeWitt, P.A. (1992) Treatment strategies for extension of levodopa effect, *Neurology Clinic*, 10: 511–26.

Li, Z. (2001) 'Recalculation of the incidence rate of alosetron-associated ischaemic colitis among women in the US' FDA memo from Zili Li, medical epidemiologist, Division of Drug Risk Evaluation II to Florence Houn, Director of Office of Drug Evaluation, 2 April.

Li Bassi L., Bertele V. and Garattini S. (2003) European regulatory policies on medicines and public health needs, *European Journal of Public Health*, 13: 246–50.

Liebenau, J. (1981) *Medical Science and Medical Industry, 1890–1929: A Study of Pharmaceutical Manufacturing in Philadelphia*. University of Pennsylvania: PhD Thesis.

Light, D.W. (ed.) (2010) *The Risks of Prescription Drugs*. New York: Columbia University Press.

Link, D. (1995) Contracting out the FDA: an Update, *GMHC Treatment Issues*, 10(5), May 1995.

Lofgren, H. and Benner, M. (2011) A global knowledge economy? Biopolitical strategies in India and the European Union, *Journal of Sociology*, 47: 163–80.

Lukes, S. (1973) On the social determinants of truth. In R. Horton and R. Finnegan (eds.), *Modes of Thought*, 230–48. London: Faber.

Lukes, S. (1982) Relativism in its Place. In M. Hollis and S. Lukes (eds.), *Rationality and Relativism*, 261–305. Oxford: Basil Blackwell.

Lukes, S. (2005) *Power: A Radical View*. Basingstoke: Palgrave Macmillan.

Lynch, T.J., Bell, D.W., Sordella, R., Gurubhagavatula, S., Okimoto, R., Brannigan, B.W. and Harris, P.L. (2004) Activating mutations in the epidermal growth factor receptor underlying responsiveness of non-small-cell lung cancer to Gefitinib, *New England Journal of Medicine*, 350: 2129–39.

Mackey, A.C. and Z. Li. (2002) 'Office of Drug Safety Postmarketing Safety Review: Alosetron' FDA Memo, March 26.

Malozowski, S. (1999a) FDA Medical Team Leader Memo, Rosiglitazone, 18 May.

Malozowski, S. (1999b) FDA Medical Team Leader Memo, Pioglitazone, 30 June.

Mannan, A. and Story, A. (2006) Abolishing the product patent: a step forward for global access to drugs. In J. Clare Cohen, P. Illingworth & U. Schuklenk (eds.), *The Power of Pills*, 179–88. London: Pluto Press.

Marciniak, T.A. (2010). FDA Medical Team Leader Memo, RECORD, 14 June.

Marks, H. (1997) *The Progress of Experiment: Science and Therapeutic Reform in the US, 1900–1990.* Cambridge: Cambridge University Press.

McCabe, C., Claxton, K. and O'Hagan, A. (2008) Why licensing authorities need to consider the net value of new drugs in assigning review priorities: Addressing the tension between licensing and reimbursement, *International Journal of Technology Assessment in Health Care*, 24: 140–5.

Medawar, C. and Hardon, A. (2004) *Medicines Out of Control? Antidepressants and the Conspiracy of Goodwill.* Amsterdam: Askant.

Medicines Control Agency (1991) *Commitment to Safety, Quality and Efficacy.* London: HMSO.

Medicines in Europe Forum (2004). The draft regulation on conditional marketing authorization for medicinal products puts patients at risk. Available at: http://www.prescrire.org/aLaUne/dossierEuropeAMMconditionnelleEn.php. Accessed 13 May 2005.

Mendelson, J. (2000) Blockade of receptors for growth factors: an anti-cancer therapy – the fourth annual Joseph H Burchenal American Association of Cancer Research Clinical Research Award Lecture, *Clinical Cancer Research*, 6: 747–53.

Meyers, A.S. (1997) Orphan drugs: the current situation in the United States, Europe, and Asia, *Drug Information Journal*, 31: 101–04.

Micek, S.T. and Ernst, M.E. (1999) Tolcapone: a novel approach to Parkinson's disease, *American Journal of Health-Systems and Pharmacy*, 56: 2195–205.

Middlemas, K. (1979) *Politics in Industrial Society: The Experience of the British System since 1911.* London: Andre Deutsch.

Miliband, R. (1983) State power and class interests, *New Left Review*, 138: 57–68.

Milne, C. (2000) The Food and Drug Administration Modernization Act and the Food and Drug Administration: Metamorphosis or Makeover? *Drug Information Journal*, 34: 681–92.

Misbin, R.I. (1999a) FDA Medical Officers Review: NDA 21–073, *Actos (Pioglitazone)*, 23 June, 1999.

Misbin, R.I. (1999b) FDA Medical Officer's Review, *Rosiglitazone*, 16 April.

Misbin R.I. (1999c) Comment on the ethics of placebo-controlled trials in patients with type 2 diabetes mellitus, *The Journal of Clinical Endocrinology & Metabolism*, 84: 823.

Misbin, R.I. (1999d) FDA Medical Officer's Review, *Rosiglitazone*, 2 April.

Moh-Jee Ng (1999) 'NDA 21071 (rosiglitazone) – Statistical Review and Evaluation', FDA memo, 22 March.

Montaner J.S.G., O'Shaughnessy M.V. and Schechter M.T. (2001) Industry-sponsored clinical research: a double-edged sword, *Lancet*, 358: 1983–95.

Morens, D.A.M., Davis, J.W. and Grandinetti, A. (1996) Epidemiological observations on Parkinson's disease: incidence and mortality in a prospective study of middle-aged men, *Neurology*, 46: 1044–50.

Mossialos, E., Mrazek, M. and Walley, T. (2004) *Regulating Pharmaceuticals in Europe: Striving for Efficiency, Equity and Quality*. Maidenhead, England: Open University Press.

Moynihan, R. and A. Cassels. (2005) *Selling Sickness*. Crows Nest: Allen & Unwin.

National Academy of Sciences (1983) *The Competitive Status of the U.S. Pharmaceutical Industry: The Influences of Technology in Determining International Industrial Competitive Advantage*, Prepared by the Pharmaceutical Panel, Committee on Technology and International Economic and Trade Issues; Office of the Foreign Secretary, National Academy of Engineering; and the Commission on Engineering and Technical Systems, National Research Council. Washington DC: National Academy Press.

National Institute for Health Care Management Foundation (2002) *Changing Patterns of Pharmaceutical Innovation*. Washington DC: NIHCM Foundation.

National Organization for Rare Diseases (2000) Statement: Prescription Drug User Fee Act (PDUFA) Public Meeting, 15 September. Available at: http://www.fda. gov/ohrms/dockets/dockets/00n_1364/ts00009.pdf. Accessed 3 August 2011.

National Organization for Rare Diseases (2002) Letter in support of increased FDA funding. Available at: http://www.rarediseases.org/nord/news/articles /fdafunding. Accessed 13 May 2005.

Nissen, S.E. & Wolski, K. (2007). Effect of rosiglitazone on the risk of myocardial infarction and death from cardiovascular causes, *New England Journal of Medicine*, 356: 2457–71.

Non-Small-Cell Lung Cancer Collaborative Group (1995) Chemotherapy in non-small-cell lung cancer: a meta-analysis using updated data on individual patients from 52 randomized clinical trials, *British Medical Journal*, 311: 899–909.

O'Donovan, O. (2007) Corporate colonization of health activism? *International Journal of Health Services*, 37: 711–33.

O'Donovan, O. (2008) The emergence of pharmaceutical industry regulation for competition in Ireland. *In* O. O'Donovan, & K. Glavanis-Grantham (eds.), *Power, Politics and Pharmaceuticals*, 61–81. Cork: Cork University Press.

Offe, C. (1973) *Structural Problems of the Capitalist States*. Frankfurt: Suhrkamp.

Office of Technology Assessment (OTA) (1993) *Costs, Risks and Rewards*, Washington DC: GPO.

Olson, M.K. (2002) Pharmaceutical policy change and the safety of new drugs, *Journal of Law and Economics*, XLV: 615–42.

Oncologic Drug Advisory Committee (2003a) Center for Drug Evaluation and Research Oncologic Drugs Advisory Committee, 74th Meeting, Transcript of the 12 March Meeting.

Onn, A., Tsuboi, M. and Thatcher, N. (2004) Treatment of non-small-cell lung cancer: a perspective on the recent advances and the experience of gefitinib, *British Journal of Cancer*, 91 (suppl. 2): S11–17.

Owen, B. and Braeutigam, R. (1978) *The Regulation Game: Strategic Use of the Administrative Process*. Cambridge, MA: Ballinger Publishing Company.

Parks, M.H. (2010). FDA Inter-Office Memo for Advisory Committee Meeting for Avandia (rosiglitazone), *16 June*.

Patients' Coalition (1996) Issues of Concern to Patients in the Debate over FDA Reform, 20 March.

Patients' Coalition (1997) Testimony by Jeff Bloom for the Patients' Coalition before the Committee on Commerce and Sub-Committee on Health and the Environment, United States House of Representatives, 23 April.

Patients and Consumer Coalition (2002) Summary of position regarding post-market safety. Press Backgrounder, 12 April.

Penn, R.G. (1982) *The Development of the Regulatory Control of the Safety, Quality, and Supply of Medicines.* Welsh National School of Medicine: MD Thesis.

Perehudoff, K. and Alves, T.L. (2010) *Patient and Consumer Organisations at the European Medicines Agency: Financial disclosures and transparency.* Netherlands: Health Action International.

Pfizer (1998) Trovan Product Label (US), September.

Phillips, C.B. (2006) Medicine goes to school: teachers as sickness brokers for ADHD, *Public Library of Science (PLoS) Medicine*, 3: 433–35.

Pines, W.L. (1999) A history and perspective on direct-to-consumer promotion, *Food and Drug Law Journal*, 54: 489–518.

Pollock, N. and Williams, R. (2010) The business of expectations: how promissory organizations shape technology and innovation, *Social Studies of Science*, 40: 525–48.

Potter, G. (1999) *The Bet: Truth in Science, Literature and Everyday Knowledges.* Aldershot: Ashgate.

Prescription Drug User Fee Amendments (2002) Section 502. Available at: http://www.fda.gov/RegulatoryInformation/Legislation/FederalFoodDrugand CosmeticActFDCAct/SignificantAmendmentstotheFDCAct/Prescription DrugAmendmentsof1992PrescriptionDrugUserFeeActof1992/default.htm. Accessed 17 September 2013.

Prizont, R. (1999) 'Alosetron Hydrochloride (Lotronex), Efficacy Review' FDA Division of Gastro-intestinal and Coagulation Drug Products Medical Officer Review, November 4.

Public Citizen (1998) Testimony before the U.S. House of Representatives' Committee on Government Reform and Oversight on inappropriate use of placebos in human experiments. 22 April. HRG Publication #1438. Available at: http://www.citizen.org/publications/release.cfm?ID=6635. Accessed 13 May 2005.

Public Citizen (1999) Comments on International Conference on Harmonization's draft guidance on choice of control group in clinical trials. Health Research Group Publication, No. 1503, 23 December. Washington DC: Public Citizen.

Public Citizen (2001) Rx R&D myths: The case against the drug industry's 'scare card', July. Washington DC: Public Citizen.

Public Citizen (2002) Statement at FDA Hearing on Risk Management of Prescription Drugs, 22 May. HRG Publication # 1620. Available at: http://www.citizen.org/publications/release.cfm?ID=7176#_Edn2. Accessed 13 May 2005.

Public Citizen (2003) Open letter to FDA against granting Iressa accelerated approval.

Ranson, M. (2002) ZD1839 (Iressa): For more than just non-small cell lung cancer, *The Oncologist*, 7 (suppl): 16–24.

Reed-Mauer, P. (1994) Restructuring the Japanese pharmaceutical industry, *Scrip Magazine* (May): 38–40.

Relman, A.S. (1980) The new medical-industrial complex, *New England Journal of Medicine*, 303: 963–70.

Renakaran, R. and Zinman, B. (2009) Thiazolidinediones and clinical outcomes in type 2 diabetes, *Lancet*, 373: 2088–90.

Roberts, T.G. and Chabner, B.A. (2004) Beyond fast-track for drug approvals, *New England Journal of Medicine*, 351: 502.

Roche Hexagon (1998) Product Label for Tasmar.

Roche Laboratories Inc. (1998) 'Dear Healthcare Professional' letter, 16 November.

Rothstein, H. (2006) The institutional origins of risk, *Health, Risk and Society*, 8: 215–21.

Sauer F. (1997) A new and fast drug approval system in Europe, *Drug Information Journal*, 31: 1–6.

Schilsky, R. (2003) Hurry up and wait: is accelerated approval of new cancer drugs in the best interests of cancer patients? *Journal of Clinical Oncology*, 21: 3718–20.

Schweitzer, S.O., Schweitzer, M.E. and Sourty-LeGuellec, M-J. (1996) Is there a US drug lag? The timing of new pharmaceutical approvals in the G-7 countries and Switzerland, *Medical Care Research and Review*, 53: 162–78.

Sculpher, M. and Claxton, K. (2005) Establishing the cost-effectiveness of new pharmaceuticals under conditions of uncertainty – when is there sufficient evidence? *Value in Health*, 8: 433–46.

Selgelid, M.J. and Sepers, E.M. (2006) Patents, profits, and the price of pills: implications for access and availability. In J. Clare Cohen, P. Illingworth & U. Schuklenk (eds.), *The Power of Pills*, 153–63. London: Pluto Press.

Senior, J.R. (1999) 'Alosetron hydrochloride (Lotronex GR68755)' FDA Division of Gastro-Intestinal and Coagulation Products: Medical Officer's New Drug Applicaton (NDA) Review, 25 October.

Shimmings, A. (2011) Comparative efficacy should have a 'formal role in drug licensing', Scrip (8 *September*).

Sirotnak, F.M. (2003) Studies with ZD1839 in pre-clinical models, *Seminars in Oncology*, 30, (suppl. 1): 12–20.

Sismondo, S. (2008) How pharmaceutical industry funding affects trial outcomes: causal structures and responses, *Social Science & Medicine*, 66: 1909–14.

Smalley, W., Shatin, D., Wysowski, D.K., Gurwitz, J., Andrade, S.E., Goodman, M., Chan, K.A., Platt, R., Schech, S.D. and Wayne, R.A. (2000) Contraindicated use of cisapride: impact of FDA regulatory action, *Journal of the American Medical Association*, 284: 3036–39.

Smith, S.R. and Kirking, D.M. (1999) Social norms and the evolution of drug regulation in the US: implications for access to medications for HIV disease, *AIDS & Public Policy Journal*, 14: 105–16.

SmithKline Beecham (1999) Printed Label. Avandia.

Sobel, B.E. and Furberg, C.D. (1997) Surrogates, semantics and sensible public policy, *Circulation*, 95: 1661–63.

Sridhara, R., Johnson, J., Justice, R., Keegan, P., Chakravarty, A. and Pazdur, R. (2010) Review of oncology and hematology drug product approvals at the US Food and Drug Administration between July 2005 and December 2007, *Journal of the National Cancer Institute*, 102: 230–43.

Stahel, R. (2003) Lessons from 'Iressa': expanded access programme, *British Journal of Cancer*, 89 (Suppl.2): 19–23.

Steigerwalt, R. (1999) 'NDA 20–073 (pioglitazone) – Review and Evaluation of Pharmacology and Toxicology Data,' FDA memo, 30 June.

Stephenson, J. (2000) Researchers describe findings for targeted cancer therapies, *Journal of the American Medical Association*, 284: 293–95.

Stigler, G.J. (1971) The theory of economic regulation. *Bell Journal of Economics and Management Science*, 2: 3–21.

Sullivan, E.J. (2003) 'Pulmonary toxicity associated with gefitinib (Iressa),' FDA Memo, Medical Officer Consultation, 30 January.

Susman, E. (2003) Accelerated approval seen as triumph and roadblock for cancer drugs, *Journal of the National Cancer Institute*, 96: 1495–6.

Takeda (1999) *Printed Label. Actos.*

Tanner, C.M. and Goldman, S.M. (1996) Epidemiology of Parkinson's disease, *Neurological Clinic*, 14: 317–35.

Temin, P. (1980). *Taking Your Medicine: Drug Regulation in the US.* Cambridge, MA: Harvard University Press.

Temple R.J. (1997) When are clinical trials of a given agent: placebo no longer appropriate or feasible?'*Controlled Clinical Trials*, 18: 613–20.

Temple, R.J. (1999) Are surrogate markers adequate to assess cardiovascular disease drugs? *Journal of the American Medical Association*, 282: 790–95.

Temple R. and Ellenberg S.S. (2000) Placebo-controlled trials and active-control trials in the evaluation of new treatments. Part 1: ethical and scientific issues, *Annals of Internal Medicine*, 133: 455–63.

Thomas K.E., McAuslane N., Parkinson C. and Walker S.R. (1998) A study of trends in pharmaceutical regulatory approval times for nine major markets in the 1990s, *Drug Information Journal*, 32: 787–801.

Tufts Center for the Study of Drug Development (2004) *Outlook 2004.* Boston, MA: Tufts CSDD.

Turner N. (2004) Pricing climate heats up in US and Europe, *Pharmaceutical Executive*, July.

Uhl, K., Li, Z. and Mackey, A.C. (2000) 'NDA 21–107: Lotronex (alosetron) Safety & Risk Management Summary' FDA Memo, 16 November.

UK House of Commons Health Committee (2005) Evidence Nos 354–56 in *Proceedings of the Health Select Committee Hearings on the Inquiry into the Influence of the Pharmaceutical Industry*, 20 January. London: TSO.

UKPDSG (1998a) Intensive blood-glucose control with sulphonylureas or insulin compared with conventional treatment and risk of complications in patients with type 2 diabetes, *Lancet*, 352: 837–53.

UKPDSG. (1998b) Effect of intensive blood-glucose control with metformin on complications in overweight patients with type 2 diabetes, *Lancet*, 352: 854–65.

Unger, E.F. (2010). FDA Office of Drug Evaluation Memo, 15 June.

US Congress (1987) *FDA's Regulation of the New Drug Merital: Fifteeenth Report of the House Committee on Government Operations*, 8 July, 100th Congress, 1st Session, House Report 100–206. Washington, DC: GPO.

US Congress (2007) Hearings on FDA's Role in the Evaluation of Avandia's Safety. Washington DC: House of Representatives Committee on Oversight and Government Reform.

US General Accounting Office (1995) *FDA Drug Approval. Review Time Had Decreased in Recent Years.* Report to Congressional Requesters. GAO/PEMD-96-1.

US General Accounting Office (2002) *Food and Drug Administration. Effect of User Fees on Drug Approval Times, Withdrawals, and Other Agency Activities.* Report to the Chairman, Committee on Health, Education, Labour, and Pensions, US Senate. GAO-02-958.

US General Accounting Office (2006) *Prescription Drugs: Improvements Needed in FDA's Oversight of Direct-to-Consumer Advertising.* Washington DC: General Accounting Office.

US Government Accountability Office (US GAO) (2009) *New Drug Approval: Report to Committee on Finance, US Senate.* GAO-09-866: GPO.

US House of Representatives (1984) *Prescription Drug Advertising to Consumers: Staff Report Prepared for the Use of the Subcommittee on Oversight and Investigations of the Committee on Energy and Commerce.* Washington DC: US GPO.

U.S. Senate (1996) *For and Drug Administration Performance and Accountability Act of 1995,* June 20, 1996, Report together with additional views to accompany S. 1477, 104th Congress, 2nd Session, Senate Report 104-284.

U.S. Senate (1997) *Food and Drug Administration Modernization Act: Report together with additional and minority reviews to accompany S.830.* July 1, 1997, 105th Congress, 1st Session. Senate Report 105-43. Washington: GPO.

Van Luijn, J.C.F., Gribnau, F.W.J. and Leufkens H.G.M. (2010) Superior efficacy of new medicines? *European Journal of Clinical Pharmacology,* 66: 445-48.

Van Zwanenberg, P. and Millstone, E. (2000) Beyond skeptical relativism: examining the social constructions of expert risk assessments, *Science, Technology & Human Values,* 25: 259-82.

Virji, S. (2011) GSK settles for US$3 billion with US government, Scrip *(3 November).*

Vogel, D. (1986) *National Styles of Regulation: Environmental Policy in Great Britain and the United* States. Ithaca, NY: Cornell University Press.

Vogel, D. (1998) The globalization of pharmaceutical regulation, *Governance,* 11: 1-22.

Vogel, D. (2003) The hare and the tortoise revisited: the new politics of consumer and environmental regulation in Europe, *British Journal of Political Science,* 33: 557-80.

Wall Street Journal (2002) 'Editorial', *Wall Street Journal,* 24 September.

Wall Street Journal (2003) A cure for cancer bureaucracy, *Wall Street Journal,* 6 May.

Waller, P. (2006) Risk management planning: time to deliver, *Pharmacoepidemiology and Drug Safety,* 15: 850-1.

Wardell, W.M. (1974) Therapeutic implications of drug lag, *Clinical Pharmacology and Therapeutics,* 15: 73-96.

Wardell, W.M. (1978) The drug lag revisited: comparison by therapeutic area of patterns of drugs marketed in the United States and Great Britain from 1972 through 1976, *Clinical Pharmacology and Therapeutics,* 24: 499-524.

Weber, M. (1949) *The Methodology of the Social Sciences.* New York: Free Press.

Wiktorowicz, M.E. (2003) Emergent patterns in the regulation of pharmaceuticals: Institutions and interests in the US, Canada, Britain, and France, *Journal of Health, Politics, Policy and Law,* 28: 615-58.

Williams, G. (2002) Team Leader Comments (Iressa), FDA Oncology Drugs Division, 24 September.

Willman, D. (2000a) How a new policy led to seven deadly drugs, *Los Angeles Times,* Wednesday, 20 December.

Willman, D. (2000b) FDA official says lessons have been learned from Rezulin, *Los Angeles Times*, 4 June.

Willman, D. (2001c) FDA moving to revive deadly drug, *The Los Angeles Times*, 30 May.

Wilson, J.Q. (ed.) (1980) *The Politics of Regulation*. New York: Basic Books.

Woodcock, J. (2002) Commentary – The re-introduction of Lotronex for diarrhoea-predominant irritable bowel syndrome: the FDA approves programme to provide patient access and manage risk, *British Medical Journal*, 2: 637–38.

Woodcock, J. (2010) 'NDA 021071 – Decision on continued marketing of rosiglitazone', FDA memo, 22 September.

World Medical Association (2000) World Medical Association Declaration of Helsinki. Ethical Principles for Medical Research Involving Human Subjects. Adopted by the 52nd WMA General Assembly, Edinburgh, Scotland, October.

Wyatt-Walter, A. (1995) Globalization, corporate identity and European technology policy, *Journal of European Public Policy*, 2: 427–46.

Yki-Jarvinen, H. (2004) Drug therapy: thiazolidinediones. *New England Journal of Medicine*, 351: 1106–18.

Yki-Jarvinen, H. (2005). The PROactive study: some answers, many questions. *Lancet*, 366: 1241–42.

Young, D. (2011a) FDA warns about Actos link to bladder cancer, but no US ban yet, Scrip *(16 June)*.

Young, D. (2011b) US FDA, GSK put Avandia pharmacy exit plan in place, Scrip *(19 May)*.

Index